The Genetic Basis of Haematological Cancers

The Genetic Basis of Haematological Cancers

EDITED BY

Sabrina Tosi
Division of Biosciences
Brunel University London
UK

Alistair G. Reid
Centre for Haematology
Imperial College London
UK

WILEY

This edition first published 2016 © 2016 by John Wiley & Sons Ltd

Registered office: John Wiley & Sons, Ltd, The Atrium, Southern Gate, Chichester, West Sussex, PO19 8SQ, UK

Editorial offices: 9600 Garsington Road, Oxford, OX4 2DQ, UK
The Atrium, Southern Gate, Chichester, West Sussex, PO19 8SQ, UK
111 River Street, Hoboken, NJ 07030-5774, USA

For details of our global editorial offices, for customer services and for information about how to apply for permission to reuse the copyright material in this book please see our website at www.wiley.com/wiley-blackwell.

Library of Congress Cataloging-in-Publication Data is applied for

ISBN: 9780470979389

A catalogue record for this book is available from the British Library.

Wiley also publishes its books in a variety of electronic formats. Some content that appears in print may not be available in electronic books.

Cover image: ©Creative Commons; Jasmijn D. E. de Rooij, C. Michel Zwaan, and Marry van den Heuvel-Eibrink, J Clin Med. 2015 Jan; 4(1): 127–149. DOI: 10.3390/jcm4010127

Typeset in 10.5/14pt MeridienLTStd by SPi Global, Chennai, India
Printed and bound in Malaysia by Vivar Printing Sdn Bhd

1 2016

Contents

List of contributors

Mary Alikian
Centre for Haematology, Imperial College London, UK

Anna Andersson
Department of Pathology, St Jude's Children's Research Hospital, USA

Philip A. Beer
Terry Fox Laboratory, BC Cancer Agency, UK

Csaba Bödör
Barts Cancer Institute, Queen Mary, University of London, UK, and MTA-SE Lendulet Molecular Oncohematology Research Group, Budapest, Hungary

Jasmijn D.E. de Rooij
Department of Pediatric Oncology/Hematology, Erasmus MC-Sophia Children's Hospital, The Netherlands

Valeria Di Battista
Hematology Unit, CREO, University of Perugia, Italy

Jude Fitzgibbon
Barts Cancer Institute, Queen Mary, University of London, UK

Christine J. Harrison
Northern Institute for Cancer Research, Newcastle University, Sir James Spence Institute, UK

Jamshid S. Khorashad
Huntsman Cancer Institute, University of Utah, USA

Philippa C. May
Centre for Haematology, Imperial College London, UK

Thomas McKerrell
Wellcome Trust Sanger Institute, University of Cambridge, UK, and Department of Haematology, Cambridge University Hospitals NHS Trust, UK

Cristina Mecucci
Hematology Unit, CREO, University of Perugia, Italy

Anthony V. Moorman
Northern Institute for Cancer Research, Newcastle University, Sir James Spence Institute, UK

Charles Mullighan
Department of Pathology, St Jude's Children's Research Hospital, USA

Valeria Nofrini
Hematology Unit, CREO, University of Perugia, Italy

Danilo Perrotti
Greenebaum Cancer Center, University of Maryland, USA

Alistair G. Reid
Centre for Haematology, Imperial College London, UK

Matthew J.J. Rose-Zerilli
Cancer Genomics, Academic Unit of Cancer Sciences, University of Southampton, UK

Matthew L. Smith
Department of Haematology, St Bartholomew's Hospital, UK

Jonathan C. Strefford
Cancer Genomics, Academic Unit of Cancer Sciences, University of Southampton, UK

Marry van den Heuvel-Eibrink
Department of Pediatric Oncology/Hematology, Erasmus MC-Sophia Children's Hospital, The Netherlands

Christian Michel Zwaan
Department of Pediatric Oncology/Hematology, Erasmus MC-Sophia Children's Hospital, The Netherlands

Preface

The haematological malignancies are a complex group of neoplastic diseases, linked by their origin in bone marrow-derived cells. Since the discovery of the Philadelphia chromosome, in the 1960s, as the pathognomonic marker of chronic myeloid leukaemia, the field of haematological malignancy has provided several important paradigms for the direct contribution of causal genetic lesions to the initiation of human cancer.

The subsequent leap in our understanding of leukaemia and lymphoma pathogenesis via a variety of molecular and cytogenetic abnormalities that disrupt normal cellular processes has challenged traditional approaches to disease classification and transformed both the diagnosis and management of patients. The characterization of tumour cells by genetic methods is now regarded as being as important as the traditional morphological approach to diagnosis. This trend is being accelerated by the introduction of monoclonal antibody therapy and by novel drugs designed to target specifically the molecular abnormalities responsible for the development of the tumour. Somatic genetic changes therefore increasingly define not just the diseases themselves, but the way in which an individual patient should best be treated and monitored.

With the following chapters, compiled by leading researchers in the field, we aim to provide a summary of current knowledge on the contribution of genetic and epigenetic lesions to the biology and management of haematological malignancies. A unifying factor of these biologically diverse diseases is the recent explosion of information on hitherto unrecognized molecular lesions arising from the application of novel next-generation sequencing technologies. In most diseases, these newly identified aberrations are already contributing to improved stratification and, in some cases, showing early promise as therapeutic targets. It is hoped that further functional analysis of recurrent lesions will permit the development of additional therapies targeted against critical oncogenic drivers. Although the majority of recurrent changes appear to have been identified, there remains scope for further refinement of this knowledge with studies of larger cohorts, the increasing use of whole genome sequencing, greater incorporation of rearrangement-based bioinformatic analysis and enhanced integration with epigenomic data. These areas, together with the investigation of the importance of sequential acquisition of mutations in the initiation of a malignant phenotype and the interaction of these lesions with the bone marrow microenvironment, are likely to keep researchers occupied for the foreseeable future. Nevertheless, as the following chapters beautifully illustrate, a comprehensive picture is emerging of the key genetic drivers of haematological malignancy, and these provide a rational basis for future research towards a complete understanding of, and effective treatment for, this complex group of diseases.

Sabrina Tosi
Alistair G. Reid

CHAPTER 1

The myelodysplastic syndromes

Cristina Mecucci, Valeria Di Battista and Valeria Nofrini

Introduction

Myelodysplastic syndromes (MDS) define neoplastic disorders with bone marrow dysplasia and insufficiency leading to one or more cytopenia in the peripheral blood. Bone marrow differentiation, although abnormal, is maintained. Despite the reduced amount of circulating blood cells, bone marrow cellularity is increased in the majority of cases. Less frequently, the bone marrow is hypoplastic, particularly in children and young adults with a predisposing genetic condition. The large majority of MDS cases affect individuals over the age of 60 years. Blast count, by definition, is less than 20%, although a minority of cases (10–20%) eventually transform to acute myeloid leukaemia (AML), defined by a blast count of 20% or more.

As bone marrow dysplasia may be induced from a variety of non-neoplastic conditions, including vitamin deficiencies, viral infections, smoking or medication, the identification of clonal genetic aberrations detected by chromosome banding or higher throughput genomic technologies plays a key role in achieving the correct diagnosis. Conventional cytogenetic analysis is able to detect abnormalities in around 40–50% of cases of *de novo* MDS, increasing to around 70–80% when integrated with whole-genome analysis detecting copy-number variations, uniparental disomy and acquired mutations.[1–3] Cytogenetic abnormalities involving partial or complete chromosome loss are more frequent than reciprocal translocations. This is in contrast to AML, which is partly subcategorized according to the presence of typical reciprocal chromosome translocations, such as t(8;21), t(15;17) and inv(16). Importantly, the latter are consistent with a diagnosis of AML, even in the presence of morphological evidence of less than 20% bone marrow blasts.[4]

The Genetic Basis of Haematological Cancers, First Edition. Edited by Sabrina Tosi and Alistair G. Reid.

The incidence of chromosome aberrations is much higher in MDS arising after chemo- or radiotherapy, including bone marrow transplantation procedures, for a prior neoplastic or non-neoplastic disease. A complex abnormal karyotype is found in more than 80% of treatment-induced MDS.

The critical role of clonal cytogenetic defects at diagnosis is underlined by the hierarchical clonal evolution and acquisition of additional chromosomal defects that often accompany disease progression. In addition to chromosomal rearrangements, newly acquired gene mutations may also mark clonal evolution and disease progression.[5–7] These changes may contribute to the development of a higher risk MDS or AML by conveying growth advantage, decreased apoptosis or avoidance of immune control.[8] The identification of driver gene mutations might also help define distinctive entities within myelodysplastic syndromes, improving classification and clinical management.[9] This chapter summarizes the current understanding of the genetic and epigenetic landscape of MDS and known predisposing conditions.

Predisposing conditions

Several inherited or congenital conditions have been associated with a predisposition to develop myelodysplasia. These conditions are characterized by the presence of inherited genetic defects and the development of MDS is often linked to additional genetic mistakes that are acquired and confined to the myeloid lineage. Table 1.1 summarizes the conditions described in this section and includes a list of constitutional genetic defects associated with the disorders.

Familial platelet disorder with propensity to myeloid malignancy (FPD/AML)

Familial platelet disorder with propensity to myeloid malignancy (FPD/AML) is an autosomal dominant disease characterized by mild to moderate bleeding tendency and modest thrombocytopenia with normal platelet size and morphology. Predisposition to develop myelodysplasia and acute leukaemia is another feature of this platelet disorder, with a leukaemic rate of approximately 35%.[10] The majority of patients exhibit impaired platelet aggregation with collagen and epinephrine, similarly to abnormalities caused by aspirin. FDP/AML is associated with alterations of *RUNX1/21q22.12*, a gene encoding for a subunit of the core binding

Table 1.1 Inherited or congenital conditions predisposing to MDS and leukaemia.

Disease	Inheritance	Gene	Locus	Other features	Incidence of MDS/AML (%)
Severe congenital neutropenia	AD	*ELANE*	19q13	None	10
	AD	*GFI1*	1p22	Monocytosis Lymphopenia	
	AR	*GSPC3*	17q24	Cardiac and urogenital malformations	
	AR	*HAX1*	1q21	Neuropsychological defects	
	XL	*WAS*	Xp11	Monocytopenia Low NK cells	
Shwachman–Diamond syndrome	AR	*SBDS*	7q11	Exocrine pancreatic insufficiency, bone marrow failure, skeletal abnormalities	10
Poikiloderma with neutropenia	AR	*C16orf57*	16q21	Poikiloderma, pachyonychia, chronic neutropenia	
Dyskeratosis congenita	XL	*DKC1*	Xq28	Mucocutaneous abnormalities, aplastic anaemia	3–5
Fanconi anaemia	AR	*FANC/BRCA*	Xp22	Multiple congenital abnormalities	50
	XL	*FANCB*			
Bloom syndrome	AR	*BLM*	15q26	Short stature, photosensitivity reactions	25
Familial platelet disorder with propensity to myeloid malignancy	AD	*RUNX1*	21q22	Dysmorphic features, intellectual disability (in cases with RUNX1 deletions)	20–60
Familial MDS/AML with GATA2 mutations	?	*GATA2*	3q21	Monocytopenia, B, NK and dendritic cell lymphopenia	
Down syndrome	N/A	*HMGN1*	21q22.2	Delayed development, learning disabilities, heart defects, vision problems, hearing loss, hypotonia	10–20-fold higher than general population

(continued)

Table 1.1 *(continued)*

Disease	Inheritance	Gene	Locus	Other features	Incidence of MDS/AML (%)
SCN	AD	*ELANE*	19q13	None	10
	AD	*GFI1*	1p22	Monocytosis Lymphopenia	
	AR	*GSPC3*	1p34	Cardiac and urogenital malformations	
	AR	*HAX1*	1q21	Neuropsychological defects	
	XL	*WAS*	Xp11	Monocytopenia Low NK cells	
SDS	AR	*SBDS*	7q11	Exocrine pancreatic insufficiency and skeletal abnormalities	10
PN	AR	*C16orf57*	16q21	Poikiloderma, pachyonychia, chronic neutropenia	
DC	XL	*DKC1*	Xq28	Mucocutaneous abnormalities, aplastic anaemia	3–5
FA	AR XL	*FANC/BRCA*			50
BS	AR	*BLM*	15q26	Short stature, photosensitivity reactions	25
FPD/AML	AD	*RUNX1*	21q22	Dysmorphic features and intellectual disability (in cases with deletions)	20–60
Familial MDS/AML with GATA2 mutations	?	*GATA2*			

AD, autosomal dominant; AR; autosomal recessive; XL, X-linked; N/A, not applicable.

factor (CBF) transcription complex. Monoallelic mutations in *RUNX1* include deletions and missense, nonsense and frameshift mutations.[11] Two functional consequences of these mutations include haploinsufficiency and a dominant negative effect.[12] Large deletions of *RUNX1* have also been described, and in these cases patients showed additional features such as short stature, malformations, dysmorphic features and intellectual disability.[13] Individuals with missense mutations have a higher risk of haematological malignancies than those carrying mutations causing haploinsufficiency.[14] However, the genetics of FPD/AML may be even more complicated; Minelli et al.[15] reported a single family with a clinical history consistent with FDP/AML in which no mutations was detected in *RUNX1* and in which linkage to chromosome 21 was excluded, implying that other genetic lesions outside this region may also cause an FDP/AML-like phenotype.

Severe congenital neutropenia (SCN)

Severe congenital neutropenia (SCN) comprises a heterogeneous group of primary immunodeficiency disorders collectively characterized by paucity of mature neutrophils, increased infections and higher risk of developing AML/MDS.[16] The majority of patients respond to treatment using recombinant human granulocyte colony-stimulating factor (rh-G-CSF) by increasing neutrophil counts and decreasing frequency and severity of infections. In recent years, progress has been made with respect to the elucidation of the genetic causes underlying syndromic and non-syndromic variants of SCN. The genes most commonly involved are the elastase gene *ELANE* (in 50–60% of cases) and the HCLS1-associated protein X-1 gene *HAX1* (in 4–30% of cases), while mutations in the growth factor-independent 1 transcription repressor gene *GFI1*, the xylanase gene *XLN* and the glucose-6-phosphatase catalytic subunit 3 *G6PC3* have been described in a smaller number of patients. Concurrent mutations have been also described (ELANE + HAX1, ELANE + G6PC3, HAX1 + G6PC3).[16,17] The majority of patients with autosomal dominant SCN bear heterozygous mutations in *ELANE*.[18] To date, more than 50 mutations have been described in *ELANE*; these mutations lead to severe neutropenia via a stress response in the endoplasmic reticulum (ER), which provokes activation of the unfolded protein response (UPR).[19] Rarely, SCN can be caused by autosomal dominant mutations in *GFI1* coding for a transcription repressor for *ELANE*.[20] In these individuals, monocytosis and leucopenia accompany the neutropenia.[16] A complex disorder characterized by SCN and developmental disorders is caused by mutations in the *G6PC3* gene. The affected individuals present with features such as cardiac and neurological malformations.[17] *HAX1* mutations were described by Klein et al.[21] as the genetic cause of Kostmann syndrome, the autosomal recessive form of SCN, associated with neurophysiological defects. In this form of SCN, *HAX1* mutations act as loss-of-function mutations, leading to increased apoptosis. Devriendt et al.[22] described X-linked neutropenia in the Wiskott–Aldrich syndrome caused by gain-of-function mutations in the *WAS* gene. These patients also have monocytopenia and very low NK cell counts.[22] Finally, acquired mutations in the granulocyte colony-stimulating factor 3 receptor gene (*CSF3R*) define a subgroup with a high risk of malignant transformation,[23] due to the concomitant presence of monosomy 7 in the myeloid cells.[24]

Poikiloderma with neutropenia

This is a rare skin condition characterized by changes in pigmentation defined as autosomal recessive inherited genodermatosis. This pathology has recently been associated with biallelic mutations in the *C16orf57* gene, located at 16q21, that encodes a U6 biogenesis 1 (USB1) protein. Mutations in this gene have also been encountered in the Rothmund–Thomson syndrome (RTS). To date, 38 PN patients have been reported, harbouring 19 different mutations that are all predicted to generate truncated protein.[25] The function of the USB1 protein is poorly characterized and the pathogenesis on PN remains obscure, but affected individuals may be predisposed to develop MDS and AML.[26]

Familial MDS/AML

The transcription factor *GATA2* has been identified as a predisposing gene in familial MDS/AML.[27] *GATA2* belongs to a family of zinc finger transcription factors that has six members in mammalians and is required in the early proliferative phase of haematopoietic development.[28] Ectopic expression of *GATA2* has yielded controversial results, promoting proliferation in some experiments and differentiation in others.[29,30] The biological functions of *GATA2* and the importance of its balanced expression have led to the suggestion that this gene might be involved in leukaemogenesis. A number of families carrying inherited heterozygous missense mutations in the *GATA2* transcription factor gene have been studied.[27] The mutations caused almost total loss of function. The MDS/AML observed in these families was characterized by various acquired chromosomal abnormalities, including trisomy 8, monosomy 7 and trisomy 21. *GATA2* mutations were also found by exome sequencing in patients with mild chronic neutropenia associated with monocytopenia and evolving to AML and/or MDS.

Shwachman–Diamond syndrome (SDS)

In the Shwachman–Diamond syndrome, a rare congenital disorder (incidence 1/75,000), 90% of patients bear Shwachman–Diamond–Bodian syndrome gene (*SBDS*) mutations. The *SBDS* gene maps to band 7q11 and encodes a protein implicated in ribosome biogenesis, mitotic spindle stability and cellular stress response.[32–34] About 10% of patients with clinical features of SDS lack *SBDS* mutations.[35] SDS is characterized by haematological abnormalities such as neutropenia, anaemia, thrombocytopenia and a high risk of developing aplastic anaemia, MDS and/or AML[36] due to defective haematopoiesis and an altered microenvironment. In these patients, the bone marrow contains few $CD34^+$ precursors which, *in vitro*, have a reduced colony-forming capacity and bone marrow stromal cells are unable to support and maintain haematopoiesis.[37] André et al.[38] showed that SDS mesenchymal stem cells have a normal karyotype and do not show the chromosomal abnormalities observed in the bone marrow of SDS patients. The cytogenetic abnormalities found in the bone marrow blasts mainly involve chromosome 7 (monosomy 7 or 7q deletion, translocations or isochromosome 7) and chromosome 20, such as del(20q).[39] The prognostic significance of the cytogenetic abnormalities in SDS is not yet resolved. Many reports suggest that chromosome 7 abnormalities were specific markers of MDS/AML evolution.[40] Nevertheless, the disease was stable in patients carrying iso(7q) and del(20q).[41,42] The mechanism of leukaemogenesis in SDS is unknown. However, oligonucleotide microarray studies showed that several genes related to other inherited marrow failure syndromes, including *FANCD2, FANCG, RUNX1, DKC1* and *MPL*, were down-regulated in SDS marrow cells, whereas several oncogenes such as *LARG, TAL1* and *MLL* were up-regulated.[43] Altered expression of the *SBDS* gene has been found in the mesenchymal cells of a *Dicer1* deleted mouse model of MDS, highlighting the importance of a healthy microenvironment for a correct haematopoiesis.[44]

Dyskeratosis congenita (DKC) and telomere syndromes

The telomere syndromes are inherited conditions in which symptoms affect different organs and tissues (skin, lung, liver and bone marrow). These syndromes are caused by dysfunctional telomerase and abnormally short telomeres. Telomeres are DNA–protein structures that protect chromosome ends from nuclease activity. Telomeric DNA is made of hundreds to thousands of repetitions of the hexanucleotide TTAGGG. The telomeric double-stranded DNA assumes a T-loop single-strand conformation at the 3′ end that protects the chromosome end from folding backwards. Several proteins form the shelterin protein complex (TERF1 and -2, TINF2, TPP1, POT1 and RAP1) and altogether their function is to stabilize the T-loop (Fig. 1.1, Table 1.2). The telomerase is the ribonucleoprotein complex that adds back additional telomeric DNA, avoiding dangerous shortening. This complex includes an RNA template, *TERC*, a reverse transcriptase enzyme encoded by the *TERT* gene, and the dyskerin proteins NHP2, GAR1, NOP10, TCAB1 and DKC1. Telomerase

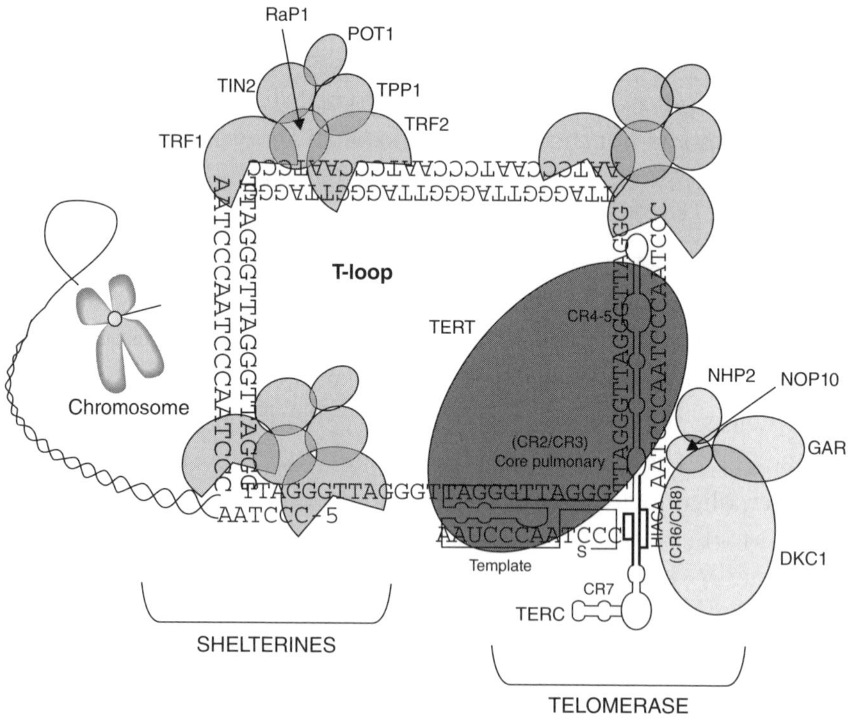

Figure 1.1 The telomere T-loop structure with shelterin complex and the elongation telomerase sub-units TERT and TERC.

Table 1.2 Telomerase components (shaded for diskerins) and shelterines.

Gene	Chromosome location	Function
Telomerase components		
TERT	5p15.33	Reverse transcriptase
TERC	3q26.2	RNA template
NHP2	5q35.3	Present in the Cajal bodies, it is the first protein that binds RNA, together with the other diskerines dials the telomerase complex
GAR1	4q25	snoRNAs belonging to H/ACA snoRNPs, specific for the pseudourilation of the RNA
NOP10	15q14	Nucleolar protein that constitutes the telomerase complex
TCAB1	17p13.1	The fitting factor between TERT, TERC and diskerine, transports the whole complex from the nucleus to the Cajal bodies
DKC1	Xq28	Part of the telomerase complex, it is specifically responsible for the first step of rRNA processing
Shelterins		
TERF1	8q21.11	Involved in a negative feedback mechanism that establishes the length of telomeres, indeed interacts with double-stranded telomeric DNA, PINX1, TINF2 and recruits POT1
TERF2	16q22.1	Telomerase inhibitor, binds to telomeric DNA double strand to limit telomere elongation
TINF2	14q12	Interconnect between the factors binding telomeric double-stranded TERF1-2 with those binding telomeric single-stranded POT1–ACD.
TPP1	16q22.1	Provides a physical link between telomerase and shelterin complex
RAP1	16q23.1	Essential for the suppression of a homologous repair system that would act on the single-stranded telomeric DNA
POT1	7q31.33	Interacts with TERF1 and ACD binds to single-stranded telomeric DNA to transmit information about the length of telomeres

enzymatic activity is repressed in the somatic cells, but activated in the stem and germinal cells and in highly proliferative tissues (skin, bone marrow, ovaries). Moreover, telomerase is up-regulated in cancer cells so that these cells can overtake the cell cycle checkpoints and escape apoptosis.[45–47] During DNA replication, there is a necessary gradual loss of telomeres due to end replication problems, but factors such as ageing, stress and mutations in telomerase complex genes accelerate progressive telomere erosion. When the telomeres are critically short (telomere length <5 kb), cells suffer defective proliferation with consequent senescence, apoptosis and genomic instability (breakage–fusion–bridge cycle, aneuploidy) that limits cell regeneration. These mechanisms underlie the development of dyskeratosis congenita (DKC), bone marrow failure, aplastic anaemia, pulmonary fibrosis and cryptogenic hepatic cirrhosis. Telomere shortening may also promote, due to genomic instability, MDS and leukaemia.[48] Telomere shortening syndromes are clinically evident relatively early in severe forms such as the premature ageing syndrome DKC whereas aplastic anaemia (AA) and idiopathic pulmonary fibrosis (IPF) are late-onset diseases that may only be recognized when MDS or AML occur. Moreover, AA and IPF are often considered complications of DKC and they are the first cause of mortality in these patients.[45]

DKC is a serious disorder characterized by nail dystrophy, lacy reticular pigmentation of the neck and upper chest and oral leukoplakia. In many patients, additional clinical manifestations such as bone marrow failure, pulmonary fibrosis, eye and dental abnormalities, oesophageal and urethral stenosis and osteopenia may be part of the clinical phenotype. The disease may be X-linked (DKCX) and is caused by mutations in the *DKC1* gene, which encodes for dyskerin, a protein necessary for the stabilization of telomerase. The autosomal dominant form (DKCA2) is caused by mutations in *TERT* and *TERC* genes coding for telomerase and RNA template, respectively. The autosomal recessive disease (DKCB1) is caused by mutations in *TINF2, NOP10* and *NHP2* genes encoding for shelterin, a protein necessary for the correct refolding of the telomeres (t-loop). owing to a complex pattern of inheritance in the DKC, disease penetrance and expressivity are highly variable and, in addition to the mutations, shortness of telomeres is required for the disease to manifest.[49]

Aplastic anaemia (AA) is a complex, heterogeneous bone marrow failure (BMF) disorder. Differential diagnosis between AA and hypoplastic forms of MDS is based on peripheral blood count and bone marrow cellularity criteria. Differential diagnosis may be difficult to establish and is sometimes arbitrary, in the absence of cytogenetic abnormalities. Correct diagnosis, however, is important for addressing prognostic stratification and treatment. Exposure to putative risk factors such as chemicals, drugs and viruses contributes to the manifestation of the acquired form. Detection of mutations in the telomerase complex and shelterin genes in patients with AA and DKC contributed to a greater understanding of the consequences of telomerase deficiency in bone marrow failure and predisposition to cancer.[46] Heterozygous mutations in the *TERT* gene impair telomerase activity by haploinsufficiency and may constitute a risk factor for marrow failure such as MDS and AML, even in patients without evidence of DKC.[50] Telomere and telomerase behaviour in T-cell deregulation has been studied in naive T-cells in MDS. Naive T-cells in MDS patients had shorter telomeres, lower TERT mRNA and reduced proliferative capacity contributing to the accumulation of senescent cells and a reduction of the naive T-cell compartment. These results suggest a mechanistic link between AA and some forms of MDS.[51]

Fanconi anaemia (FA)

Fanconi anaemia (FA) is a rare autosomic recessive and X-linked disease characterized by multiple somatic malformations, haematological abnormalities and predisposition to a variety of cancers.[52–54] The classical diagnostic test for FA is based on an assessment of cellular hypersensitivity to DNA interstrand crosslinking agents, such as diepoxybutane (DEB) and mitomycin (MMC).[55]

Haematological abnormalities represent the most prominent pathological manifestation of FA; 75–90% of FA patients develop bone marrow failure during the first decade of life and, in addition, patients develop aplastic anaemia, MDS or AML.[56,57] A recent report revealed a common pattern of specific chromosomal abnormalities in FA patients with MDS or AML, which include gain of 1q or 3q, monosomy 7 or deletion of 7q and 11q loss.[58] Moreover, cryptic *RUNX1* lesions such as translocations, deletions or mutations have also been reported.[59] Interestingly, translocations and/or duplications of 1q can be seen at all stages in the BM, including 'normal' or hypoplastic BM without signs of transformation. This suggests that these abnormalities are not predictive of MDS–AML development.[59,60] Fifteen *FANC* genes have been identified to date (*FANCA, FANCB, FANCC, FANCD1, FANCD2, FANCE, FANCF, FANCG, FANCI, FANCJ, FANCL, FANCM, FANCN, FANCO* and *FANCP*), mutations of which give rise to FA. Their products are components of a common cellular pathway, the Fanconi anaemia signalling pathway, involved in controlling multiple functions related to DNA repair and cellular response to stress.[61] There are some clear genetic–phenotype correlations in the FA patients: 'hypomorphic' mutations are associated with mild phenotype whereas FANCD2 patients usually have a more severe phenotype.[62] Identification the of breast cancer susceptibility gene *BRCA2* as an additional FA gene suggests a close relationship between the DNA repair mechanisms of the FA and BRCA1/2 pathways.[63–65]

Down syndrome

The increased risk of leukaemia in children with Down syndrome (DS) is estimated at 50-fold. The World Health Organization (WHO) in 2008 proposed a unique biological entity for DS-related myeloid leukaemia to include both MDS and AML.[4] Among AML patients, acute megakaryocytic leukaemia (AMkL) is the most common subtype identified. Cytogenetic abnormalities in myeloid leukaemia of DS showed trisomy 8 in 13–44% of patients whereas monosomy 7 is very rare.[66,67] In addition to the constitutional trisomy 21, somatic mutations of the gene encoding the transcription factor *GATA1* have also been reported.[68–70] Younger DS patients showed a better response to chemotherapy than non-DS AML children,[71] whereas older DS patients with *GATA1* mutations had a poorer prognosis comparable to that of AML patients without DS.[72] Recent research using mouse models showed that of the many genes present on chromosome 21 that are over-represented in Down syndrome patients, *HMGN1* is particularly important as it is responsible for switching off *PRC2* with the consequence of increasing B-cell proliferation.[73] *HMGN1* is a plausible candidate for regulating genome-wide chromatin modification, with potential impact on other forms of cancer.[74] This discovery opens up new avenues for targeted therapy in individuals with Down syndrome and haematological malignancies.

Cytogenetics

Standard cytogenetic methods based on chromosome banding techniques allow the detection of chromosomal abnormalities in about 50% of patients with *de novo* MDS and 80% of those with treatment-related MDS. In the remaining cases, karyotypes are normal or non-informative.[75] These figures are based on the definition of cytogenetic clone that requires the presence of a chromosomal aberration in at least two metaphases (in the case of structural aberration or trisomy) or three metaphases (in the case of monosomy).

Single chromosomal aberrations are typical of primary MDS whereas complex karyotypes are more common in therapy-related MDS (t-MDS), although there is considerable overlap. Overall, deletions predominate over reciprocal translocations, suggesting that tumour suppressor gene inactivation/haploinsufficiency plays a pivotal role in MDS development.[75] Translocations are shared with AML and/or MPD. Besides one distinct MDS type defined by its chromosome change, i.e. del(5q),[4] other recurrent cytogenetic findings are associated with typical morphological features and clinical course of disease.[76,77] Cytogenetic findings play a major role in determining prognostic stratification of MDS addressing treatment choice and experimental therapies. Conventional cytogenetics, i.e. karyotyping, provided the most relevant biological markers for MDS diagnosis (Figure 1.2, Table 1.3). Further refinements have been introduced by higher resolution technologies such as fluorescence *in situ* hybridization (FISH) and single nucleotide polymorphism (SNP) array analysis. FISH allows the detection of abnormalities in non-dividing cells, whereas SNP array technology allows the detection of regions of loss of heterozygosity (LOH) that are cryptic at the level of G-banding, either because they are caused by deletions too small to discern microscopically or because they occur without genomic loss via a phenomenon known as acquired uniparental disomy (UPD). The latter phenomenon is relatively common in MDS and has been shown to result in the unmasking of mutated tumour suppressor genes.[78,79]

Loss of Y chromosome (–Y) and del(11q)

According to the recent international prognostic scoring,[80] both abnormalities are associated with even better prognosis than normal karyotype, which itself confers a favourable outcome.[81,82] The role of –Y in MDS pathogenesis is not clear as it is also found in the bone marrow of healthy

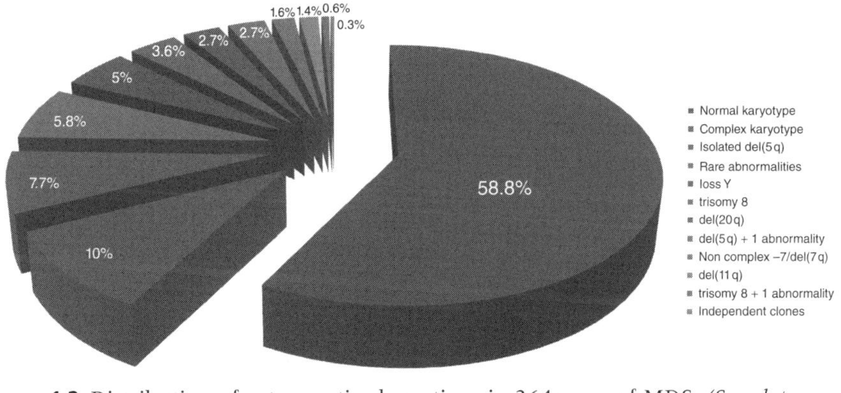

Figure 1.2 Distribution of cytogenetic aberrations in 364 cases of MDS. *(See plate section for color representation of this figure.)*

Table 1.3 List of chromosomal abnormalities identified by conventional cytogenetics relevant for MDS diagnosis.

MDS chromosome abnormalities	***Structural***
	t(11;21)(q23.3; t;(2;11)
	del(7q)
	del(5q)
	del(20q)
	del(12p)
	del(11q)
	del(17p)
	iso(X)(q13)
	del(18q)
	Numerical
	monosomy 7
	trisomy 6
	–Y
	trisomy 8
	Structural/numerical
	1q unbalanced translocations
MDS/AML chromosome abnormalities	*Structural*
	t(6;9)(p23;q34)/DEK-CAN
	t(3;5)(q25;q34)/NPM1-MLF1
	inv(3)(q21q26/t(3;3)(q21;q26)/EVI1, MDS1
	t(1;3)
	Numerical
	trisomy 4
	trisomy 8
	trisomy 13
	trisomy 14
	trisomy 21
MDS/MPD chromosome abnormalities	i(17)(q13)
	trisomy 14/i(14)(q11)
	del(13)(q14)
	trisomy 8
	del(17)(p13)
	del(20)(q13)

elderly males. Consequently, –Y, like del(20q), is not a diagnostic marker of MDS without additional morphological evidence. However, detection of –Y at diagnosis and its disappearance at remission indicate that, at the very least, it represents a marker of clonality in confirmed MDS cases.

Del(11q) describes interstitial deletions of chromosome 11 that occur with variable breakpoints between q14 and q23 in MDS. Ring sideroblasts are a frequent morphological feature accompanying this aberration.[83,84] Importantly, given its location, the *MLL* gene is not involved in this chromosomal change.[84]

Del(20q)

Common to MDS and MPD, isolated 20q deletions are also sporadically found in cases without evidence of BM dysplasia and transient cytopenia. Both interstitial and terminal deletions have been detected by conventional cytogenetics. Molecular cytogenetics has established that virtually all deletions are interstitial and have a common deleted region at band 20q13. A single 20q deletion may result in the complete loss of expression of two imprinted genes, i.e. *L3MBTL1* and *SGK2*, whose concomitant loss is responsible for dysregulation of both erythro- and megakaryopoiesis.[85] Del(20q) is included among cytogenetic abnormalities with a relative favourable course.[86] When del(20q) is found in bone marrow without typical morphological signs of bone marrow dysplasia, it is insufficient to support diagnosis of MDS.[4,87]

idic(X)(q13)

A dicentric isochromosome composed of two copies of the short (p) arm of the X chromosome, which may involve either the active or inactive X, is a recurrent finding in elderly women with MDS and is frequently associated with the presence of ringed sideroblasts. Breakpoints involving Xq13 may be also found in MDS associated with translocations, such as t(X;11)(q13;q24), t(X;19)(q13;p11), in which ring sideroblasts are not common.[88]

Del(17)(p13)/i(17q)

Unbalanced 17p translocations, monosomy 17 or i(17q), all resulting in 17p13 deletion, have been found in roughly 5% of MDS.[89] Loss of *TP53* usually occurs as a result of these abnormalities, often accompanied by deletion or mutation affecting the second allele. Often therapy related, these cases usually have excess blasts in the bone marrow. The most frequent translocations that, in an unbalanced state, lead to loss of 17p are t(5;17)(p11;p11) and t(7;17)(p11;p11).[90,91] The majority of patients with del(17p) have additional chromosomal changes, with more than 75% displaying a peculiar type of dysgranulopoiesis, i.e. pseudo-Pelger–Huët hypolobulation of the nucleus and small vacuoles in neutrophils. Interestingly, isolated isochromosome 17q, which also results in loss of 17p, is a distinct clinico-pathological entity with myelodysplastic and myeloproliferative features, a high risk of leukaemic transformation and a wild-type remaining *TP53* allele.[92]

Del(12p)

Abnormalities of the short arm of chromosome 12 (12p) are found in 1–3% of primary MDS as an isolated change or with additional anomalies. Deletions at 12p are more frequent in t-MDS with complex karyotype. The deletions vary in size with a common deleted region within band 12p13, between *ETV6* (distally) and *CDKN1B* (proximally).[93] It has been reported that 12p deletions of smaller size often occur as a sole abnormality and appear to confer a relatively good clinical outcome.[94] Indeed, according to the IPSS-R, cases with isolated del(12p) fall within the good risk group with del(5q).[80]

Trisomy 8

In primary MDS, the incidence of +8, whether alone or associated with other abnormalities, is about 10–15%. Together with monosomy 7, it is the most frequent numerical aberration and is also commonly found in other myeloid malignancies, namely AML and MPN.[76,77] In MDS, a trisomy 8 constitutional mosaicism may underlie trisomic cell growth in bone marrow.[95] Independently of treatment or clinical–haematological variations, trisomy 8 may be involved in the so-called 'clonal fluctuation,' i.e. spontaneous trisomic clone disappearance and re-expansion during disease follow-up.[96–99] The prognostic impact of trisomy 8 is not well defined since patients have a wide ranging survival, hence the aberration is grouped in the 'intermediate' prognostic category. Phenotype/genotype correlations suggest that trisomy 8 in MDS mainly affects an early myeloid progenitor at the level of CFU–GEMM, downstream of the totipotent myeloid–lymphoid stem cell. Gene expression profiling of $CD34^+$ cells with trisomy 8 detected the aberrant expression of genes regulating immune and inflammatory responses and apoptosis.[100]

Rare trisomies: +6, +13, +14, +15, +16, +19, +21

These trisomies have been sporadically reported in MDS as isolated numerical changes. As trisomy 6 is typically associated with hypoplastic MDS, it distinguishes MDS from true aplastic anaemia. The neoplastic

+6 clone involves both myeloid and erythroid lineages. Although very few cases have been reported to date, trisomy 6 does not appear to be associated with an aggressive clinical course.[101]

Trisomy 14 appears to predict poor prognosis in MDS, MPD and AML. It is associated with advanced age, male gender, thrombocytosis and morphological abnormalities in red blood cells. Similar clinical–haematological features were observed in cases with an isochromosome 14q.[102,103]

Trisomy 15 is predominantly found in low-grade MDS, such as refractory anaemia, and is often associated with –Y. Its significance has not yet been completely defined. It is prevalent in males and, like –Y, has been related to ageing. In fact, the largest analysis of patients with trisomy 15 did not find frank BM dysplasia and showed that PB cytopenia resolved spontaneously in many cases.[104,105]

Trisomy 19 is closely associated with *de novo* MDS and MDS-derived AML, suggesting that it plays a role in disease progression.[106]

Trisomy 21, the second most common trisomy in patients with MDS/AML, is rarely isolated but, when it is, it appears to predict poor prognosis.[107] It is frequently associated with –5/5q–, –7/7q– and +8, which determine outcome. A recent study designed to identify putative accompanying genetic lesions, such as imbalances, uniparental disomy and gene mutations, and to define associated clinical features found that +21 positive myeloid malignancies were clinically highly variable and had a heterogeneous pattern of cryptic copy-number variations and gene mutations.[108]

Monosomy 7 and del(7q)

Loss of whole chromosome 7 or partial deletions at its long arm [–7/del(7q)], whether isolated or as part of a complex karyotype,[109] occur in ~10% of MDS with unfavourable prognosis. In a minority of cases, an underlying familial monosomy 7 syndrome with more than one affected sibling may emerge. Monosomy 7 is associated with susceptibility to serious infections. MDS with monosomy 7 responds well to demethylating agents.[110] Monosomy 7 is often observed in emerging MDS/AML clones from hypo/aplastic bone marrow in congenital conditions such as Fanconi anaemia, Shwachman–Diamond syndrome, Kostmann syndrome and neurofibromatosis 1, and also in acquired aplastic anaemia after benzene exposure. Insights into selective G-CSF stimulation of myeloid cells bearing monosomy 7 have been obtained *in vitro* and *in vivo*.[24,111] The gene expression profile in CD34$^+$ cells with monosomy 7 revealed down-regulation of genes associated with differentiation. Partial 7q deletions are large in size and relevant genes are still elusive,[112,113] although by analogy with del(5q), haploinsufficiency for one or more critical gene(s) is likely.[114] At least three commonly deleted regions (CDRs) have been identified, at 7q22, 7q32–33 and 7q35–36 (Fig. 1.3).[115–119] Interesting candidate genes are located at the 7q22 CDR, although deletion of a 2 Mb synthenic region did not induce myeloid malignancies in mice.[114]

Three contiguous genes located immediately upstream of 7q22, i.e. *SAMD9*/7q21.2, *SAMD9L*/7q21.2 and *MIKI*/7q21.3, have been proposed as candidate genes. In particular, down-regulation of *MIKI* gene, located at the mitotic spindle and centrosome, produced mitotic abnormalities and a nuclear morphology similar to that observed in MDS.[120] *CUX1* (*CUTL1*, 7q22), encoding a homeodomain DNA binding transcription factor, has also been implicated.[116,119] Normally highly expressed in multipotent haematopoietic progenitors, *CUX1* was down-regulated in CD34$^+$ cells from patients with –7/del7q. Haploinsufficiency of its orthologue in *Drosophila* resulted in increased haemocyte proliferation and melanotic tumour proliferation in developing larvae.[121] Another candidate at 7q22 is *MLL5*, a member of the *MLL* family thought to regulate stable transcriptional states during the developmental processes. In loss-of-function mouse models, *MLL5* behaved as a key regulator of normal haematopoiesis; however, its inactivation did not result in overt myeloid diseases.[122–124] Moving downstream, a recent candidate is the *DOCK4* gene at 7q31. This gene encodes a GTPase regulator

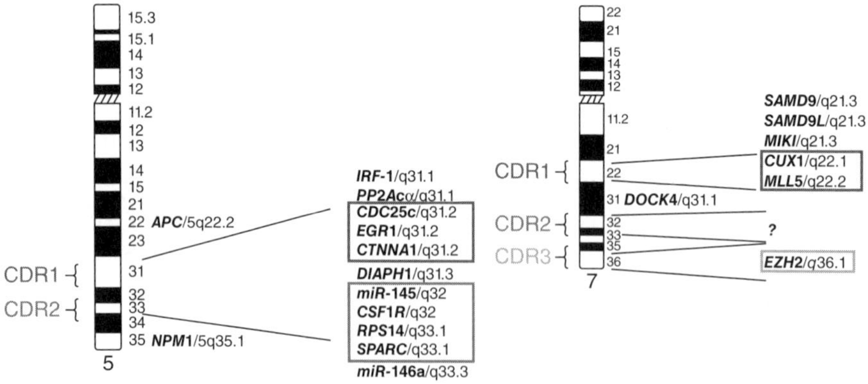

Figure 1.3 Chromosomes 5 and 7 common deleted regions (CDRs) and examples of deleted genes.

and was identified in a methylation profiling study on peripheral blood leukocytes of 21 MDS patients with either normal or abnormal karyotypes (including monosomy 7).[125] *DOCK4* was significantly hypermethylated and weakly expressed in MDS. Genetic and epigenetic events such as promoter methylation and 7q loss may underlie this deep down-regulation. *DOCK4* knockdown in primary marrow $CD34^+$ stem cells reduced erythroid colony formation and increased apoptosis, recapitulating MDS bone marrow failure.[125] Whole-genome analysis identified *EZH2* gene mutations in patients with –7/del(7q) by cytogenetics or UPD on SNP analysis).[118] *EZH2* maps at 7q36.1, encodes for a methyltransferase protein in the PRC2 complex and serves an essential function in maintaining transcriptional silencing through specific post-translational histone modifications.[126] A diverse range of missense, nonsense and frameshift mutations in *EZH2* were identified in ~6.4% of MDS cases. Interestingly *EZH2* mutations are found in monoallelic and biallelic states and are more frequently associated with 7q UPD and 7q36.1 microdeletions than with –7/del(7q).[118,127] These observations suggest that *EZH2* acts as a tumour suppressor. In contrast, however, an activating mutation increasing *EZH2* methylation activity was found in lymphomas[128] and *EZH2* over-expression has been reported in various malignancies.[126]

Rare monosomies

Monosomy 5 is rarer than estimated on the basis of karyotype; FISH reveals that a proportion of cases are in fact long arm deletions or unbalanced translocations, with the p-arm and proximal q-arm retained in abnormal marker chromosomes.[129] It is typically found in t-MDS arising after alkylating agents and predicts a poor outcome.[130] Monosomy 18, and also other monosomies, such as –13, –17, –20 and –21, are usually found as part of complex karyotypes.[131–133] When present as a sole abnormality, monosomy 21 is considered a good prognostic marker.[132]

Unbalanced translocations involving 1q

These translocations involve a supernumerary 1q that rearranges with a variety of chromosome loci (Yq12, 6p21, 6p24, 7q10, 9p10, 10q11, 13q10, 14q10, 15q10, 16q11, 16q24, etc.), resulting in partial trisomy of 1q. In the majority of cases the 1q breakpoint occurs within the heterochromatic region. The most frequent unbalanced 1q translocation, t(1;7)(q10;p10), occurs in both MDS and AML, particularly in therapy-related cases. der(6)t(1;6)(q21–25;p21–23), another recurrent unbalanced translocation producing a partial trisomy 1q, has been reported in MDS, AML and chronic myeloproliferative neoplasms.[134,135]

t(17;18)(p10;q10)

The rare t(17;18) whole arm chromosome translocation is recurrent in AML and MDS, even in cases with ringed sideroblasts. Rarely found as an isolated abnormality, it results in loss of the short arms of chromosomes 17 and 18. FISH has shown that the derivative bears both centromeres.[96,136]

Rare or sporadic balanced translocations

Sporadic balanced translocations are seen in ~2–3% of low-grade MDS. Unlike most translocations in MDS/AML, a recurrent breakpoint at 11q23.3, telomeric to *MLL* in the t(11;21)(q23.3;q11.2) and the t(2;11)(p21;q23.3), does not produce a fusion gene. Remarkably, the t(2;11)(p21;q23.3) is associated with strong miR-125b up-regulation which, in *in vitro* experiments, interfered with primary human $CD34^+$ cell differentiation and inhibited terminal (monocytic and granulocytic) differentiation in leukaemic cell lines. In some patients, the normal chromosome 11 was lost and the derivative was duplicated, reinforcing the critical role of the rearranged sequences in disease pathogenesis. These translocations are frequently seen as isolated abnormalities. However, the most commonly associated changes are del(5q) and/or –7/del(7q).[137,138] Remarkably, several new reciprocal translocations have been reported recently in MDS and AML,[139,140] but the molecular counterpart(s) still remain to be determined.

t(3;5)(q25.1;q34)

t(3;5)(q25.1;q34) produces a fusion between *NPM1*, encoding for a nucleocytoplasmic shuttle protein, and *MLF1*, encoding for a cytoplasmic protein. The *NPM1-MLF1* fusion has been linked to increased BM apoptosis.[141,142]

3q21–q26 rearrangements

These consist of inv(3)(q21q26) and t(3;3)(q21;q26) and other 3q26 rearrangements. Typically associated with rapidly evolving MDS, they are frequently found in therapy-related cases. Both the former rearrangements deregulate expression of at least two genes, i.e. MECOM

(*MDS1 and EVI1* complex locus) at 3q26 and ribophorin I (*RPN1*) at 3q21. 3q21 rearrangements predict poor response to chemotherapy and short survival and are associated with marked dysmegakaryopoiesis (micromegakaryocytes) in 90% of cases. The platelet count is normal in 50% of cases and even increased in 20% of 3q21 abnormalities.[143,144] Both 3q21 and 3q26 breakpoints may be involved in other recurrent rearrangements in MDS. Among them, the t(1;3)(p36;q21) typically occurs in MDS with trilineage dysplasia, especially dysmegacaryocytopoiesis and poor prognosis. *RPN1* is involved at 3q21 and the putative oncogene *MEL1* (MDS/EVI1-like gene 1) at 1p36.3. Since *MEL1* and *MECOM* are highly similar and t(1;3) and inv(3)/t(3;3) both result in RPN1-driven ectopic *MEL1* or *MECOM* expression, they are hypothesized to share a common molecular mechanism. The t(3;21)(q26;q22), generating the *RUNX1-MECOM* fusion, is typically associated with t-MDS/t-AML arising after exposure to radiation, epipodophyllotoxins and anthracyclines. Cases usually show multilineage dysplasia but do not have thrombocytosis.[145] Like other 3q26 rearrangements, t(2;3)(p15–22;q26), another recurrent translocation involving 3q26, causes *MECOM* ectopic expression. The translocation fuses *MDS1* and the 5′ region of *EVI1* from the *MECOM* locus at 3q26 with *BCL11A, THADA* and other loci at 2p16.1–21. In a few cases, accompanying cryptic deletions distal to the 3q26 breakpoint were detected. As with any other 3q26 rearrangement and *MDS1/EVI1* ectopic expression, clinical outcome is severe.[146,147]

AML-associated translocations

Typical balanced translocations associated with eponymous AML subtypes such as t(8;21)(q22;q22), (15;17)(q22;q21) and inv(16)(p13q22) can be found in cases with less than 20% of bone marrow blasts. Nevertheless, these cases should be diagnosed as AML rather than MDS.[4] Other chromosome rearrangements, such as the t(6;9)(p23;q34), *MLL*/11q23 translocations and *NUP98* translocations, when found in cases with <20% blasts, predict a rapid evolution towards a frank leukaemia. The t(6;9)(p23;q34) was associated with a younger age of patients whereas, unlike AML, it did not appear to impact on prognosis if compared with the overall outcome of RAEB and RAEB-T.[148] *MLL* translocations are typically associated with t-MDS/t-AML, which develops after a relatively short period from treatment with topoisomerase II inhibitors for a previous neoplasia.[149]

NUP98/11p15 translocations

At least 30 different partners have been reported for the *NUP98* gene (11p15) in MDS (Fig. 1.4), AML, MPN and T-ALL. Despite the variety of *NUP98* partner genes, they fall into two main categories: homeodomain (*HD*) and *non-HD* genes. The chimeric transcripts encode for fusion proteins that juxtapose the NUP98 N-terminal GLFG repeats to the partner gene C-terminus. Remarkably, several translocations were found in patients with t-MDS/t-AML, suggesting that genotoxic chemotherapeutic agents can play a role in generating chromosomal rearrangements involving *NUP98*. *NUP98-NSD1* is a cryptic change frequent in paediatric AML but rare in adult MDS.[150–153] In NUP98-positive cases, events cooperating with NUP98 in leukaemogenesis were identified as *FLT3*-ITD and *WT1* and *KIT* mutations.[154]

Complex karyotypes

A karyotype with multiple abnormalities, termed complex, is seen in up to 20% of MDS and more than 50% of t-MDS and is associated with poor prognosis and a median survival time of less than 1 year. Most current

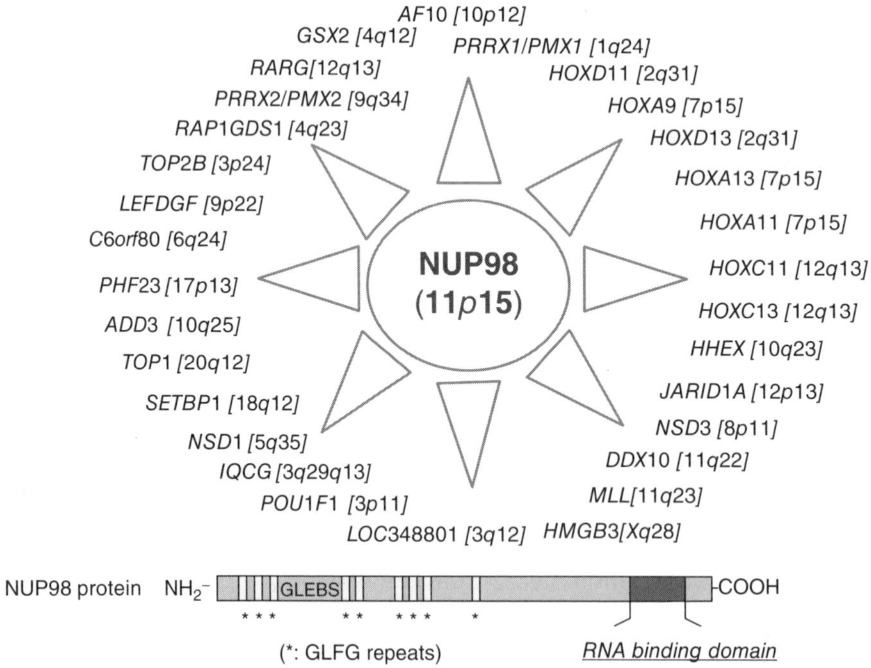

Figure 1.4 *NUP98* partner genes (top) and schematic representation of the NUP98 protein (bottom).

clinical guidelines agree that the presence of more than three numerical and/or structural changes is the threshold for poor prognosis.[75] Exposure to mutagens and/or radiation can be found in the medical history of MDS patients with complex karyotypes,[155] which often include –7/7q– and –5/5q–. In around 40% of MDS/AML with –5/5q– and complex karyotypes NPM1, mapping at 5q35, is deleted, although it is never lost in isolated 5q–.[156] TP53 deletions and/or mutations, seen in more than 90% of cases, characterize patients with extremely poor prognosis.[156,157] However, a monosomal karyotype, i.e. at least two clonal autosomal monosomies or one monosomy associated with a structural change, may possibly be prognostically worse than a complex karyotype in MDS.[158–161]

Chromosome 5q deletions

Chromosome 5q deletions, with or without additional karyotypic abnormalities, vary in size. In *de novo* MDS, 10–15% of patients display a 5q loss, making it the most frequent recurrent cytogenetic abnormality in MDS.[162] As the most specific aberration in MDS, an isolated 5q deletion is recognized as a distinct entity, MDS del(5q) or '5q– syndrome,' in the WHO 2008 classification[4] (described below) (Fig. 1.5). In MDS induced by cancer treatment with alkylating agents or radiotherapy, a 5q deletion also is seen in ~80% of patients with complex karyotypes, with or without chromosome 7 abnormalities.[163] Outcomes in MDS patients with 5q deletion vary greatly in terms of overall survival (OS) and risk of transformation to acute myeloid leukaemia (AML), depending on additional associated changes.[162] It is well established that patients with del(5q) plus two or more chromosomal abnormalities (i.e. complex karyotypes) have poor outcomes. OS ranges from 27 to 145 months in isolated del(5q) compared with 6–33 months for del(5q) in a complex karyotype. Risk of AML transformation is greater in the latter category, with almost all patients developing AML within 5 years of diagnosis of MDS.[132,162,164–167] The prognostic value of a single chromosomal abnormality (excluding chromosome 7 anomalies) in addition to del(5q) is still a matter of debate. Some evidence suggests that patients with del(5q) plus one additional abnormality have significantly worse prognosis than patients with isolated del(5q)[164] (Fig. 1.3), whereas other studies found no significant differences in OS.[132,162,164–167] Response rates to treatment indicate the profound biological and clinical differences in MDS with isolated 5q– and MDS with 5q– in complex karyotypes. Lenalidomide successfully elicits

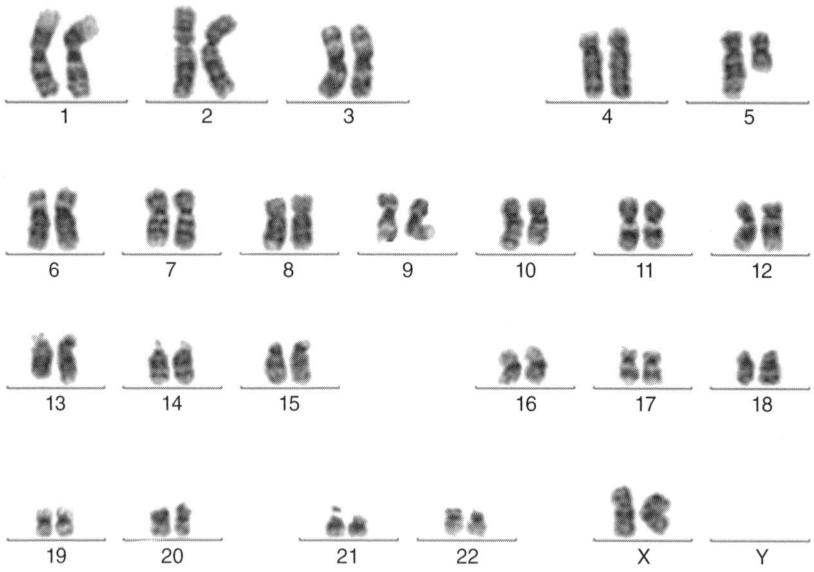

Figure 1.5 Example of G-banded karyotype showing a del(5q).

erythroid response and transfusion independence in up to 70% of cases of 5q– syndrome, but in only 20% of cases with 5q– in complex karyotype.

Isolated 5q deletion: 5q– syndrome and other del(5q) MDS

The '5q– syndrome,' the only MDS to be recognized as a separate category in the WHO 2008 classification,[4] was first reported by Van den Berghe et al. in 1974.[168] They described the isolated interstitial deletion at the long arm of chromosome 5 as a distinct entity that was characterized by refractory macrocytic anaemia, normal or high platelet count, hypolubulated micromegakaryocytes, hypoplastic erythropoiesis, female gender preponderance and relatively good prognosis. Subsequent reports confirmed a female preponderance (7:3 females:males) and good prognosis (median survival of 145 months) with ~10% of patients transforming to acute myeloid leukaemia.[4,169–171] Remarkably, the 5q– syndrome can often be predicted on the basis of bone marrow morphology alone.[171] It is worth noting that some MDS with del(5q) as the sole anomaly e.g. refractory anaemias (RA) and refractory anaemia with excess of blasts (RAEB)[169] do not satisfy 5q– syndrome criteria. An isolated del(5q) has occasionally been reported in myeloproliferative neoplasms (MPN) and acute myeloid leukaemia (AML). Recent evidence suggests that an isolated del(5q) other than 5q– syndrome is an adverse prognostic factor in myeloid malignancies, regardless of the clinical-pathological disease phenotype.[172]

When the del(5q) is the sole karyotypic abnormality, cryptic genomic imbalances are very rare.[78,173,174] Both centromeric and telomeric boundaries are different in isolated del(5q) compared with those in 5q– in complex karyotypes.[175,176] Mutations in *JAK2* and *MPL* genes have rarely been reported, in 6% and 3% of 5q– syndrome, respectively, without consequences on phenotype or prognosis.[172] The del(5q) and the JAK2^{V617F}, however, were shown to be independent clones from a single patient.[177] IDH mutations are more frequent in other myeloid malignancies with del(5q).[172] Interestingly, 18% of patients with low-risk, early-stage MDS and isolated del(5q) carried small subclones bearing p53 mutation which subsequently expanded with leukaemic evolution.[178] The p53-positive small clones may expand at time of disease progression.[179]

The 5q– syndrome is well established as a disorder of the human haematopoietic stem cell (HSC). In a subset of patients with 5q–, B-cells may be involved, suggesting HSCs with combined lympho-myeloid potentialities.[180,181] Investigation into del(5q) MDS patients who received lenalidomide showed that virtually all CD34$^+$, CD38$^-$/low, CD90$^+$ stem cells and CD34$^+$, CD38$^+$ progenitor cells had the 5q deletion before treatment. Despite complete remission, however, a fraction of the minor, quiescent CD34$^+$, CD38$^-$/low, CD90$^+$ del(5q) stem cells remained resistant and expanded considerably at the time of disease progression. This CD34$^+$ HSC subset plays a role in leukaemia initiation.[182]

Decades of research has been focused on analysing chromosome 5q deletions in 5q– syndrome and other MDS with del(5q) to define a commonly deleted region (CDR) and to identify the gene(s) responsible for the disease. Two distinct CDRs were identified and linked to differences in disease behaviour (Fig. 1.3).[170] The proximal CDR (CDR1) at 5q31, associated with advanced MDS and AML, includes a cytokine cluster (i.e. *IL3, IL4, IL5, IL13, GM-CSF*), *CDC25c, PP2Acalfa, CTNNA1* and *ERG1* (see below).[183,184a] A more distal CDR (CDR2) in 5q– syndrome was narrowed to an ~1.5 Mb interval at 5q32–q33 flanked by the DNA marker D5S2413 and the *GLRA1* gene.[184b] CDR2 encompasses around 40 genes including *SPARC, mir-143, mir-145, mir-146* and *RPS14*. An absence of detectable mutations in any of the 40 genes is in keeping with haploinsufficiency (i.e. the dosage effect resulting from loss of a single allele) as a major pathogenetic event.[171]

Whether phenotypic heterogeneity in MDS patients with del(5q) is the inevitable consequence of the number and type of haploinsufficent genes inside the deleted region of individual patients has not been established. Additional genetic and epigenetic abnormalities, and also microenvironment interactions with clonal cells, have been hypothesized.[185] However, *in vitro* and/or *in vivo* models provided functional evidence of the role of individual genes, located at 5q, in haematopoietic stem cell function and leukaemic progression (Table 1.4).[185] Next-generation sequencing in MDS with del(5q) evolving to AML showed mutations of *RYR1* and *TP53* only at progression to AML.[186] The critical role of *TP53*, however, is better illustrated by the finding of clonal hyperexpression and mutation of the gene in around 20% of low-risk MDS evolving to acute myeloid leukaemia who fail to respond to lenalidomide.[178,187]

Candidate genes at 5q

APC

The adenomatous polyposis coli (*APC*) tumour suppressor gene is located at chromosome band 5q23. A single *APC* allele is lost in over 95% of patients with myeloid neoplasms and del(5q). *APC* encodes for a multifunctional protein involved in the regulation of, for example, the Wnt signalling pathway via its ability to control β-catenin degradation, cell migration, cell adhesion, spindle assembly and chromosome segregation. Since *APC* loss of function underlies colorectal cancer initiation and progression[199] it may therefore contribute to MDS by haploinsufficiency. In two murine models, *APC* haploinsufficiency led to ineffective haematopoiesis and altered haematopoietic stem cell function, inducing an MDS/myeloproliferative phenotype.[188,189]

IRF-1

The *IRF1* gene maps to 5q31.1, upstream of the proximal del(5q) CDR. It encodes the interferon regulatory factor-1, a transcription factor controlling functions such as immune response regulation, cell growth, cytokine signalling and haematopoietic development. Interestingly, in $IRF1^{-/-}$ mice, immature granulocytic precursors were increased, suggesting a role in early myelopoiesis.[190]

PP2Acα and *CDC25c*

Protein phosphatase 2a (*PP2A*) dephosphorylates many substrates and plays a role in the broad cellular regulatory functions of many signalling and metabolic pathways. A holoenzyme, it is composed of diverse

Table 1.4 List of 5q genes with a role in haematopoietic stem cell function and leukaemic progression.

Gene	Locus	CDR1	CDR2	Murine model phenotype
APC	5q22	No	No	Anaemia, macrocytosis, monocytosis[188,189]
IRF-1	5q31.1	No	No	Impaired granulocytic differentiation and maturation, reduction in granulo-monocytic progenitors[190]
*EGR1**	5q31.2	Yes	No	Development of T-cell lymphoma or MDS/MPD phenotype characterized by leukocytosis, anaemia, thrombocytopenia, with dysplastic neutrophils[191]
DIAPH1	5q31.3	Yes	No	Development of age-dependent myeloproliferative defects: splenomegaly, fibrotic and hypercellular bone marrow, extramedullary haematopoiesis in both spleen and liver, presence of immature myeloid progenitor cells with high nucleus-to-cytoplasm ratio, dysplastic erythrocytes[192]
CSF1R	5q32	No	Yes	Mononuclear phagocytic deficiency, increased primitive progenitor cell frequency[193]
miR-145/miR-146a	5q32/5q33.3	No	Yes/adjacent to distal border of CDR	Thrombocytosis, neutropenia, megakaryocytic dysplasia[194]
SPARC	5q33	No	Yes	Thrombocytopenia, reduced erythroid colony formation[195]
RPS14 and *RBN22*	5q33	No	Yes	Development of a 5q– syndrome phenotype: macrocytic anaemia, monolobulated megakaryocytes[196]
NPM1	5q35.1	No	No	MPD/AML phenotypes,[197] MDS phenotype[198]

*No haematological phenotype; T-cell lymphoma or MPD developed after treatment with alkylating agents.

subunits. The catalytic subunit alpha isoform (*PP2ACα)* and its regulatory subunit (beta isoform PR 52) are located at 5q31.1 and 5q32, respectively. PP2A activates the G2-M cell cycle transition by dephosphorylating an inhibitory serine residue on CDC25c, a serine/threonine and tyrosine phosphatase which the gene also maps at 5q31.2. *CDC25c*, in turn, dephosphorylates cyclin-B/cyclin-dependent kinase (CDK)-1, allowing mitotic start.[200] Lenalidomide directly inhibits *CDC25c* activity and indirectly suppresses *PP2A* activity, inducing G2/M arrest in del(5q) myeloid progenitors. Thus, PP2A and CDC25c inhibition emerge as key effectors of del(5q) MDS sensitivity to lenalidomide.[201]

EGR1

The early growth response 1 (*EGR1*) gene maps to 5q31.1 and encodes for a transcription factor with a zinc finger DNA binding domain.[170] Although not itself transforming, $EGR1^{+/-}$ and $EGR1^{-/-}$ mice developed a, MPN-like syndrome when treated with *N*-ethylnitrosourea (ENU) with shorter latencies and higher frequencies than wild-type *EGR1*.[191]

CTNNA1

CTNNA1 maps at 5q31.2, downstream to the proximal del(5q) CDR. Known as alpha-catenin, *CTNNA1* encodes for a transmembrane glycoprotein that is involved in establishing tight epithelial cell connections via intracellular domain anchorage to the actin cytoskeleton. The tumour suppressor role of CTNNA1 was reported in patients with intestinal tumours.[202] *CTNNA1* expression was reduced in del(5q) leukaemic cells and in the HL-60 cell line harbouring a chromosome 5q interstitial deletion. Interestingly, methylation and histone deacetylation may suppress the retained allele in MDS with del(5q) and in high-risk MDS and AML.[203]

DIAPH1

DIAPH1 encodes for the mammalian diaphanous (mDia)-related formin mDia-1 protein. It maps at 5q31.3 immediately upstream of the del(5q) distal CDR and participates in many cytoskeletal remodelling events, including cytokinesis, vesicle trafficking and filopodia assembly. Knockdown experiments in Drf1, the murine homologue of *DIAPH1*, resulted in haematopoietic progenitor defects mimicking a myeloproliferative disorder.[192]

miR-145 and *miR-146*

miR-145 and miR-146a, which are abundant in haematopoietic/stem progenitor cells, are implicated in features of the 5q– syndrome phenotype, namely thrombocytosis and hypolobulated megakaryocytes.[194] Both are deleted in virtually all 5q– syndromes as miR-145 maps within the typical 5q– syndrome CDR whereas miR-146a is located adjacent to its distal boundary. Indeed, miR-145 and miR-146a were both down-regulated in the $CD34^+$ cells of three patients with 5q– syndrome.[194] miR-145 targets Toll–interleukin-1 receptor domain-containing adaptor protein (*TIRAP*) and miR-146a the tumour necrosis factor receptor-associated factor 6 (*TRAF6*). Knockdown of both miR-145 and miR-146a, or enforced *TRAF6* expression, in mouse haematopoietic stem cells resulted in thrombocytosis, mild neutropenia and megakaryocytic dysplasia, as observed in patients with 5q– syndrome.[194] However, miR-145 expression levels are discrepant in patients with 5q– syndrome.[204–206] Notably, p53 up-regulates miR-145.[207,208]

SPARC

The secreted protein acidic and rich in cysteine/osteonectin/BM40 (*SPARC*) gene is located on 5q31.3–q32. It encodes for a protein that regulates extracellular matrix components. Deletion of one allele of *SPARC* may provide a malignant clone with a stromal advantage. *SPARC* expression is up-regulated in del(5q) erythroblasts after treatment with lenalidomide.[209] In addition, homozygous *SPARC* knock-out mice developed several features, including anaemia and thrombocytopenia, suggesting a cooperative effect on erythropoiesis and anaemia as seen in del(5q) MDS.[195]

RPS14

Mapping at 5q32, the ribosomal protein S14 gene encodes a 40S ribosome subunit protein. In yeast its homologue is essential for endonucleolytic cleavage, which removes 200 nucleotides from the 3′ end of 20S pre-rRNA and generates mature 18S rRNA and functional 40S ribosomal subunits.[210] In 2008, Ebert et al.[211] used RNA-mediated interference (RNAi) to reproduce a disease phenotype in normal haematopoietic progenitor cells. Forced expression rescued the phenotype in patient-derived bone marrow cells. A block in pre-ribosomal RNA processing in *RPS14*-deficient cells was identified. Interestingly, the block in ribosome biogenesis in 5q– syndrome cells was functionally equivalent to the defect in Diamond–Blackfan anaemia (DBA), a congenital disorder characterized by anaemia, erythroblastopenia and increased malignancy caused by mutations in several ribosomal protein genes. The role of *RPS14* in del(5q) MDS pathogenesis was reinforced by evidence from a murine model of the human 5q– syndrome. Haploinsufficiency in mice induced macrocytic anaemia with dysplastic features, monolobulated megakaryocytes and hypocellular bone marrow that was rescued by p53 inactivation.[196] This study first suggested that a p53-dependent mechanism underlay the 5q– syndrome and a link between the typical 5q– syndrome ribosome biogenesis dysregulation and the p53 checkpoint was later demonstrated. Indeed, disruption of ribosome assembly released free ribosomal proteins. These interacted with the product of the human homologue of the mouse double minute 2 gene (*MDM2*). MDM2 is an E3–ubiquitin ligase that directly binds TP53, leading to its proteasomal degradation. Therefore, interaction of free ribosomal proteins with MDM2 prevented TP53 binding, inhibited ubiquitination and promoted protein stabilization.[212,213]

Consistent with the 5q– syndrome phenotype, RNAi-mediated *RPS14* gene silencing in human haematopoietic stem cells was followed by selective p53 activation in erythroid progenitor cells, resulting in erythroid-specific p21 accumulation, cell cycle arrest and apoptosis.[214]

CSNK1A1

Strong evidence has been accumulated for an important pathogenetic role of this gene in del(5q).[215] *CSNK1A1* mutations were identified in ~7% of del(5q) cases. Interestingly, mutant expression induced cell cycle progression of haematopoietic stem cells in mice.

NPM1

The *NPM1* gene maps at chromosome 5q35.1 and encodes a highly conserved phosphoprotein of around 37 kDa. NPM protein resides in nucleoli, although it shuttles rapidly between nucleus and cytoplasm, participating in many cellular processes. These include pre-ribosomal particle transport and ribosome biogenesis, response to stress stimuli such as UV irradiation and hypoxia, maintenance of genomic stability through control of cellular ploidy, DNA repair processes and DNA transcription regulation through chromatin condensation/decondensation modulation. NPM regulates the activity and stability of crucial tumour suppressors such as p53 and ARF.[216] Bone marrow dysplastic features and chromosome instability were found in mouse models with *NPM1* haploinsufficiency.[197,198] This is in keeping with human high-risk MDS and AML in which *NPM1* loss is significantly associated with karyotype complexity (i.e. chromosome markers and aneuploidies).[156,217,218]

Somatic mutations

In addition to cytogenetic abnormalities, a range of genetic and epigenetic lesions are found in patients with MDS. To date, only one mutation has been associated with a particular MDS subtype, and this involves *SF3B1* in MDS with ring sideroblasts.[219] The majority of lesions are shared with acute myeloid leukaemia and myeloproliferative neoplasms. Several somatic mutations are predictive of overall survival and progression to acute myeloid leukaemia, although they are yet to be included in any of the risk stratification systems. Commonly mutated gene types in MDS include

1 oncogenes and tumour suppressor gene;
2 genes involved in epigenetic modification (CpG island methylation or histone modification);
3 genes encoding for spliceosome machinery.

Oncogenes and tumour suppressor genes

RUNX1

The human runt-related transcription factor 1 gene (*RUNX1*, previously termed *CBFA2*, *AML1* or *PEBP2aB*) is located on chromosome 21q22, spans 260 kb and is encoded by 10 exons that produce three transcript variants by alternative splicing. The encoded protein is one of the conserved core binding transcription factor (CBF) alpha subunits. This transcription factor is a heterodimeric protein composed of three distinct DNA-binding CBFα subunits (*RUNX1*, *RUNX2* and *RUNX3*) and a common invariable non-DNA-binding subunit (CBFβ). The RUNX1 protein consists of two well-defined domains: a highly conserved N-terminal RUNT domain mediates DNA binding and heterodimerization with the non-DNA binding regulatory CBFβ subunit and a transactivation domain (TAD) followed by a highly conserved five amino acid sequence (VWRPY) at the C-terminus.[220] RUNX1 plays an essential role in specifying the haematopoietic stem cell and is required for haematopoiesis. Molecular studies in $runx1^{-/-}$ mice showed arrested development and death at an early embryonic stage due to blocked haematopoiesis.[221]

Two promoters regulate RUNX1 expression transcription: RUNX1c, the longest isoform, is transcribed from a distal promoter in exon 2, whereas RUNX1a and RUNX1b are transcribed from the proximal promoter in exon 4.[222] Functional studies in mice suggested that relative expression of RUNX1a and RUNX1b splice isoforms affects cell fate because RUNX1a expression increases the self-renewal capacity of haematopoietic stem and progenitor cells, whereas RUNX1b expression promotes differentiation of these cell types. The balance between these antagonist effects underlies physiological regulation of haematopoietic ontogeny.[223]

RUNX1 is a DNA-binding transcription factor which, in the myeloid lineage, directly binds several genes related to myeloid growth factor signalling, such as *IL-3*, *GM-CSF*, the *M-CSF* receptor and *c-Mpl*,[224–227] thus regulating expression in immature myeloid cells.[228,229] Chromosomal rearrangements involving the *RUNX1* gene, such as t(8;21) *RUNX1-RUNX1T1*, t(3;21) *RUNX1/EAP* and t(3;21) *RUNX1/MECOM(EVI1)* are typically associated with acute myeloid leukaemia and rarely described in MDS. Nevertheless, RUNX1 is a promiscuous gene recombining with multiple genomic sites. A translocation t(X;21)(p22.3;q22.1) with *RUNX1/FOG2* rearrangement was described in one case of refractory anaemia with excess of blasts.[230] Conversely, loss-of-function *RUNX1* mutations are far more common in MDS.[231–233]

RUNX1 gene mutations were detected in about 10–20% of MDS, particularly in radiation-associated and therapy-related MDS/AML. At the protein level, these mutations are distributed throughout the full length of the protein. N-terminal mutations involve, almost exclusively, the RUNT domain; half are missense mutations replacing amino acid residues in direct contact with DNA. Other frameshift or nonsense mutations abolish the physiological function of the RUNT domain. Although these N-terminal mutations are associated with low or no DNA-binding ability, they can still bind the CBF complex B subunit. N-terminal mutations predominantly affect younger patients with a history of radiation and/or chemical exposure, who present with hypocellular bone marrow.

C-terminal mutations are seldom missense. The majority are the so-called 'chimera-like,' frameshift mutations that result in truncation of the original RUNX1 protein followed by a stretch of additional amino acids originating from the new reading frame. *RUNX1* C-terminal mutations display a high cellularity in bone marrow and peripheral blood and increased dysplastic megakaryocytes, resulting in a myelodysplastic/myeloproliferative disease.[234] Recent studies have shown that haploinsufficiency of *RUNX1* activity leads to thrombocytopenia, whereas an almost complete loss of *RUNX1* activity leads to increased genomic instability and predisposes to leukaemia.[235]

Germinal heterozygous *RUNX1* loss-of-function mutation causes the familial platelet disorder with predisposition to acute myeloid leukaemia (AML) (FPD/AML). Patients have moderate thrombocytopenia and impaired platelet function, usually show progressive pancytopenia and dysplastic features with age and share a markedly increased risk of developing AML during their lifetime with MDS patients.[236] All families showed loss-of-function mutations affecting the RUNT domain. Evidence that RUNX1 haploinsufficiency causes FPD/AML came from observations in families with a *RUNX1* large intragenic deletion and a fully functional residual allele. However, current evidence indicates that complete RUNX1 loss of function might be required for or, at least, contribute to AML progression, as a significant proportion of sporadic AML cases have biallelic *RUNX1* mutations. *RUNX1* gain of function by amplification events was detected in primary paediatric ALL.

A multivariate analysis that included risk stratification using IPSS showed that *RUNX1, ASXL1, TP53, EZH2* and *ETV6* mutations were independent predictors of poor overall survival and were mainly associated with high-risk MDS subtypes (RAEB, RAEB-T, CMML).[237] *RUNX1, TP53* and *NRAS* mutations correlated with severe thrombocytopenia and elevated blast count, but not with neutropenia or anaemia.[237,238] Loss-of-function *RUNX1* mutations in mice affect megakaryocyte but not granulocyte or erythroid differentiation, which is consistent with the MDS phenotype in humans with *RUNX1* mutations. The close association of *RUNX1* mutations with radiation suggests that either the *RUNX1* gene is particularly sensitive to DNA damage following radiation or that pre-existing *RUNX1* mutations may predispose patients to MDS following DNA damage.[239] Patients with *RUNX1* mutations at diagnosis also have a higher frequency of −7/7q deletion than those without.[237]

NRAS and *KRAS*

The RAS family, namely the proto-oncogenes with intrinsic GTPase activity *H-ras, K-ras* and *N-ras,* is the most extensively studied gene family in MDS. Ras proteins are signal switch molecules that cycle between active GTP-bound and inactive GDP-bound conformations (Ras-GTP and Ras-GDP). Once activated by a cell surface receptor, RAS proteins mediate signal transduction across the plasma membrane and also the intracellular pathways that underlie proliferation and differentiation. Two protein families regulate the Ras protein activation status. Negative regulators are GTPase-activating proteins (p120GAP, neurofibromin and GAP1) that stimulate hydrolysis of bound GTP to GDP. Positive regulators are guanine nucleotide exchange factors (Sos and Ras-GRF) that stimulate the exchange of Ras-bound GDP for fresh GTP from the cytosol.[240]

In the active conformation, GTP-bound Ras binds to and activates effector proteins such as Raf kinases, PI-3 kinase and Ral-GDS. In one well-defined Ras signalling pathway, activation of the serine–threonine kinase Raf protein family activates the ERK MAP kinase protein kinase cascade. The RAS-to-MAPK pathway appears to be essential for mitogenic signalling.[241] Ras proteins shuttle between the Golgi apparatus and plasma membrane. Produced as cytoplasmic proteins, they require several post-translational modifications such as prenylation, proteolysis, carboxymethylation and palmitoylation to acquire full biological activity.[242]

In RAS family molecules, primary amino acid sequence alignment indicates four domains: (i) at the N-terminus, 85 amino acids are identical in H-, K- and N-ras; (ii) another 80 amino acids, showing 70–80% conservation within RAS proteins; (iii) the rest of the molecule, except for the last four amino acids, is a hypervariable region; (iv) the highly conserved C-terminal motif CAAX (C = cysteine, A = any aliphatic residue, X = any uncharged amino acid) is the consequence of post-translational modifications.

The human neuroblastoma RAS viral oncogene homolog gene (*NRAS*, NM_002524.4) maps on 1p13.2, spans 8 kb and consists of seven exons encoding for a 189 amino acid protein. The v-Ki-ras2 Kirsten rat sarcoma viral oncogene homologue gene (*KRAS*) maps on 12p12, is spread over 35 kb and consists of six exons that can be alternatively spliced to produce a 188 or 189 amino acid protein (NM_004985, 5312 bp, and NM_033360, 5436 bp, respectively) according to whether exon 5 is retained or not.

KRAS defects are implicated in AML and juvenile myelomonocytic leukaemia (JMML). Mutations give rise to abnormal protein products that retain the active GTP-bound form, promoting constitutive activation. *RAS* mutations were detected in haematological malignancies and *NRAS* was particularly implicated in AML, chronic myeloid leukaemia and acute lymphoblastic leukaemia.[243–245] The most frequent mutation, a single base change at codon 12 (G→A transition) results in aspartic acid being substituted for glycine in the protein. Codons 13 and 61 are also frequently mutated. Codons 12 and 13 are located on the pocket that binds GTP and mutant proteins have decreased phosphatase activity. *NRAS* is mutated twice as often as *KRAS*.[246]

A study focusing on point mutations in 439 cases of myelodysplastic syndromes identified 3.6% with *NRAS* mutations and 0.9% with *KRAS* mutations. Both mutations were associated with adverse clinical features and a high risk of transformation to AML.[1]

ETV6

ETS variant gene 6 (*ETV6*, previously known as *TEL*) maps at 12p13, consists of eight exons spanning 8 kb and is a member of the ETS (E-26 transforming specific) family of transcriptional repressors that bind to DNA sequence 5′-CCGGAAGT-3′ factors. The protein (NM_001987.4, 452 amino acids) has two specific domains. At the N-terminus, the HLH domain (encoded by exons 3 and 4) is also referred to as the pointed or sterile alpha motif (SAM) domain. It underlies hetero- and homo-dimerization with other ETV6 proteins and possibly other ETS family members. At the C-terminus, the ETS domain encoded by exons 6–8 underlies sequence specific DNA binding. The ubiquitously expressed ETV6 protein is found in the nucleus. Mice carrying a targeted *ETV6* gene disruption demonstrated a role for ETV6 in early embryonic angiogenesis and adult haematopoiesis.[247] *ETV6* knock-out mice display profound embryonic lethal phenotypes. Rescue of embryonic lethality in ES cell–embryo chimeras revealed that *ETV6* was required for adult haematopoiesis, specifically erythropoiesis, myelopoiesis and lymphopoiesis.[248]

ETV6 accelerates erythroid but represses megakaryocytic differentiation.[249,250] *ETV6* is reported to be involved in over 40 translocations with various chromosome partners[251] (Fig. 1.6). However, translocations involving *ETV6* are rare in myelodysplastic syndromes and mainly concentrated in myelodysplastic/myeloproliferative disorders.[251] Moreover, *ETV6* is a typical partner of *PDGFRB*/5q33 in a distinct myeloid neoplasm.[4]

Although its role in fusion genes supports a mechanism of oncogene activation, there is mounting evidence that *ETV6* may also function as a tumour suppressor gene. In a large MDS study, *ETV6* mutations were present in 2.7% of patients and were associated with decreased overall survival.[1]

Germline mutations in *ETV6* have been reported in a number of unrelated individuals, defining a new hereditary syndrome. Functional studies have shown that these mutations have an impact in haematopoiesis and represent predisposing conditions for cytopenia and malignant transformation.[252]

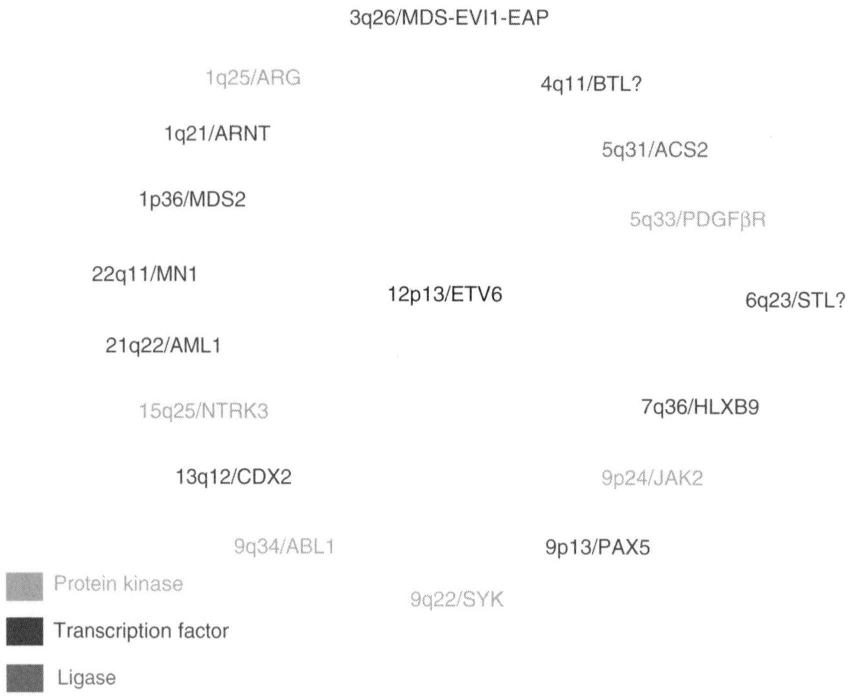

Figure 1.6 Example of *ETV6* partner genes shaded according to their function.

TP53

Human tumour protein 53 (TP53), at chromosome 17p13.1, is an 11-exon gene spanning over 19,149 kb (GenBank Accession Number: NC_000017). This gene encodes for 12 different isoforms because of alternative splicing, alternative promoter usage and alternative initiation of translation.[253] TP53 is a 53 kDa phosphoprotein consisting of four main functional domains (UniProtKB P04637 – P53_human; www.uniprot.org/uniprot/P04637) (Fig. 1.7):

1 an N-terminal transactivation domain (TAD) that is responsible for TP53 transcriptional function (amino acids 17–56) and a proline-rich SH3 target region;

2 a DNA-binding domain (DBD) (amino acids 102–292), also known as the 'core' domain that is responsible for the specific DNA-binding ability of p53;

3 an oligomerization domain (OD) (amino acids 325–356) that is responsible for TP53 quaternary structure;

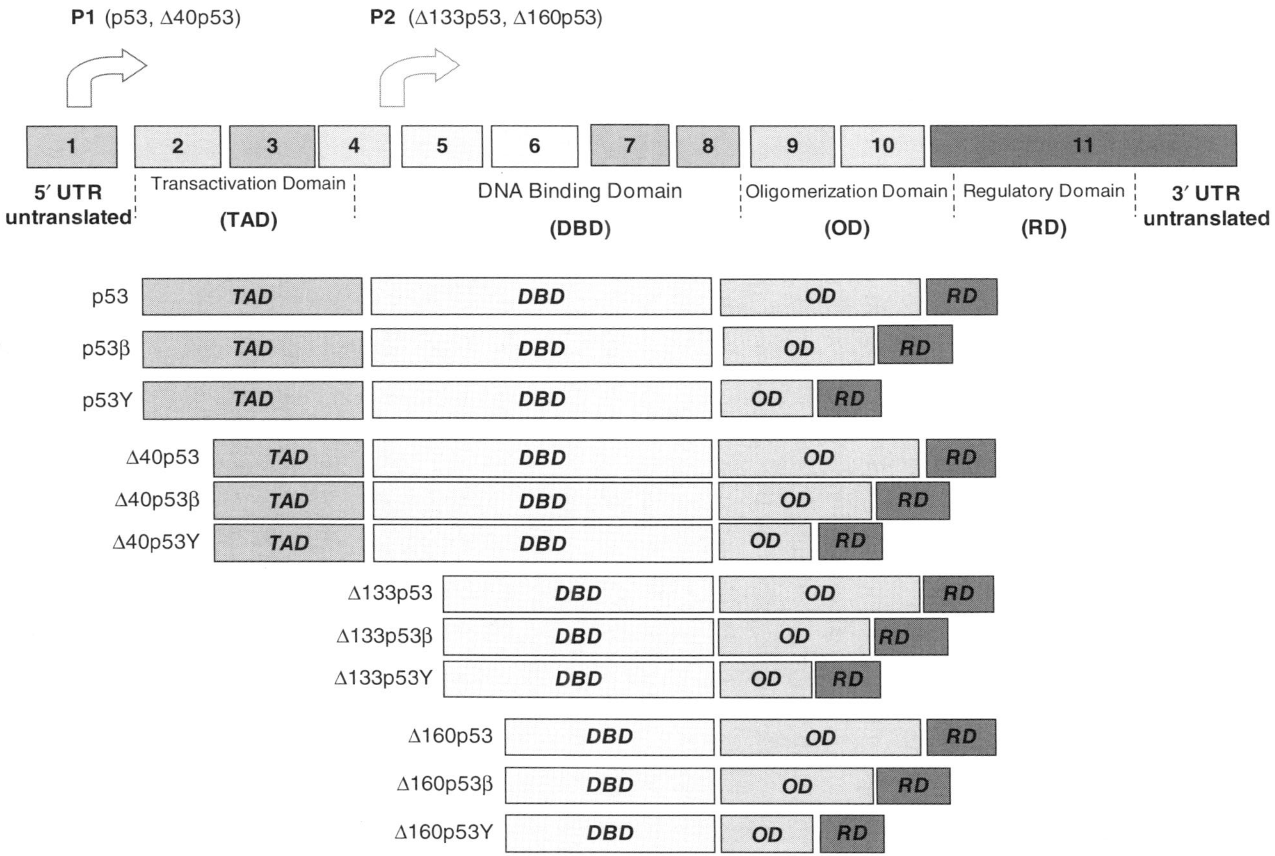

Figure 1.7 Structure of the TP53 protein showing the four main functional domains: (i) N-terminal transactivation domain (TAD), (ii) DNA-binding domain (DBD), (iii) oligomerization domain (OD) and (iv) C-terminal regulatory domain (RD).

4 a C-terminal regulatory domain (RD) (amino acids 368–387) that seems to have a negative effect on specific DNA binding of the full-length protein. It is connected to the tetramerization domain through a basic linker region that contains a nuclear localization sequence.

TP53 was originally isolated as a cellular partner of simian virus 40 (SV-40)-derived tumour antigens[254,255] and later shown to be an important tumour suppressor.[256,257] The encoded protein is a transcription factor[258,259] that binds directly and specifically as a tetramer to target sequences of DNA through p53-responsive elements (p53REs)[259,260] to transactivate several genes involved in p53 tumour suppressor activities such as p21[261] (cell-cycle arrest), BCL2 binding component 3 (PUMA)[262] and SHISA5 (SCOTIN)[263] involved in apoptosis.

It mediates DNA damage responses to a variety of cellular stresses, inducing cell-cycle arrest,[264] senescence[265] and apoptosis[266,267] in order to maintain genomic stability.[268]

TP53 is often described as the 'guardian of the genome' because of its key role in cell-cycle arrest and in inducing apoptosis when DNA becomes extensively damaged. Five functional pathways involving p53 have been schematically grouped.[269] There are various physical and chemical causes of DNA damage that is detected by a dedicated set of proteins and then repaired by specialized enzymes. The p53 protein seems to be involved in multiple types of DNA damage that involve an active repair system.[270,271] The levels of the p53 protein are predominantly regulated by its proteolytic turnover. In normal unstressed cells, its level is down-regulated via the binding of proteins such as MDM2, COP1, PIRH2 or JNK that promote p53 degradation via the ubiquitin/proteasome pathway.[272] Once p53 is activated in response to a stress signal, it gains the ability to bind to p53-responsive DNA sequence elements in the genome (RRRCWWGYYY, where R = purine, W = A or T and Y = pyrimidine).[259] TP53 regulates a set of genes that are clearly involved in cell-cycle arrest (p21, 14-3-3 sigma, GADD-45), apoptosis (intrinsic and extrinsic apoptotic pathways) and senescence.

Mutations of TP53 are often deleterious for the cell and favour tumorigenesis. Germline mutations were found in Li–Fraumeni syndrome, a rare disease in which half of the affected patients are predicted to get cancer before their thirties.[273,274] Somatic TP53 mutations are associated with a variety of human cancers. They are distributed along all the coding sequence of the gene, although several hot-spot mutations are frequently described. The hot-spot amino acids R273 and R248 directly contact the major and minor grooves of DNA. Four other hot-spot mutations (R175, G245, R249, R282) are involved in stabilizing the structure through a hydrogen-bond network. There are three major classes of mutations that prevent DNA binding: (i) mutations involved in direct DNA contact, (ii) mutations that destabilize the structural integrity of the DNA binding region and result in loss of DNA-binding affinity and (iii) mutations in the helix region that prevent cooperative binding.[275]

TP53 mutations had been extensively described in MDS and AML before the era of large-scale sequencing, although the advent of extremely sensitive techniques for whole-genome analysis allowed the detection of a higher number of cases affected. The presence of TP53 mutations confers a poor prognosis in MDS.[276] They have been significantly associated with complex karyotypes including –7/7q–, –5/5q– and 17p deletions.[1,277,278] A range of different mutations has been reported, although the majority are missense mutations in the sequence-specific DNA-binding domain of p53, between residues 102 and 292.[276]

In 30/318 (9.4%) consecutive MDS patients, TP53 mutations concentrated in isolated del(5q) (19%) and in complex karyotypes (CK) with –5/5q– (72%), with an intermediate-2 prognostic index, a low level of p53 protein expression by immunohistochemistry, high blast count and risk of leukaemic progression.[278] Jadersten et al.[178] detected mutations in bone marrow progenitors in almost one-fifth of patients with low-risk del(5q) MDS (10/55, 18%) and showed that small clones with mutations may occur at an early stage of disease, even years before disease progression, and were associated with an increased risk of leukaemic evolution.

Mutations of genes involved in epigenetic modulation

TET2

Molecular and cytogenetic approaches identified the *TET2* gene in a common 500 kb minimal deleted region in myeloid malignancies.[279] *TET2*, ten-eleven translocation 2, maps at chromosome 4q24 and encodes for two isoforms of which the longer (isoform A, NM 001127208.2) includes 11 exons and encodes a 2002 amino acid protein. *TET2* belongs to the TET family of proteins, TET1–2–3, which are Fe(II)- and α-ketoglutarate-dependent enzymes that modify DNA by hydroxylating 5-methylcytosine (5-mC) to 5-hydroxymethylcytosine (5-hmC).[279,280] 5-hmC blocks the activity of methyl-DNA-binding proteins that normally confer transcriptional silencing.[282] Additionally, 5-hmC may lead

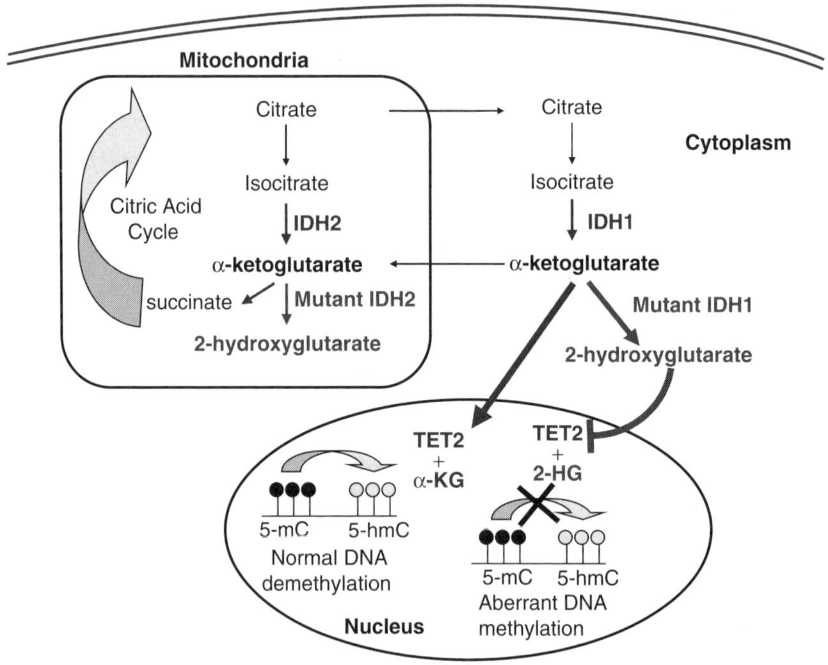

Figure 1.8 Mechanisms of action for IDH1, IDH2 and TET2.

to passive demethylation as DNMT1 is unable to recognize 5-hmC. Methylation marks are therefore lost in succeeding DNA replication cycles, as newly incorporated cytosines in the daughter strand fail to become methylated[283] (Fig. 1.8).

TET2 somatic deletions and inactivating mutations were identified in MPN and MDS. Loss of heterozygosity and microdeletions were mapped within a minimal region of chromosome 4q24.[279,284] *TET2* sequencing identified missense, frameshift and nonsense somatic mutations in 11–26% of patients with MDS, 37–44% of patients with MDS/MPN and 11–24% of patients with sAML.[283,285–288] Importantly, most *TET2* mutations are heterozygous in leukaemia and wild-type allele expression is retained. These data suggest that *TET2* can function as a haploinsufficient tumour suppressor in most patients. Biallelic *TET2* inactivation occurs in <10% of patients.[279,289]

TET2 mutations in MDS are thought to result in a loss of catalytic activity. Loss-of-function mutations are predicted to decrease global 5-hmC levels and increase 5-mC levels.[290] Several missense mutations are associated with impaired 5-hmC production and reduced 5-hmC levels *in vitro* and *in vivo*, respectively.[290] *Tet2* knock-out mice developed a disease with MDS-like features including erythroid progenitor expansion.[291] Notably, in murine models a considerable latency to overt myeloid leukaemias suggests that secondary events are required for leukaemogenesis.[126]. A study performed on a large series of patient samples showed that progression to acute myeloid leukaemia was accompanied by evolution of a novel clone or development of minor pre-existing clones often carrying additional genetic lesions or biallelic *TET2* mutations.[292]

Multiple studies have reported conflicting results on the impact of *TET2* mutations on survival[279,287,293] and on response to demethylating agents, such as 5-azacytidine. To date, no consistent impact of *TET2* mutations on survival has been observed in MDS.[287,293]

DNMT3A

DNA cytosine-5-methyltransferase 3A (*DNMT3A*) is a 26-exon gene mapping at chromosome 2p23. It encodes three isoforms: the longest is A (NM_022552.4 and NM_175629.2, which differ in the 5′ UTR), B (NM_153759.3) and C (NM_175630.1). They consist of 23, 19 and 4 exons, respectively, and encode for corresponding proteins of 912, 723 and 166 amino acids. DNMT3A, a member of the mammalian family of DNA methyltransferases (DNMTs), catalyses the addition of a methyl group to the cytosine residue of CpG dinucleotides. Since DNMT3A initiates *de novo* DNA methylation, it affects the expression of a variety of genes and also affects genome stability.[294] The DNMT3A N-terminus contains PWWP and ADD domains that recognize histone H3 lysine modifications and recruit the protein to specific gene targets (e.g. promoter region),[295] subsequently modifying gene expression.[296,297]

DNMT3A mutations were initially identified in *de novo* AML with poor prognosis, at a frequency of 22%,[298,299] and later in a subset of MDS patients (<10%), about 30% of whom developed secondary AML.[300] In high-risk MDS, 3.3% of patients bear heterozygous mutations in the DNMT3A methyltransferase domain. Identifying *DNMT3A* mutations in MDS might be relevant in prognosis as they may precede AML development.[299,300] The most frequent mutation in AML and MDS is a heterozygous missense mutation that converts arginine to histidine at position 882 (R882H). This mutation reduces DNMT3A methyltransferase activity *in vitro*.[301] It is not clear whether *DNMT3A* mutations result in a function loss or gain.[126,301,302] Frameshift and nonsense mutations occur upstream of the methyltransferase domain and are likely to be loss-of-function mutations,[300] but as missense mutations involve highly conserved residues within the *DNMT3A* methyltransferase domain, they may not be simple loss-of-function mutations but may confer a novel protein function. *DNMT3A* loss causes a defect in gene silencing that is necessary for haematopoietic stem cell renewal. Impairing *de novo* DNA methylation has been shown to block haematopoietic differentiation.[303]

The prognostic significance of *DNMT3A* mutation status in *de novo* MDS needs to be validated in large clinical studies as patients with mutations could be candidates for more intensive treatment such as allogeneic transplant[300] early in their disease course. Further functional analysis is also required; mechanistically, partial loss of DNMT3A activity is thought to lead to hypomethylation, but the role of the loci targeted by these changes in *DNMT3A*-mutant or *DNMT3A*-haploinsufficient cells remains to be delineated.

A large study performed on more than 17,000 individuals unaffected by haematological disorders showed that detectable somatic gene mutations affecting the *DNMT3A*, *TET2* and *ASXL1* genes are rare in the younger population (age <40 years). However, they become more frequent in the older population (age >70 years) and increase in a manner that is directly proportional to age. These findings support the notion that age-related clonal haematopoiesis is associated with the risk of developing haematological disorders.[338]

IDH1/IDH2

The *IDH1* gene, mapping at chromosome 2q33.3, encodes for a single 414 amino acid protein (NM_005896.2) whereas the *IDH2* gene, mapping at chromosome 15q26.1, encodes for a 452 amino acid protein (NM_002168.2). Both IDH1 and IDH2 catalyse an essential step in the Krebs cycle: conversion of isocitrate to α-ketoglutarate in an $NADP^+$-dependent manner. IDH1 is active in the cytoplasm and peroxisomes and IDH2 in the mitochondria (Fig. 1.8).

First described in metastatic colon cancer and glyomas,[304,305] *IDH1* and *IDH2* mutations were later reported in *de novo* and secondary AML (15–30%), MDS (~5%) and MDS/MPN (~9%).[306–312] Mutations involved exon 4 of both the *IDH1* and *IDH2* genes, which encodes three arginine residues (R100, R109 and R132 in IDH1 and R140, R149 and R172 in IDH2).[313] *IDH1* and *IDH2* mutations were always heterozygous and were mutually exclusive, suggesting aberrant gain of function. Distinct mutations in both enzymes (IDH1: R132 → C, H, L or S; IDH2: R172 → C, H, L or S; R140 → Q) reduced enzymatic function with respect to α-ketoglutarate (α-KG) production but increased 2-hydroxyglutarate (2-HG) production 50-fold.[311,314,315] Increased 2-HG might result in DNA hypermethylation through inhibition of multiple enzymes, such as the large family of Jumonji domain-containing histone demethylases[314] and TET enzymes[317] that require α-KG as a cofactor.[318,319] hence mutations in *IDH1* or *IDH2* seem to deregulate widely DNA and histone demethylation processes.

IDH1 mutations seem to be associated with shorter overall survival and a higher rate of transformation into acute myeloid leukaemia,[320] although a prospective clinical trial is needed to confirm these data.

ASXL1

ASXL1 (additional sex combs-like protein 1) is a 12-exon gene (81 kb) mapping at band 20q11.21. Alternative splicing results in two transcript variants (NM_015338.5 and NM_001164603.1). The longer isoform, ASXL1, a 1541 amino acid protein (170 kDa), is ubiquitously expressed and, like ASXL2 and ASXL3, is a mammalian homologue of the *Drosophila* gene *ASX.* Structurally, ASXL1 has two highly conserved regions, the N-terminal ASX homology domain (ASXN and ASXM) and the C-terminal plant homeodomain (PHD). ASXL1 belongs to the polycomb group and trithorax complex family and regulates gene expression by conformational chromatin changes (epigenetic mechanisms). At the cellular level, it interacts with many different molecules [histone acetyltransferase (SRC-1), LSD1 protein, etc.], which accounts for its pleiotropic role as either a transcriptional activator or repressor.[321,322] Although expressed in many human tissues, ASXL1 activity is of major importance for haematopoiesis and bone marrow homeostasis. *ASXL1*-deficient human $CD34^+$ stem and progenitor cells showed impaired granulo-monocyte differentiation.[323] A mouse model showed that *ASXL1* silencing cooperates with other genetic aberrations in inducing myeloid leukaemia.[322]

ASXL1 is often mutated in MDS (10–15%), MPN (10–15%), AML (40%) and CMML (40%) and in the congenital Bohring–Opitz syndrome,[324,325] but rearrangements are rare in lymphoid neoplasms.[326] In myelodysplastic syndromes, heterozygous base exchanges leading to stop codons (nonsense mutations) are frequently seen. Frameshift mutations via nucleotide insertion or deletion are also common. Mutations are usually concentrated in exon 12, around the Gly-rich domain, but there are reports of mutations in a more proximal portion to the C-terminal end.[327–329] The clinical outcome is unfavourable, with patients harbouring *ASXL1* mutations showing significantly shorter overall survival.[1,330,331]

EZH2

EZH2 (enhancer of zeste homolog 2) on chromosome 7q35–36 transcribes five splice variants; the longest isoform (isoform a, NM_004456.4) encodes a 751 amino acid protein (NP_004447.2).

The EZH2 protein is a catalytic subunit of the polycomb repressive complex-2 (PRC2), involved in transcriptional silencing. PRC2 is a highly conserved histone methyltransferase that targets lysine 27 on histone 3 (H3K27me3). It is commonly associated with the silencing of genes that are implicated in cellular development, differentiation and fundamental processes, such as fate decision, cell-cycle regulation, senescence, cell differentiation and cancer.[332] Its core enzymatic subunit mediates methylation activity by interacting with SUZ12 and EED, which are two auxiliary subunits. EZH2 contains a highly conserved C-terminal SET domain with an unusual pseudoknot structure[333] and a cysteine-rich CXC domain that is required, together with the SET domain, for histone methyltransferase activity. The N-terminus contains an EID domain that binds to the PRC2 subunit EED and two SANT domains. Domain II is required for binding SUZ12, a non-catalytic protein.[334]

The gene is mutated in ~6% of patients with MDS.[126] Missense mutations mostly target highly conserved residues located in the SANT domain II and CXC–SET domain, whereas frameshift mutations are dispersed throughout the gene. All mutations are clearly inactivating and are classified as loss of function as they reduce EZH2 catalytic activity. Whether EZH2 acts as a co-dominant tumour suppressor for myelopoiesis is still under debate. In lymphoma, a recurrent missense mutation of a single tyrosine in the EZH2 protein SET domain (Y641) appears to lead to gain of function rather than reduced H3K27 methylation. Heterozygous Y641 mutants seem to work in combination with wild-type EZH2, to increase H3K27 trimethylation, which is the functional equivalent of EZH2 over-expression. Even though a variety of EZH2 mutations, including missense, nonsense and premature stop codon, were described in MDS, the lymphoma-associated Y641 mutation, was never reported. Moreover, many other co-existing factors on chromosome 7 may contribute to leukaemogenesis in MDS patients, thus contributing to their poor prognosis. EZH2 mutations are found in both MDS and MDS/MPN and are frequently associated with uniparental disomy (UPD) of 7q or 7q36.1 microdeletions.[127,335–337] *EZH2* mutations are strongly associated with decreased overall survival (OS) in patients with MDS.[1,238]

Mutations of genes involved in the spliceosome machinery

Spliceosome proteins are essential for spliceosome complex assembly and function and provide a basic cellular mechanism for gene expression. Alternative splicing generates a large diversity of proteins from a limited set of genes. The RNA splicing complex is made of multiple snRNP protein and other protein factors which together remove introns from newly transcribed pre-mRNA. Spliceosome recognition of exon–intron boundaries, transesterification reactions and intron sequence splicing generate mature and truncated mRNA. Splicing starts with recognition of a 5′ splice site by a U1 snRNP complex followed by recruitment of U2AF1/2 (35/65) heterodimer, ZRSR2 and an SR protein, such as SRSF1 or -2, that recognized the 3′ splice site. Finally, a U2 snRNP complex replaces SF1 bound to the branch point sequence with one of its components, SF3B1, to establish a splicing complex.[339]

In MDS, mutually exclusive spliceosome mutations mostly affect *SF3B1*, *U2AF1*, *SRSF2* and *ZRSR2* in the E/A splicing complex[340] (Table 1.5). Compromised function of the E/A complex, the hallmark of this unique category of myeloid neoplasms, plays a central role in the pathogenesis of myelodysplasia. The close relationship between the mutation type and unique disease subtypes is evidence of a pivotal role in MDS, but the functional link between the abnormal RNA species splicing and MDS phenotype is still unclear.[341]

U2AF1

U2AF1 (U2 small nuclear RNA auxiliary factor 1; 21q22.3) belongs to the SR family of splicing factors and has nine exons; alternative splicing results in multiple transcript variants. U2AF1 plays a critical role in constitutive and enhancer-dependent RNA splicing.[342] Some cases of MDS, AML and MDS/MPN were reported to carry heterozygous missense mutations, affecting codon 34 (S34) or codon 157 (Q157) of the *U2AF1*

Table 1.5 Spliceosome components mutated in MDS.

Gene	Chromosome location	Mutations in MDS (%)
U2AF1	21q22.3	6–20
ZRSR2	Xp22.1	3
SRSF2	17q25.1	8–14 (75 MDS-RS, 40 CMML)
SF3B1	2q33.1	14–28 (75–80 MDS-RS, 68–82 RSRS, 57–76 RCMD-RS)

gene. The S34 and Q257 residues are located within zinc finger domains that may be important for RNA binding activity, as they recruit the RNA 3′ splicing site. Pre-mRNAs may influence U2AF1 function and specificity. The nucleotide sequence of this immature messenger RNA may be important in determining the genes altered in cells expressing mutant U2AF1. These observations suggest that mutations are gain of function, with the mutant protein showing increased splicing activity.[342] The frequency of *U2AF1* mutations in MDS is estimated to be 6–20% but they are not significantly associated with haematological findings, FAB subtypes or karyotype and their impact on clinical outcome is controversial.

The role of *U2AF1* mutations and the exact trigger mechanisms in cancer pathogenesis are not known, but it is clear that they induce abnormal global RNA splicing.[343] *U2AF1* is implicated in 3′ SS recognition. Mutations involving this site would lead to incorrect splicing, resulting in increased production of transcript with unspliced intronic sequences. A loss-of-function mutation (*S35F*) was also identified.[341] This mutation probably contributes to MDS-related ineffective haematopoiesis and cytopenia.[341] *U2AF1* and *SRSF2* mutations are more frequent in CMML and higher risk MDS.[344]

ZRSR2

ZRSR2 (small nuclear ribonucleoprotein auxiliary factor 35 kDa subunit-related protein 2), mapping at Xp22.1, is involved in development of the splicing complex and is specifically required for pre-spliceosome assembly. The *ZRSR2* gene has 11 exons and generates only one transcript that encodes for a 428 amino acid protein. Rare mutations were found in 3% of MDS patients. Mutations were prevalent in MDS without ring sideroblasts. Specific association with IPSS risk profile or cytogenetic aberrations are unknown. Thol et al.[340] identified recurrent mutations in this gene: a frameshift mutation in codon R448, heterozygous nonsense mutations in codons R126 and R295 and a missense mutation in codon C181. These mutations involved splicing of the donor/acceptor site, which caused either premature truncation or gross structural changes, leading to loss of protein function. *ZRSR2* mutation occurs mainly in males, as *ZRSR2* appears to act as an X-linked recessive tumour suppressor gene. Many missense mutations that were found in both males and females were functional somatic changes.[341]

SRSF2

The serine/arginine-rich splicing factor 2 (*SRSF2)* gene on chromosome 17q25.1 has five exons and generates two transcript variants that differ in the 3′ UTR. The gene encodes a 221 amino acid protein, which is localized in the cell nucleus. The protein is a member of the serine/arginine (SR)-rich family of pre-mRNA splicing factors. *SRSF2* mutations were found in 8–14% of patients with MDS, in 75% of MDS with ring sideroblasts and 40% of CMML.[345] RNA splicing regulation is important for normal cell functions and, because of its role in splicing complex assembly and splice-site recognition, *SRSF2* genetic alterations may be crucial to the pathogenesis of MDS and other haematopoietic neoplasms.[346] Missense or deletion *SRSF2* mutations involve the P95 codon in a sequence between the RNA recognition motif (RRM) and the arginine–serine-rich domains[341,347] and are frequently associated with mutations of other genes (*ASXL1, RUNX1, IDH*) in MDS. *SRSF2* gene mutations are associated with older age and male gender and predict poor overall and leukaemia-free survival.[340] In a study of a large series of *SRSF*-mutated patients, the mutations were shown to be present at diagnosis rather than being acquired during disease progression.[346]

SF3B1

SF3B1 (splicing factor 3B, subunit 1) on chromosome 2q33.1 has 26 exons; alternative splicing results in multiple transcript variants encoding different isoforms. Isoform 1 is the longest, comprising all 26 exons. Isoform 2 uses an alternate splice site in the 3′ coding region to encode a shorter isoform with a distinct C-terminus that lacks exon 5. *SF3B1* encodes subunit 1 of the splicing factor 3b protein complex and is a 1304 amino acid protein that forms nuclear speckles. Splicing factor 3b, splicing factor 3a and a 12S RNA unit form the U2 small nuclear ribonucleoproteins complex (U2 snRNP). Splicing factor 3b is required for 'A' complex assembly and the 3b/3a splicing factor complex binds pre-mRNA upstream of the intron's branch site in a sequence-independent manner and may anchor U2 snRNP to pre-mRNA.[348]

SF3B1 sequencing has revealed mutations in 14–28% of MDS cases, 75–80% of MDS with ring sideroblasts (MDS-RS), 68–82% of refractory anaemia with ring sideroblasts (RARS) and 57–76% of refractory cytopenia with multilineage dysplasia with ring sideroblasts (RCMD-RS). Ring sideroblasts (RS) are recurrent findings in MDS, MPN and MDS/MPN overlap syndromes in which erythroid precursors show a perinuclear ring of iron-laden mitochondria.[345] *SF3B1* mutations are strongly associated with increased sideroblasts, with one study finding a positive predictive value of 97.7%.[349] The association between *SF3B1* mutations and distinct RS and erythroid lineage defects suggests that *SF3B1* plays a unique role in the pathogenesis of these MDS subgroups.[350] In particular the perturbation of the iron metabolism in SF3B1 mutant was linked to deregulation of SCL25A37.[351] *SF3B1* mutation is associated with better overall survival and lower risk of progression to AML.[345,352] *SF3B1* may have the potential to become a novel predictor of favourable clinical outcome in the diagnosis of RARS.[350,353,354]

Rare gene mutations in myelodysplastic syndromes

JAK2

Janus kinase 2 (*JAK2*) is a 25-exon gene (NG_009904.1) of 142,939 kb mapping at 9p24.1. The transcript (NM_004963.1) is translated into the 1132 amino acid JAK2 protein (130.7 kDa) belonging to Janus kinase subfamily, which also includes JAK1, JAK3 and TYK2 proteins (NP_004963.1). JAK2 is a non-receptor tyrosine kinase associated with the intracellular domain of many cytokine receptors and ubiquitously expressed in human tissue, including haematopoietic organs. The critical role of *JAK2* was demonstrated *in vivo* as germline deletion in mice resulted in embryonic lethality and no haematopoiesis.[355,356] A somatic point mutation of JAK2 at nucleotide 1849 (exon 14) involving a G to T transversion results in phenylalanine substitution for a valine at codon 617 (V617F), leading to a gain of kinase activity function and constitutive activation.[357] *JAK2 V617F* is frequently involved in the pathogenesis of *BCR-ABL1*-negative myeloproliferative neoplasms (MPN), particularly polycythemia vera. *JAK2* mutations were described in about 50% of rare MDS/MPN subgroups, such as refractory anaemia with ring sideroblasts and thrombocytosis (RARS-T).[358] This is a haemopathy with some characteristics of myelodysplasia, such as ring sideroblasts, but also increased platelet count and hypercellular marrow typical of MPN. In these cases, the association of *JAK2* and *SF3B1* mutations has been established (reviewed by Musto et al.[359]).

c-CBL

c-CBL (also named *CBL2, c-CBL, RNF55*, etc.), a 101,873 kb human gene, maps at 11q23.3. It has 16 exons (NM_005188.3) and encodes a 906 amino acid protein (115 kDa). Its expression is ubiquitous but especially predominant in haematopoietic cells. c-CBL belongs to a highly conserved mammalian protein family[360] and has a dual function. As an E3 ubiquitin ligase, by ubiquitination and intracellular degradation, it negatively regulates the activity of receptor and non-receptor tyrosine kinases[361] and non-kinase cell surface receptors.[362,363] In its adaptor function, c-CBL interacts with a plethora of proteins including kinases and structural proteins.[360] These multiple c-CBL activities are associated with its particular domain structure, which, in keeping with other Cbl proteins, includes a conserved N-terminal portion consisting of two important domains: the tyrosine kinase-binding domain and the RING finger domain.[364]

c-Cbl mutations were widely described in solid tumours[365,366] and myeloid neoplasms.[367] Missense intragenic insertion–deletion mutations revealed that *c-CBL* has a dual role, acting as a tumour suppressor gene with loss of function causing abrogation of negative regulation of E3 ubiquitin ligase activity and as an oncogene with gain-of-function mutations causing enhanced adaptor activity.[368] In mice expressing a *BCR-ABL1* transgene, loss of *c-Cbl* accelerated evolution to blast crisis.[364]

Mutations of *c-CBL* have been reported in 15% of juvenile myelomonocytic leukaemia (JMML) and 13% of chronic myelomonocytic leukaemia (CMML) and also in AML and MDS, mostly in RAEB and RCMD. In MDS, relatively few *c-Cbl* missense and insertion mutations (2–5% incidence) are found in early-stage disease, but they increase during AML transformation.[369] *c-CBL* mutations are associated with loss of heterozygosity via acquired uniparental disomy (aUDP) at region 11q.[370] Transgenic mice expressing the fusion gene NUP98-HOXD13 acquired the Cbl aberration with progression from MDS to AML.[371] This suggests that *c-Cbl* mutations might be associated with a more aggressive type of MDS and might cooperate with other genomic abnormalities in the development of AML.

Epigenetics

Epigenetic changes influence chromatin remodelling or post-transcriptional regulation without affecting DNA sequence. Well-characterized epigenetic mechanisms include methylation of CpG-enriched promoter regions, chromatin remodelling via histone modification and post-transcriptional regulation such as that that caused by variation in

Table 1.6 Molecules and events related to epigenetics.

DNA and chromatin	RNA
CpG islands methylation	Micro RNA
Histone modifications	Piwi RNA
Methylation	Transfer RNA
Acetylation	Ribosomal RNA

the expression of non-coding RNA (Table 1.6). Dynamic CpG methylation throughout cell differentiation correlates with cell type-specific gene regulation and expression levels.[372] It has been proposed that transcriptional and post-transcriptional regulation via epigenetic changes play a key role in the conserved differentiation programme of the haematopoietic system[373] and could therefore plausibly be hijacked by cancer. Notably, epigenetic changes are clonally inherited but reversible. Indeed, the role of epigenetic events in MDS pathogenesis and progression is emphasized by the efficacy of DNA hypomethylating agents and histone deacetylase inhibitors.[374,375] The importance of epigenetic changes in MDS pathogenesis is further emphasized by the frequent mutation of epigenetic regulator genes, namely *TET2, IDH1, IDH2, DNMT3A, EZH2* and *ASXL1,* discussed in the previous section.

DNA methylation

DNA methylation involves the conversion of cytosine residues to 5-methylcytosine with subsequent changes in protein–DNA interactions. Methylation of CpG islands, typically located at gene promoters, can inhibit gene expression by two mechanisms: directly by precluding the recruitment of DNA-binding proteins to their target sites, or indirectly by recruitment of histone and chromatin remodelling complexes to methylated sites, which in turn physically inhibits normal DNA–protein interactions.

Genome-wide aberrant hypermethylation of promoter regions has been shown in MDS cells compared with normal CD34 positive progenitors.[376] Interestingly, a profile of hypermethylated genes differentiating MDS and secondary AML from *de novo* AML has been reported,[376] allowing the identification of distinct biological pathways critical in MDS and secondary AML, such as transcription factors, NOTCH signalling, DNA repair and the WNT pathway. Methylation of the DAP-kinase 1 promoter has also been reported to differentiate *de novo* from treatment-induced MDS.[377] Aberrantly hypermethylated genes may contribute to the development of new diagnostic tools; for example, transcription factor genes such as *KLF11*, *MAFB* and *KLF5* are hypermethylated in 15, 7 and 1.7%, respectively, of cases of MDS other than the 5q– syndrome.[378] Global hypermethylation can occur early in MDS, independent of cytogenetic changes, and influences prognosis, allowing the identification of patients at risk of rapid transformation.[376,379]

The paradigm for epigenetic deregulation in MDS is silencing of the *p15INK4B* cell cycle-regulating gene by aberrant promoter methylation, found in 10–30% of MDS cases and associated with disease progression. Inactivation of *p15INK4B* in a murine model results in MDS/MPN phenotype with frequent progression to AML.[380] Methylation of other specific genes, namely *HIC1*, *CDH1*, *ER* and *p73*, has been also related to adverse outcome.[381–383] A list of single genes involved in epigenetic events in MDS is provided in Table 1.7.

Overall, the genome-wide methylation level seems to have an impact on MDS progression, with high risk MDS showing higher average methylation levels than low-risk cases.[384] Hypermethylation of *GSTM5*, *BIK* and *ANGPTL2* promoters, in particular, was observed in bone marrow samples from patients with refractory anaemia with excess blast (RAEB), and correlate with high risk, as does hypermethylation of *ID4*, which is seen with higher frequency in advanced stages.[385,386] The gene *FHIT* encompassing a common fragile site at 3p14.2 was methylated in 47% of MDS, particularly higher risk cases.[387] In contrast, hypermethylation

Table 1.7 List of single genes involved in epigenetic events in MDS.

Gene	Location	Gene	Location
PRDM2	1p36.21	*DOCK4*	7q31.1
INPP5D	2q37.1	*CDKN2B*	9p21
FHIT	3p14.2	*DAPK1*	9p21.33
ZMYND1	3p21.3	*MEG3*	14q32.2
RASSF1	3p21.3	*SOCS1*	16p13.13
PLK2	5q12.1	*CDH1*	16q22.1
RIL	5q31.1	*HIC1*	17p13.3
CTNNA1	5q31.2	*PLCB1*	20p12
ID4	6p22.3	*GATA1*	Xp11.23
ESR1	6q25.1		

and under-expression of *IL27RA* and *DICER1* were consistently found in low-risk MDS.[388]

Although the precise molecular mechanisms underlying the role of most recurrent epigenetic alterations are not fully understood, some findings support their role in pathogenesis of MDS. For example, CpG hypermethylation of the *GATA1* erythropoietic gene promoter seems to contribute to the ineffective erythropoiesis that characterizes MDS.[389]

Recent genome-wide studies of the methylome have emphasized the importance of evaluating DNA methylation in a range of genomic regions such as transcriptional start sites, gene bodies, regulatory regions and repeat sequences.[390] Recent data also support the existence of regions of intermediate methylation that seem to contradict the previously accepted bimodal model of DNA methylation that implies a binary methylated–unmethylated state.[391] Although DNA methylation mainly occurs in the CpG dinucleotide context in mammals, non-CG methylation has been described in human stem cells.[392] The latter findings therefore suggest that further research is required to define the precise nature and role of disease-specific methylation in MDS.

Histone modifications

Histones are composed of an octamer of proteins that include dimers of H2A, H2B, H3 and H4. These eight-protein complexes are the basic structure of nucleosomes around which DNA is coiled. Histone-based alterations include methylation, acetylation, ubiquitination, phosphorylation, sumoylation and ADP-ribosylation.[393] Post-translational modifications of histone tails are closely linked to gene expression patterns and have been proposed as an epigenetic code.[394] Histone lysine (K) or arginine (R) mono-, di- or trimethylation does not direct affect nucleosomal compaction but serves as a binding site for a range of reader proteins that participate in histone mobility and stability, influencing chromatin conformation and DNA transcription. Trimethylations of lysine 4 (H3K4me3) and lysine 27 (H3K27me3) are typically active and inactive histone modifications, respectively. However, methylation of specific residues can be associated with either activation or silencing of transcription.[395] Lysine (K) and arginine (R) residues of histone H3 and H4 were shown to be aberrantly methylated in MDS.[396] Histone modifications in MDS are a major effect of recurrent somatic mutations involving the *EZH2* and *ASXL1* genes (see the previous section).

RNA

The importance of the RNA machinery emerged relatively recently with the application of deep sequencing technologies to the human transcriptome. Although the most relevant information to date came from microRNA (miRNA), the human genome may encode over 1400 miRNAs, which frequently target many genes related to cancer development or prevention.[397] The relationship between miRNAs and components of the epigenetic machinery is twofold: miRNAs can be deregulated by epigenetic silencing/activation or can themselves regulate the epigenetic machinery. The relative amounts of all components may undergo significant variation in MDS.

An important role is played by miRNA in the control of normal haematopoietic stem cell function and their deregulation may correspond to specific features of MDS.[398] A diagnostic signature allowing discrimination of MDS from normal haematopoiesis has been reported, including increased expression of hsa-miR-378, hsa-miR-632 and hsa-miR-636.[399] Assessment of miRNA signature was also shown to be helpful in distinguishing MDS entities with chromosomal alterations from those with a normal karyotype.[400] In MDS with del(5q), miR-150 was markedly increased compared with normal haematopoiesis.[401,402] In addition, a specific signature of 17 up-regulated and four down-regulated miRNAs has been identified in the 5q– syndrome.[400] It is important to note, however, that down-regulation of miRNAs in MDS is not simply a result of deletion of the corresponding coding region.[403] Increased miR125b-1 was found in MDS with t(2;11)(p21;q24.1), with or without del(5q).[138] The level of expression of miR181 significantly influenced survival in MDS of lower risk according to the International Prognostic Scoring System (IPSS)[404] and increased expression of miR-155 and miR-210 was reported to be associated with disease progression.[405]

The pathogenic function of some miRNAs is likely to be explained by the critical role of their molecular targets, which frequently involve established MDS pathways. For instance, binding of miR21 to the 3′-UTR of the *SMAD7* gene (located at chromosome 18q21) in haematopoietic cells reduces the expression of SMAD7 protein, thus potentiating TGF-β signalling, which is responsible for decreased erythroid colony formation.[406] Members of the miR-17-92 family (miR-17-5p and miR-20a) which down-regulate *E2F1* and let-7a, which in turn reduces the expression of *KRAS*, are under-expressed in high-risk compared with low-risk MDS.[407] At least two predicted targets of microRNA, the *MYB* gene, critical in haematopoiesis, and the *Sufu* gene, involved in the Hedgehog signalling pathway, emerged from a 13 miRNA signature diagnostic for MDS.[399]

Aside from miRNA, other species of sRNA exist that, when deregulated, could potentially impact on oncogenesis via epigenetic means. A recent RNA-seq study of MDS revealed a significantly higher level of transfer RNA (tRNA) in RAEB2 than in RA.[408] tRNA is involved in chromatin organization via post-translation editing.[409] Interestingly, tRNAs are able to inhibit cytochrome *c*-activated apoptosis, which has been shown to be decreased in high-grade MDS).[410,411] In low-grade MDS, the same study also reported extensive post-transcriptional regulation via the recently discovered Piwi interacting RNAs (piRNA).[408]

Conclusion

Our understanding of the contribution of somatic alterations to the development of the myelodysplastic syndromes is developing rapidly with the increased use of post-genomic techniques. Important links have recently been made between genetic and epigenetic factors and clinical and biological features of MDS that have advanced our understanding of disease pathogenesis and provided rational targets for future therapeutic use. It is hoped that a molecular approach to clinical management, in combination with established methodology, will hasten improved outcomes in MDS.

References

1 Bejar, R., Stevenson, K., Abdel-Wahab, O., et al. Clinical effect of point mutations in myelodysplastic syndromes. *N Engl J Med* 2011; 364: 2496–2506.

2 Papaemmanuil, E., Gerstung, M., Malcovati, L., et al. Clinical and biological implications of driver mutations in myelodysplastic syndromes. *Blood* 2013; 122: 3616–3627.

3 Tiu, R., Gondek, L., O'Keefe, C., Maciejewski, J.P. Clonality of the stem cell compartment during evolution of myelodysplastic syndromes and other bone marrow failure syndromes. *Leukemia* 2007; 21: 1648–1657.

4 Vardiman, J.W., Thiele, J., Arber, D.A., et al. The 2008 revision of the World Health Organization (WHO) classification of myeloid neoplasms and acute leukemia: rationale and important changes. *Blood* 2009; 114: 937–951.

5 Shih, L.Y., Huang, C.F., Wang, P.N., et al. Acquisition of FLT3 or N-ras mutations is frequently associated with progression of myelodysplastic syndrome to acute myeloid leukemia. *Leukemia* 2004; 18: 466–475.

6 Horiike, S., Kita-Sasai, Y., Nakao, M., Taniwaki, M. Configuration of the TP53 gene as an independent prognostic parameter of myelodysplastic syndrome. *Leuk Lymphoma* 2003; 44: 915–922.

7 Xu, L., Gu, Z.H., Li, Y., et al. Genomic landscape of $CD34^+$ hematopoietic cells in myelodysplastic syndrome and gene mutation profiles as prognostic markers. *Proc Natl Acad Sci U S A* 2014; 111: 8589–8594.
8 Dan, C., Chi, J., Wang, L. Molecular mechanisms of the progression of myelodysplastic syndrome to secondary acute myeloid leukaemia and implication for therapy. *Ann Med* 2015; 47: 209–217.
9 Malcovati, L., Papaemmanuil, E., Ambaglio, I., et al. Driver somatic mutations identify distinct disease entities within myeloid neoplasms with myelodysplasia. *Blood* 2014; 124: 1513–1521.
10 Balduini, C.L., Savoia, A. Genetics of familial forms of thrombocytopenia. *Hum Genet* 2012; 131: 1821–1832.
11 Song, W.J., Sullivan, M.G., Legare, R.D., et al. Haploisufficiency of CBFA2 causes familial thrombocytopenia with propensity to develop acute myeloid leukemia. *Nat Genet* 1999; 23: 166–175.
12 Cohen, M.M., Jr., Perspectives on RUNX1 genes: an update. *Am J Med Genet A* 2009; 149A: 2629–2646.
13 Berì-Dexheimer, M., Latger-Cannard, V., Philippe, C., et al. Clinical phenotype of germline RUNX1 haploinsufficiency: from point mutations to large genomic deletions. *Eur J Hum Genet* 2008; 16: 1014–1018.
14 Matheny, C.J., Speck, M.E., Cushing, P.R., et al. Disease mutations in RUNX1 and RUNX2 create nonfunctional, domina-negative, or hypo-morphic alleles. *EMBO J* 2007; 26: 1163–1175.
15 Minelli, A., Maserati, E., Rossi, G., et al. Familial platelet disorder with propensity to acute myelogenous leukemia: genetic heterogeneity and progression to leukemia via acquisition of clonal chromosome anomalies. *Genes Chromosomes Cancer* 2004; 40: 165–171.
16 Vandenberghe, P., Beel, K. Severe congenital neutropenia, a genetically heterogeneous disease group with an increased risk of AML/MDS. *Pediatr Rep* 2011; 3: 21–24.
17 Boztug, K., Appaswamy, G., Ashikov, A., et al. A syndrome with congenital neutropenia and mutations in G6PC3. *N Engl J Med* 2009; 360: 32–43.
18 Germeshausen, M., Deerberg, S., Peter, Y., Reimer, C., Kratz, C.P., Ballmaier M. The spectrum of ELANE mutations and their implications in severe congenital and cyclic neutropenia. *Hum Mutat* 2013; 34: 905–914.
19 Grenda, D.S., Murakami, M., Ghatak, J., et al. Mutations of the ELA2 gene found in patients with severe congenital neutropenia induce the unfolded protein response and cellular apoptosis. *Blood* 2007; 110: 4179–4187.
20 Badolado, R., Fontana, S., Notarangelo, S.D., Savoldi, G. Congenital neutropenia: advances in diagnosis and treatment. *Curr Opin Allergy Clin Immunol* 2004; 4: 513–521.
21 Klein, C., Grudzien, M., Appaswamy, G., et al. HAX1 deficiency causes autosomal recessive severe congenital neutropenia (Kostmann disease). *Nat Genet* 2007; 39: 86–92.
22 Devriendt, K., Kim, A.S., Mathijs, G., et al. Constitutively activating mutation in WASP causes X-linked severe congenital neutropenia. *Nat Genet* 2001; 27: 313–317.

23 Beel, K., Vandenberghe, P. G-CSF receptor (CSF3R) mutations in X-linked neutropenia evolving to acute myeloid leukemia or myelodysplasia. *Haematologica* 2009; 94: 1449–1452.
24 Sloand, E.M., Yong, A.S., Ramkissoon, S., et al. Granulocyte colony-stimulating factor preferentially stimulates proliferation of monosomy 7 cells bearing the isoform IV receptor. *Proc Natl Acad Sci U S A* 2006; 103: 14483–14488.
25 Colombo, E.A., Bazan, J.F., Negri, G., et al. Novel C16orf57 mutations in patients with poikiloderma with neutropenia: bioinformatic analysis of the protein and predicted effects of all reported mutations. *Orphanet J Rare Dis* 2012; 7: 7.
26 Walne, A.J., Vulliamy, T., Beswick, R., Kirwan, M., Dokal, I. Mutations in C16orf57 and normal-length telomeres unify a subset of patients with dyskeratosis congenita, poikiloderma with neutropenia and Rothmund–Thomson syndrome. *Hum Mol Genet* 2010; 19: 4453–4461.
27 Hahn, C.N., Chong, C.E., Carmichael, C.L., et al. Heritable GATA-2 mutations associated with familial myelodysplastic syndrome and acute myeloid leukemia. *Nat Genet* 2011; 43: 1012–1017.
28 Tsai, F.Y., Keller, G., Kuo, F.C., et al. An early haematopoietic defect in mice lacking the transcription factor GATA-2. *Nature* 1994; 371: 221–226.
29 Briegel, K., Lim, K.C., Plank, C., Beug, H., Engel, J.D., Zenke, M. Ectopic expression of a conditional GATA-2/estrogen receptor chimera arrests erythroid differentiation in a hormone-dependent manner. *Genes Dev* 1993; 7: 1097–1109.
30 Persons, D.A., Allay, J.A., Allay, E.R., et al. Enforced expression of the GATA-2 transcription factor blocks normal hematopoiesis. *Blood* 1999; 93: 488–499.
31 Pasquet, M., Bellanné-Chantelot, C., Tavitian, S., et al. High frequency of GATA2 mutations in patients with mild chronic neutropenia evolving to MonoMac syndrome, myelodysplasia, and acute myeloid leukemia. *Blood* 2013; 121: 822–829.
32 Austin, K.M., Leary, R.J., Shimamura, A. The Shwachman–Diamond SBDS protein localizes to the nucleolus. *Blood* 2005; 106: 1253–1258.
33 Austin, K.M., Gupta M.L., Jr.,, Coats, S.A., et al. Mitotic spindle destabilization and genomic instability in Shwachman–Diamond syndrome. *J Clin Invest* 2008; 118: 1511–1518.
34 Ball, H.L., Zhang, B., Riches, J.J., et al. Shwachman–Bodian Diamond syndrome is a multi-functional protein implicated in cellular stress responses. *Hum Mol Genet* 2009; 18: 3684–3695.
35 Boocock, G.R.B., Morrison, J.A., Popovic, M., et al. Mutations in SBDS are associated with Shwachman–Diamond syndrome. *Nat Genet* 2003; 33: 97–101.
36 Smith, O.P., Hann, I.M., Chessells, J.M., Reeves, B.R., Milla, P. Haematological abnormalities in Shwachman–Diamond syndrome. *Br J Haematol* 1996; 94: 279–284.
37 Dror, Y., Freedman, M.H. Shwachman–Diamond syndrome: an inherited preleukemic bone marrow failure disorder with aberrant hematopoietic progenitors and faulty marrow microenviroment. *Blood* 1999; 94: 3048–3054.

38 André, V., Longoni, D., Bresolin, S., et al. Mesenchymal stem cells from Shwachman–Diamond syndrome patients display normal functions and do not contribute to hematological defects. *Blood Cancer J* 2012; 2(10): e94.
39 Dror, Y., Freedman, M.H. Shwachman–Diamond syndrome. *Br J Haematol* 2002; 118: 701–713.
40 Dror, Y., Squire, J., Durie, P., Freedman, M.H. Malignant myeloid transformation with isochromosome 7q in Shwachman–Diamond syndrome. *Leukemia* 1998; 12: 1591–1595.
41 Cunningham, J., Sales, M., Pearce, A., et al. Does isochromosome 7q mandate bone marrow transplant in children with Shwachman–Diamond syndrome? *Br J Haematol* 2002; 119: 1062–1069.
42 Crescenzi, B., La Starza, R., Sambani, C., et al. Totipotent stem cells bearing del(20q) maintain multipotential differentiation in Shwachman–Diamond syndrome. *Br J Haematol* 2009; 144: 116–119.
43 Rujkijyanont, P., Beyene, J., Wei, K., Khan, F., Dror, Y. Leukaemia-related gene expression in bone marrow cells from patients with the preleukaemic disorder Shwachman–Diamond syndrome. *Br J Haematol* 2007; 137: 537–544.
44 Raaijmakers, M.H., Mukherjee, S., Guo, S., et al. Bone progenitor dysfunction induces myelodysplasia and secondary leukaemia. *Nature.* 2010; 464: 852–857.
45 Armanios, M., Blackburn, E.H. The telomere syndromes. *Nat Rev Genet* 2012; 13: 693–704.
46 Savage, S.A., Bertuch, A.A. The genetics and clinical manifestations of telomere biology disorders. *Genet Med* 2010; 12: 753–764.
47 Young, N.S. Bone marrow failure and the new telomere diseases: practice and research. *Hematology* 2012; 17(Suppl 1): S18–S21.
48 Calado, R.T., Young, N.S. Telomere maintenance and human bone marrow failure. *Blood* 2008; 111: 4446–4455.
49 Du, H.Y., Pumbo, E., Manley, P., et al. Complex inheritance pattern of dyskeratosis congenita in two families with 2 different mutations in the telomerase reverse transcriptase gene. *Blood* 2008; 111: 1128–1130.
50 Calado, R.T. Telomeres and marrow failure. *Hematology Am Soc Hematol Educ Program* 2009: 338–343.
51 Yang, L., Mailloux, A., Rollison, D.E., et al. NaiveT-cells in myelodysplastic syndrome display intrinsic human telomerase reverse transcriptase (hTERT) deficiency. *Leukemia* 2013; 27: 897–906.
52 Shimamura, A., Alter, B.P. Pathophysiology and management of inherited bone marrow failure syndromes. *Blood Rev* 2010; 24: 101–122.
53 Auerbach, A.D. Fanconi anemia and its diagnosis. *Mutat Res* 2009; 668: 4–10.
54 Alter, B.P. Cancer in Fanconi anemia, 1927–2001. *Cancer* 2003; 97: 425–440.
55 Auerbach, A.D. Fanconi anemia diagnosis and the diepoxybutane (DEB) test. *Exp Hematol* 1993; 21: 731–733.
56 Butturini, A., Gale, R.P., Verlander, P.C., Adler-Brecher, B., Gillio, A.P., Auerbach, A.D. Hematologic abnormalities in Fanconi anemia: an International Fanconi Anemia Registry study. *Blood* 1994; 84: 1650–1655.
57 Kutler, D.I., Singh, B., Satagopan, J., et al. A 20-year perspective on the International Fanconi Anemia Registry (IFAR). *Blood* 2003; 101: 1249–1256.

58 Soulier, J. Fanconi anemia. *Hematology Am Soc Hematol Educ Program* 2011; 492–497.
59 Quentin, S., Cuccuini, W., Ceccaldi, R., et al. Myelodysplasia and leukemia of Fanconi anemia are associated with a specific pattern of genomic abnormalities that includes cryptic RUNX1/AML1 lesions. *Blood* 2011; 117: e161–e170.
60 Mehta, P.A., Harris, R.E., Davies, S.M., et al. Numerical chromosomal changes and risk of development of myelodysplastic syndrome – acute myeloid leukemia in patients with Fanconi anemia. *Cancer Genet Cytogenet* 2010; 203: 180–186.
61 Kee, Y., D'Andrea, A.D. Expanded roles of the Fanconi anemia pathway in preserving genomic stability. *Genes Dev* 2010; 24: 1680–1694.
62 Kalb, R., Neveling, K., Hoehn, H., et al. Hypomorphic mutations in the gene encoding a key Fanconi anemia protein, FANCD2, sustain a significant group of FA-D2 patients with severe phenotype. *Am J Hum Genet* 2007; 80: 895–910.
63 Garcia-Higuera, I., Taniguchi, T., Ganesan, S., et al. Interaction of the Fanconi anemia proteins and BRCA1 in a common pathway. *Mol Cell* 2001; 7: 249–262.
64 Howlett, N.G., Taniguchi, T., Olson, S., et al. Biallelic inactivation of BRCA2 in Fanconi anemia. *Science* 2002; 297: 606–609.
65 Xia, B., Dorsman, J.C., Ameziane, N., et al. Fanconi anemia is associated with a defect in the BRCA2 partner PALB2. *Nat Genet* 2007; 39: 159–161.
66 Hasle, H., Niemeyer, C.M., Chessells, J.M., et al. A pediatric approach to the WHO classification of myelodisplastic and myeloproliferative disease. *Leukemia* 2003; 17: 277–282.
67 Henderson, R., Spence, L. Down syndrome with myelodysplasia of megakaryoblastic lineage. *Clin Lab Sci* 2006; 19: 161–164.
68 Greene, M.E., Mundschau, G., Wechsler, J., et al. Mutations in GATA1 in both transient myeloproliferative disease and acute megakaryoblastic leukemia of Down syndrome. *Blood Cell Mol Dis* 2003; 31: 351–356.
69 Hitzler, J.K., Cheung, J., Li, Y., Scherer, S.W., Zipursky, A. GATA1 mutations in transient leukemia and acute megakaryoblastic leukemia of Down syndrome. *Blood* 2003; 101: 4301–4304.
70 Magalhaes, I.Q., Splendore, A., Emeranciano, M., Figueiredo, A., Ferrari, I., Pombo-de Oliveira, M.S. GATA1 mutations in acute leukemia in children with of Down syndrome. *Cancer Genet Cytogenet* 2006; 166: 112–116.
71 Lange, B.J., Kobrinsky, N., Barnard, D.R., et al. Distinctive demography, biology, and outcome of acute myeloid leukemia and myelodysplastic syndrome in children with Down syndrome: Children's Cancer Group Studies 2861 and 2891. *Blood* 1998; 91: 608–615.
72 Gamis, A.S., Woods, W.G., Alonzo, T.A., et al. Increased age at diagnosis has a significantly negative effect on outcome in children with Down syndrome and acute myeloid leukemia: a report from the Children's Cancer Group Study 2891. *J Clin Oncol* 2003; 21: 3415–3422.
73 Lane, A.A., Chapuy, B., Lin, C.Y., et al. Triplication of a 21q22 region contributes to B cell transformation through HMGN1 overexpression and loss of histone H3 Lys27 trimethylation. *Nat Genet* 2014; 46: 618–623.
74 Letourneau, A., Santoni, F.A., Bonilla, X., et al. Domains of genome-wide gene expression dysregulation in Down's syndrome. *Nature* 2014; 508: 345–350.

75 Haase, D. Cytogenetic features in myelodysplastic syndromes. *Ann Hematol* 2008; 87: 515–526.
76 Mecucci, C. Molecular features of primary MDS with cytogenetic changes. *Leukemia Res* 1998; 22: 293–302.
77 Vallespi, T., Imbert, M., Mecucci, C., Preudhomme, C., Fenaux, P. Diagnosis, classification, and cytogenetics of myelodysplastic syndrome. *Haematologica* 1998; 83: 258–275.
78 Mohamedali, A., Gäken, J., Twine, N.A., et al. Prevalence and prognostic significance of allelic imbalance by single-nucleotide polymorphism analysis in low-risk myelodysplastic syndromes. *Blood* 2007; 110: 3365–3373.
79 Mohamedali, A.M., Smith, A.E., Gaken, J., et al. Novel TET2 mutations associated with UPD4q24 in myelodysplastic syndrome. *J Clin Oncol* 2009; 27: 4002–4006.
80 Greenberg, P.L., Tuechler, H., Schanz, J., et al. Revised international prognostic scoring system for myelodysplastic syndromes. *Blood* 2012; 120: 2454–2465.
81 Solé, F., Luno, E., Sanzo, C., et al. Identification of novel cytogenetic markers with prognostic significance in a series of 968 patients with primary myelodysplastic syndromes. *Haematologica* 2005; 90: 1168–1178.
82 Wong, A.K., Fang, B., Zhang, L., Guo, X., Lee, S., Schreck, R. Loss of the Y chromosome: an age-related or clonal phenomenon in acute myelogenous leukemia/myelodysplastic syndrome? *Arch Pathol Lab Med* 2008; 132: 1329–1332.
83 Mecucci, C., Van Orshoven, A., Vermaelen, K., et al. 11q– chromosome is associated with abnormal iron stores in myelodysplstica syndromes. *Cancer Genet Cytogenet* 1987; 27: 39–44.
84 Wang, S.A., Abruzzo, L.V., Hasserjian, R.P., et al. Myelodysplastic syndromes with deletions of chromosome 11q lack cryptic MLL rearrangement and exhibit characteristic clinicopathologic features. *Leukemia Res* 2011; 35: 351–357.
85 Aziz, A., Baxter, E.J., Edwards, C., et al. Cooperativity of imprinted genes inactivated by acquired chromosome 20q deletions. *J Clin Invest* 2013; 123: 2169–2182.
86 Braun, T., de Botton, S., Taksin, A.L.S., et al. Characteristics and outcome of myelodysplastic syndromes (MDS) with isolated 20q deletion: a report on 62 cases. *Leukemia Res* 2011; 35: 863–867.
87 Mhawech, P., Saleem, A. Myelodysplastic syndrome: review of the cytogenetic and molecular data. *Crit Rev Oncol Hematol* 2001; 40: 229–238.
88 Michaux, L., Wlodarska, I., Mecucci, C., et al. Characterization by chromosome painting of balanced and unbalanced X chromosome translocations in myelodysplastic syndromes. *Cancer Genet Cytogenet* 1995; 82: 17–22.
89 Soenen, V., Preudhomme, C., Roumier, C., Daudignon, A., Lai, J.L., Fenaux, P. 17p deletion in acute myeloid leukemia and myelodysplastic syndrome. Analysis of breakpoints and deleted segments by fluorescence in situ. *Blood* 1998; 91: 1008–1015.
90 Laï, J.L., Zandecki, M., Fenaux, P., et al. Translocations (5;17) and (7;17) in patients with de novo or therapy-related myelodysplastic syndromes or acute nonlymphocytic leukemia. A possible association with acquired pseudo-Pelger–Huët anomaly and small vacuolated granulocytes. *Cancer Genet Cytogenet* 1990; 46: 173–183.

91 Wang, P., Spielberger, R.T., Thangavelu, M., et al. dic(5;17): a recurring abnormality in malignant myeloid disorders associated with mutations of TP53. *Genes Chromosomes Cancer* 1997; 20: 282–291.
92 Kanagal-Shamanna, R., Bueso-Ramos, C.E., Barkoh, B., et al. Myeloid neoplasms with isolated isochromosome 17q represent a clinicopathologic entity associated with myelodysplastic/myeloproliferative features, a high risk of leukemic transformation, and wild-type TP53. *Cancer* 2012; 118: 2879–2888.
93 Sato, Y., Suto, Y., Pietenpol, J., et al. TEL and KIP1 define the smallest region of deletions on 12p13 in hematopoietic malignancies. *Blood* 1995; 86: 1525–1533.
94 Streubel, B., Sauerland, C., Heil, G., et al. Correlation of cytogenetic, molecular cytogenetic, and clinical findings in 59 patients with ANLL or MDS and abnormalities of the short arm of chromosome 12. *Br J Haematol* 1998; 100: 521–533.
95 Secker-Walker, L., Fitchett, M. Commentary: constitutional and acquired trisomy 8. *Leukemia Res* 1995; 19: 737–740.
96 Mecucci, C., Rege-Cambrin, G., Michaux, J.L., Tricot, G., Van den Berghe, H. Multiple chromosomally distinct cell populations in myelodysplastic syndromes and their possible significance in the evolution of the disease. *Br J Haematol* 1986; 64: 699–706.
97 Mecucci, C., Tricot, G., Boogaerts, M., Van den Berghe, H. An identical translocation between chromosome 1 and 15 in two patients with myelodysplastic syndromes. *Br J Haematol* 1986; 62: 439–445.
98 Anastasi, J., Feng, J., Le Beau, M.M., Larson, R.A., Rowley, J.D., Vardiman, JW. Cytogenetic clonality in myelodysplastic syndromes studied with fluorescence in situ hybridization: lineage, response to growth factor therapy, and clone expansion. *Blood* 1993; 81: 1580–1585.
99 Iwabuchi, A., Ohyashiki, K., Ohyashiki, J.H., et al. Trisomy of chromosome 8 in myelodysplastic syndrome. Significance of the fluctuating trisomy 8 population. *Cancer Genet Cytogenet* 1992; 62: 70–74.
100 Paulsson, K., Johansson, B. Trisomy 8 as the sole chromosomal aberration in acute myeloid leukemia and myelodysplastic syndromes. *Pathol Biol (Paris)* 2007; 55: 37–48.
101 Mohamed, A.N., Varterasian, M.L., Dobin, S.M., et al. Trisomy 6 as a primary karyotypic aberration in hematologic disorders. *Cancer Genet Cytogenet* 1998; 106: 152–155.
102 Vasef, M.A., Murata-Collins, J.L., Alsabeh, R., Medeiros, L.J. Trisomy 14 in myelodysplastic syndromes: report of two cases and review of the literature. *Arch Pathol Lab Med* 1998; 122: 77–83.
103 Horton, Y.M., Johnson, P.R. Trisomy 14 in myeloid malignancies: report of two cases and review of the literature. *Cancer Genet Cytogenet* 2001; 124: 172–174.
104 Baumgartner, B.J., Shurafa, M., Terebelo, H., Tapazoglu, E., Van Dyke, D.L. Trisomy 15, sex chromosome loss, and hematological malignancy. *Cancer Genet Cytogenet* 2000; 117: 132–135.
105 Natelson, E. Myelodisplasia with isolated trisomy 15: a 15 year follow-up without specific therapy. *Am J Med Sci* 2006; 331: 157–158.

106 Johansson, B., Billstrom, R., Mauritzon, N., Mitelman, F. Trisomy 19 as the sole chromosomal anomaly in haematological neoplasms. *Cancer Genet Cytogenet* 1994; 74: 62–65.
107 Wan, T.S., Au, W.Y., Chan, J.C.W., Chan, L.C., Ma, S.K. Trisomy 21 as the sole acquired karyotypic abnormality in acute myeloid leukemia and myelodysplastic syndrome. *Leukemia Res* 1999; 23: 1079–1083.
108 Larsson, N., Lilljebjorn, H., Lassen, C., Johansson, B., Fioretos, T. Myeloid malignancies with acquired trisomy 21 as the sole cytogenetic change are clinically highly variable and display a heterogeneous pattern of copy number alterations and mutations. *Eur J Haematol* 2012; 88: 136–143.
109 Ademà, V., Hernández, J.M., Abáigar, M., et al. Application of FISH 7q in MDS patients without monosomy 7 or 7q deletion by conventional G-banding cytogenetics: does –7/7q– detection by FISH have prognostic value? *Leukemia Res* 2013; 37: 416–421.
110 Raj, K., John, A., Ho, A., et al. CDKN2B methylation status and isolated chromosome 7 abnormalities predict responses to treatment with 5-azacytidine. *Leukemia* 2007; 21: 1937–1944.
111 Imashuku, S., Hibi, S., Bessho, F., et al. Detection of myelodysplastic syndrome/acute myeloid leukemia evolving from aplastic anemia in children, treated with recombinant human G-CSF. *Haematologica* 2003; 88: ECR31.
112 Kratz, C.P., Emerling, B.M., Donovan, S., et al. Candidate gene isolation and comparative analysis of a commonly deleted segment of 7q22 implicated in myeloid malignancies. *Genomics* 2001; 77: 171–180.
113 Curtiss, N.P., Bonifas, J.M., Lauchle, J.O., et al. Isolation and analysis of candidate myeloid tumor suppressor genes from a commonly deleted segment of 7q22. *Genomics* 2005; 85: 600–607.
114 Wong, J.C., Zhang, Y., Lieuw, K.H., et al. Use of chromosome engineering to model a segmental deletion of chromosome band 7q22 found in myeloid malignancies. *Blood* 2010; 115: 4524–4532.
115 Le Beau, M.M., Espinosa R., III, Davis, E.M., Eisenbart, J.D., Larson, R.A., Green, E.D. Cytogenetic and molecular delineation of a region of chromosome 7 commonly deleted in malignant myeloid diseases. *Blood* 1996; 88: 1930–1935.
116 Fischer, K., Fröhling, S., Scherer, S.W., et al. Molecular cytogenetic delineation of deletions and translocations involving chromosome band 7q22 in myeloid leukemias. *Blood* 1997; 89: 2036–2041.
117 Döhner, K., Brown, J., Hehmann, U., et al. Molecular cytogenetic characterization of a critical region in bands 7q35–q36 commonly deleted in malignant myeloid disorders. *Blood* 1998; 92: 4031–4035.
118 Jerez, A., Sugimoto, Y., Makishima, H., et al. Loss of heterozygosity in 7q myeloid disorders: clinical associations and genomic pathogenesis. *Blood* 2012; 119: 6109–6117.
119 Tosi, S., Scherer, S.W., Giudici, G., et al. Delineation of multiple deleted regions in 7q in myeloid disorders. *Genes Chromosomes Cancer* 1999; 25: 384–392.
120 Asou, H., Matsui, H., Ozaki, Y., et al. Identification of a common microdeletion cluster in 7q21.3 subband among patients with myeloid leukemia and myelodysplastic syndrome. *Biochem Biophys Res Commun* 2009; 383: 245–251.

121 McNerney, M.E., Brown, C.D., Wang, X., et al. CUX1 is a haploinsufficient tumor suppressor gene on chromosome 7 frequently inactivated in acute myeloid leukemia. *Blood* 2013; 121: 975–983.
122 Zhang, Y., Wong, J., Klinger, M., Tran, M.T., Shannon, K.M., Killeen, N. MLL5 contributes to hematopoietic stem cell fitness and homeostasis. *Blood* 2009; 113: 1455–1463.
123 Heuser, M., Yap, D.B., Leung, M., et al. Loss of MLL5 results in pleiotropic hematopoietic defects, reduced neutrophil immune function, and extreme sensitivity to DNA demethylation. *Blood* 2009; 113: 1432–1443.
124 Madan, V., Madan, B., Brykczynska, U., et al. Impaired function of primitive hematopoietic cells in mice lacking the Mixed-Lineage-Leukemia homolog MLL5. *Blood* 2009; 113: 1444–1454.
125 Zhou, L., Opalinska, J., Sohal, D., et al. Aberrant epigenetic and genetic marks are seen in myelodysplastic leukocytes and reveal Dock4 as a candidate pathogenic gene on chromosome 7q. *J Biol Chem* 2011; 286: 25211–25223.
126 Shih, A.H., Abdel-Wahab, O., Patel, J.P., Levine, R.L. The role of mutations in epigenetic regulators in myeloid malignancies. *Nat Rev Cancer* 2012; 12: 599–612.
127 Ernst, T., Chase, A.J., Score, J., et al. Inactivating mutations of the histone methyltransferase gene EZH2 in myeloid disorders. *Nat Genet* 2010; 42: 722–726.
128 Morin, R.D., Johnson, N.A., Severson, T.M., et al. Somatic mutations altering EZH2(Tyr641) in follicular and diffuse large B-cell lymphomas of germinal-center origin. *Nat Genet* 2010; 42: 181–185.
129 Galván, A.B., Mallo, M., Arenillas, L., et al. Does monosomy 5 really exist in myelodysplastic syndromes and acute myeloid leukemia? *Leukemia Res* 2010; 34: 1242–1245.
130 Tasaka, T., Tohyama, K., Kishimoto, M., et al. Myelodysplastic syndrome with chromosome 5 abnormalities: a nationwide survey in Japan. *Leukemia* 2008; 22: 1874–1881.
131 Berger, R., Le Coniat, M., Derré, J., Flexor, M.A., Hillion, J. Abnormalities of chromosome 18 in myelodysplastic syndromes and secondary leukemia. *Cancer Genet Cytogenet* 1992; 63: 97–99.
132 Haase, D., Germing, U., Schanz, J., et al. New insights into the prognostic impact of the karyotype in MDS and correlation with subtypes: evidence from a core dataset of 2124 patients. *Blood* 2007; 110: 4385–4395.
133 Raza, S., TaherNazerHussain, F., Patnaik, M., Knudson, R., Van Dyke, D., Tefferi, A. Autosomal monosomies among 24,262 consecutive cytogenetic studies: prevalence, chromosomal distribution and clinicopathologic correlates of sole abnormalities. *Am J Hematol* 2011; 86: 353–356.
134 La Starza, R., Aventin, A., Falzetti, D., et al. Regions of juxtaposition in unbalanced 1q rearrangements of malignant hemopathies. *Leukemia* 2001; 15: 861–863.
135 Caramazza, D., Hussein, K., Siragusa, S., et al. Chromosome 1 abnormalities in myeloid malignancies: a literature survey and karyotype–phenotype associations. *Eur J Haematol* 2013; 84: 191–200.

136 Jonveaux, P., Derré, J., Berger, R. Whole arm translocation t(17;18): a non-random abnormality of myeloid cell proliferation. *Leukemia* 1993; 7: 1987–1989.
137 Wlodarska, I., Selleri, L., La Starza, R., et al. Molecular cytogenetics localizes two new breakpoints on 11q23.3 and 21q11.2 in MDS with t(11;21) translocation. *Genes Chromosomes Cancer* 1999; 24: 199–206.
138 Bousquet, M., Quelen, C., Rosati, R., et al. Myeloid cell differentiation arrest by miR-125b-1 in myelodysplastic syndrome and acute myeloid leukemia with the t(2;11)(p21;q23) translocation. *J Exp Med* 2008; 205: 2499–2506.
139 Costa, D., Munoz, C., Carrio, A., et al. Reciprocal Translocations in Myelodysplastic Syndromes and Chronic Myelomonocytic Leukemias: Review of 5,654 Patients with an Evaluable Karyotype. *Genes Chromosomes Cancer* 2013; 52: 753–763.
140 Walker, A., Mrozek, K., Kohlschmidt, J., et al. New recurrent balanced translocations in acute myeloid leukemia and myelodysplastic syndromes: cancer and leukemia group B 8461. *Genes Chromosomes Cancer* 2013; 52: 385–401.
141 Yoneda-Kato, N., Fukuhara, S., Kato, J. Apoptosis induced by the myelodysplastic syndrome-associated NPM-MLF1 chimeric protein. *Oncogene* 1999; 18: 3716–3724.
142 Arber, D.A., Chang, K.L., Lyda, M.H., Bedell, V., Spielberger, R., Slovak, M.L. Detection of NPM/MLF1 fusion in t(3;5)-positive acute myeloid leukemia and myelodysplasia. *Hum Pathol* 2003; 34: 809–813.
143 Jotterand Bellomo, M., Parlier, V., Mühlematter, D., Grob, J.P., Beris, P. Three new cases of chromosome 3 rearrangement in bands q21 and q26 with abnormal thrombopoiesis bring further evidence to the existence of a 3q21q26 syndrome. *Cancer Genet Cytogenet* 1992; 59: 138–160.
144 Cui, W., Sun, J., Cotta, C.V., Medeiros, L.J., Lin, P. Myelodysplastic syndrome with inv(3)(q21q26.2) or t(3;3)(q21;q26.2) has a high risk for progression to acute myeloid leukemia. *Am J Clin Pathol* 2011; 136: 282–288.
145 Li, S., Yin, C.C., Medeiros, L.J., Bueso-Ramos, C., Lu, G., Lin, P. Myelodysplastic syndrome/acute myeloid leukemia with t(3;21)(q26.2;q22) is commonly a therapy-related disease associated with poor outcome. *Am J Clin Pathol* 2012; 138: 146–152.
146 Trubia, M., Albano, F., Cavazzini, F., et al. Characterization of a recurrent translocation t(2;3)(p15–22;q26) occurring in acute myeloid leukaemia. *Leukemia* 2006; 20: 48–54.
147 Poppe, B., Dastugue, N., Vandesompele, J., et al. EVI1 is consistently expressed as principal transcript in common and rare recurrent 3q26 rearrangements. *Genes Chromosomes Cancer* 2006; 45: 349–356.
148 Shapira, M.Y., Hirshberg, B., Amir, G., Rund, D. 6;9 translocation in myelodysplastic syndrome. *Cancer Genet Cytogenet* 1999; 112: 57–59.
149 Super, H.J., McCabe, N.R., Thirman, M.J., et al. Rearrangements of the MLL gene in therapy-related acute myeloid leukemia in patients previously treated with agents targeting DNA-topoisomerase II. *Blood* 1993; 82: 3705–3711.
150 Jaju, R.J., Fidler, C., Haas, O.A., et al. A novel gene, NSD1, is fused to NUP98 in the t(5;11)(q35;p15.5) in de novo childhood acute myeloid leukemia. *Blood* 2001; 98: 1264–1267.

151 La Starza, R., Gorello, P., Rosati, R., et al. Cryptic insertion producing two NUP98/NSD1 chimeric transcripts in adult refractory anemia with an excess of blasts. *Genes Chromosomes Cancer* 2004; 41: 395–399.

152 Shiba, N., Ichikawa, H., Taki, T., et al. NUP98-NSD1 gene fusion and its related gene expression signature are strongly associated with a poor prognosis in pediatric acute myeloid leukemia. *Genes Chromosomes Cancer* 2013; 52: 683–693.

153 Thol, F., Kölking, B., Hollink, I.H., et al. Analysis of NUP98/NSD1 translocations in adult AML and MDS patients. *Leukemia* 2013; 27: 750–754.

154 Gough, S.M., Slape, C.I., Aplan, P.D. NUP98 gene fusions and hematopoietic malignancies: common themes and new biologic insights. *Blood* 2011; 118: 6247–6257.

155 Pedersen-Bjergaard, J., Andersen, M.K., Christiansen, D.H., Nerlov, C. Genetic pathways in therapy-related myelodysplasia and acute myeloid leukemia. *Blood* 2002; 99: 1909–1912.

156 La Starza, R., Matteucci, C., Gorello, P., et al. NPM1 deletion is associated with gross chromosomal rearrangements in leukemia. *PLoS One* 2010; 5(9): e12855.

157 Lindvall, C., Nordenskjöld, M., Porwit, A., Björkholm, M., Blennow, E. Molecular cytogenetic characterization of acute myeloid leukemia and myelodysplastic syndromes with multiple chromosome rearrangements. *Haematologica* 2001; 86: 1158–1164.

158 Patnaik, M.M., Hanson, C.A., Hodnefield, J.M., Knudson, R., Van Dyke, D.L., Tefferi, A. Monosomal karyotype in myelodysplastic syndromes, with or without monosomy 7 or 5 is prognostically worse than an otherwise complex karyotype. *Leukemia* 2011; 25: 266–270.

159 Belli, C.B., Bengio, R., Negri Aranguren, P., et al. Partial and total monosomal karyotypes in myelodysplastic syndromes: comparative prognostic relevance among 421 patients. *Am J Hematol* 2011; 86: 540–545.

160 van Gelder, M., de Wreede, L.C., Schetelig, J., et al. Monosomal karyotype predicts poor survival after allogeneic stem cell transplantation in chromosome 7 abnormal myelodysplastic syndrome and secondary acute myeloid leukemia. *Leukemia* 2013; 27: 879–888.

161 Valcárcel, D., Ademà, V., Solé, F., et al. Complex, not monosomal, karyotype is the cytogenetic marker of poorest prognosis in patients with primary myelodysplastic syndrome. *J Clin Oncol* 2013; 31: 916–922.

162 Mallo, M., Cervera, J., Schanz, J., et al. Impact of adjunct cytogenetic abnormalities for prognostic stratification in patients with myelodysplastic syndrome and deletion 5q. *Leukemia* 2011; 25: 110–120.

163 Smith, S.M., Le Beau, M.M., Huo, D., et al. Clinical–cytogenetic associations in 306 patients with therapy-related myelodysplasia and myeloid leukemia: the University of Chicago series. *Blood* 2003; 102: 43–52.

164 Holtan, S.G., Santana-Davila, R., Dewald, G.W., et al. Myelodysplastic syndromes associated with interstitial deletion of chromosome 5q: clinicopathologic correlations and new insights from the pre-lenalidomide era. *Am J Hematol* 2008; 83: 708–713.

165 Germing, U., Lauseker, M., Hildebrandt, B., et al. Survival, prognostic factors and rates of leukemic transformation in 381 untreated patients with MDS and del(5q): a multicenter study. *Leukemia* 2012; 26: 1286–1292.
166 Giagounidis, A.A., Germing, U., Haase, S., et al. Clinical, morphological, cytogenetic, and prognostic features of patients with myelodysplastic syndromes and del(5q) including band q31. *Leukemia* 2004; 18: 113–119.
167 Giagounidis, A.A., Germing, U., Aul, C. Biological and prognostic significance of chromosome 5q deletions in myeloid malignancies. *Clin Cancer Res* 2006; 12: 5–10.
168 Van den Berghe, H., Cassiman, J.J., David, G., Fryns, J.P., Michaux, J.L., Sokal, G. Distinct haematological disorder with deletion of long arm of no. 5 chromosome. *Nature* 1974; 251: 437–438.
169 Van den Berghe, H., Michaux, L. 5q–, twenty-five years later: a synopsis. *Cancer Genet Cytogenet* 1997; 94: 1–7.
170 Padron, E., Komrokji, R., List, A.F. The 5q– syndrome: biology and treatment. *Curr Treat Options Oncol* 2011; 12: 354–368.
171 Boultwood, J., Pellagatti, A., McKenzie, A.N., Wainscoat, J.S. Advances in the 5q– syndrome. *Blood* 2010; 116: 5803–5811.
172 Patnaik, M.M., Lasho, T.L., Finke, C.M., et al. Isolated del(5q) in myeloid malignancies: clinicopathologic and molecular features in 143 consecutive patients. *Am J Hematol* 2011; 86: 393–398.
173 Crescenzi, B., La Starza, R., Romoli, S., et al. Submicroscopic deletions in 5q– associated malignancies. *Haematologica* 2004; 89: 281–285.
174 Wang, L., Fidler, C., Nadig, N., et al. Genome-wide analysis of copy number changes and loss of heterozygosity in myelodysplastic syndrome with del(5q) using high-density single nucleotide polymorphism arrays. *Haematologica* 2008; 93: 994–1000.
175 Nofrini, V., La Starza, R., Crescenzi, B., Pierini, V., Barba, G., Mecucci, C. Different boundaries characterize isolated and non-isolated 5q deletions in myelodysplastic syndromes and acute myeloid leukemias. *Haematologica* 2012; 97: 792–794.
176 Douet-Guilbert, N., De Braekeleer, E., Basinko, A., et al. Molecular characterization of deletions of the long arm of chromosome 5 (del(5q)) in 94 MDS/AML patients. *Leukemia* 2012; 26: 1695–1697.
177 Sokol, L., Caceres, G., Rocha, K., Stockero, K.J., Dewald, D.W., List, A.F. JAK2(V617F) mutation in myelodysplastic syndrome (MDS) with del(5q) arises in genetically discordant clones. *Leukemia Res* 2010; 34: 821–823.
178 Jadersten, M., Saft, L., Smith, A., et al. TP53 mutations in low-risk myelodysplastic syndromes with del(5q) predict disease progression. *J Clin Oncol* 2011; 29: 1971–1979.
179 Jädersten, M., Saft, L., Pellagatti, A., et al. Clonal heterogeneity in the 5q– syndrome: p53 expressing progenitors prevail during lenalidomide treatment and expand at disease progression. *Haematologica* 2009; 94: 1762–1766.
180 Jaju, R.J., Jones, M., Boultwood, J., et al. Combined immunophenotyping and FISH identifies the involvement of B-cells in 5q– syndrome. *Genes Chromosomes Cancer* 2000; 29: 276–280.

181 Nilsson, L., Edén, P., Olsson, E., et al. The molecular signature of MDS stem cells supports a stem-cell origin of 5q myelodysplastic syndromes. *Blood* 2007; 110: 3005–3014.

182 Tehranchi, R., Woll, P.S., Anderson, K., et al. Persistent malignant stem cells in del(5q) myelodysplasia in remission. *N Engl J Med* 2010; 363: 1025–1037.

183 Horrigan, S.K., Arbieva, Z.H., Xie, H.Y., et al. Delineation of a minimal interval and identification of 9 candidates for a tumor suppressor gene in malignant myeloid disorders on 5q31. *Blood* 2000; 95: 2372–2377.

184a Lai, F., Godley, L.A., Joslin, J., et al. Transcript map and comparative analysis of the 1.5-Mb commonly deleted segment of human 5q31 in malignant myeloid diseases with a del(5q). *Genomics* 2001; 71: 235–245.

184b Boultwood, J., Fidler, C., Strickson, A.J., et al. Narrowing and genomic annotation of the commonly deleted region of the 5q– syndrome. *Blood* 2002; 99: 4638–4641.

185 Ebert, B.L., Molecular dissection of the 5q deletion in myelodysplastic syndrome. *Semin Oncol* 2011; 38: 621–626.

186 Pellagatti, A., Fernandez-Mercado, M., Di Genua, C., et al. Whole-exome sequencing in del(5q) myelodysplastic syndromes in transformation to acute myeloid leukemia. *Leukemia* 2014; 28: 1148–1151.

187 Caceres, G., McGraw, K., Yip, B.H., et al. P53 suppression promotes erythropoiesis in del(5q) MDS, suggesting a targeted therapeutic strategy in lenalidomide-resistant patients. *Proc Natl Acad Sci U S A* 2013; 110: 16127–16132.

188 Lane, S.W., Sykes, S.M., Al-Shahrour, F., et al. The Apc(min) mouse has altered hematopoietic stem cell function and provides a model for MPD/MDS. *Blood* 2010; 115: 3489–3497.

189 Wang, J., Fernald, A.A., Anastasi, J., Le Beau, M.M., Qian Z. Haploinsufficiency of Apc leads to ineffective hematopoiesis. *Blood* 2010; 115: 3481–3488.

190 Testa, U., Stellacci, E., Pelosi, E., et al. Impaired myelopoiesis in mice devoid of interferon regulatory factor 1. *Leukemia* 2004; 18: 1864–1871.

191 Joslin, J.M., Fernald, A.A., Tennant, T.R., et al. Haploinsufficiency of EGR1, a candidate gene in the del(5q), leads to the development of myeloid disorders. *Blood* 2007; 110: 719–726.

192 Peng, J., Kitchen, S.M., West, R.A., et al. Myeloproliferative defects following targeting of the Drf1 gene encoding the mammalian diaphanous related formin mDia1. *Cancer Res* 2007; 67: 7565–7571.

193 Dai, X.M., Ryan, G.R., Hapel, A.J., et al. Targeted disruption of the mouse colony-stimulating factor 1 receptor gene results in osteopetrosis, mononuclear phagocyte deficiency, increased primitive progenitor cell frequencies, and reproductive defects. *Blood* 2002; 99: 111–120.

194 Starczynowski, D.T., Kuchenbauer, F., Argiropoulos, B., et al. Identification of miR-145 and miR-146a as mediators of the 5q– syndrome phenotype. *Nat Med* 2010; 16: 49–58.

195 Lehmann, S., O'Kelly, J., Raynaud, S., Funk, S.E., Sage, E.H., Koeffler, H.P. Common deleted genes in the 5q– syndrome: thrombocytopenia and reduced erythroid colony formation in SPARC null mice. *Leukemia* 2007; 21: 1931–1936.

196 Barlow, J.L., Drynan, L.F., Hewett, D.R., et al. A p53-dependent mechanism underlies macrocytic anemia in a mouse model of human 5q– syndrome. *Nat Med* 2010; 16: 59–66.
197 Sportoletti, P., Grisendi, S., Majid, S.M., et al. Npm1 is a haploinsufficient suppressor of myeloid and lymphoid malignancies in the mouse. *Blood* 2008; 111: 3859–3862.
198 Grisendi, S., Bernardi, R., Rossi, M., et al. Role of nucleophosmin in embryonic development and tumorigenesis. *Nature* 2005; 437: 147–153.
199 Fodde, R., Smits, R., Clevers, H. APC, signal transduction and genetic instability in colorectal cancer. *Nat Rev Cancer* 2001; 1: 55–67.
200 Bulavin, D.V., Higashimoto, Y., Popoff, I.J., et al. Initiation of a G2/M checkpoint after ultraviolet radiation requires p38 kinase. *Nature* 2001; 411: 102–107.
201 Wei, S., Chen, X., Rocha, K., et al. A critical role for phosphatase haplodeficiency in the selective suppression of deletion 5q MDS by lenalidomide.*Proc Natl Acad Sci U S A* 2009; 106: 12974–12979.
202 Shibata, H., Takano, H., Ito, M., et al. Alpha-catenin is essential in intestinal adenoma formation. *Proc Natl Acad Sci U S A* 2007; 104: 18199–18204.
203 Ye, Y., McDevitt, M.A., Guo, M., et al. Progressive chromatin repression and promoter methylation of CTNNA1 associated with advanced myeloid malignancies. *Cancer Res* 2009; 69: 8482–8490.
204 Kumar, M.S., Narla, A., Nonami, A., et al. Coordinate loss of a microRNA and protein-coding gene cooperate in the pathogenesis of 5q– syndrome. *Blood* 2011; 118: 4666–4673.
205 Boultwood, J., Pellagatti, A., Cattan, H., et al. Gene expression profiling of $CD34^+$ cells in patients with the 5q– syndrome. *Br J Haematol* 2007; 139: 578–589.
206 Votavova, H., Grmanova, M., Dostalova Merkerova, M., et al. Differential expression of microRNAs in $CD34^+$ cells of 5q– syndrome. *J Hematol Oncol* 2011; 4: 1.
207 Suzuki, H.I., Yamagata, K., Sugimoto, K., Iwamoto, T., Kato, S., Miyazono, K. Modulation of microRNA processing by p53. *Nature* 2009; 460: 529–533.
208 Sachdeva, M., Zhu, S., Wu, F., et al. p53 represses c-Myc through induction of the tumor suppressor miR-145. *Proc Natl Acad Sci U S A* 2009; 106: 3207–3212.
209 Pellagatti, A., Jädersten, M., Forsblom, A.M., et al. Lenalidomide inhibits the malignant clone and up-regulates the SPARC gene mapping to the commonly deleted region in 5q– syndrome patients. *Proc Natl Acad Sci U S A* 2007; 104: 11406–11411.
210 Jakovljevic, J., de Mayolo, P.A., Miles, T.D., et al. The carboxy-terminal extension of yeast ribosomal protein S14 is necessary for maturation of 43S preribosomes. *Mol Cell* 2004; 14: 331–342.
211 Ebert, B.L., Pretz, J., Bosco, J., et al. Identification of RPS14 as a 5q– syndrome gene by RNA interference screen. *Nature* 2008; 451: 335–339.
212 Riley, T., Sontag, E., Chen, P., Levine, A. Transcriptional control of human p53-regulated genes. *Nat Rev Mol Cell Biol* 2008; 9: 402–412.
213 Zhou, X., Hao, Q., Liao, J., Zhang, Q., Lu, H. Ribosomal protein S14 unties the MDM2–p53 loop upon ribosomal stress. *Oncogene* 2013; 32: 388–396.

214 Dutt, S., Narla, A., Lin, K., et al. Haploinsufficiency for ribosomal protein genes causes selective activation of p53 in human erythroid progenitor cells. *Blood* 2011; 117: 2567–2576.
215 Schneider, R.K., Ademà, V., Heckl, D., et al. Role of casein kinase 1A1 in the biology and targeted therapy of del(5q) MDS. *Cancer Cell* 2014; 26: 509–520.
216 Grisendi, S., Mecucci, C., Falini, B., Pandolfi, P.P. Nucleophosmin and cancer. *Nat Rev Cancer* 2006; 6: 493–505.
217 Ammatuna, E., Panetta, P., Agirre, X, et al. NPM1 gene deletions in myelodysplastic syndromes with 5q– and complex karyotype. *Haematologica* 2011; 96: 784–785.
218 Pellagatti, A., Cazzola, M., Giagounidis, A., et al. Marked down-regulation of nucleophosmin-1 is associated with advanced del(5q) myelodysplastic syndrome. *Br J Haematol* 2011; 155: 272–274.
219 Malcovati, L., Karimi, M., Papaemmanuil, E., et al. SF3B1 mutation identifies a distinct subset of myelodysplastic syndrome with ring sideroblasts. *Blood* 2015; 126: 233–241.
220 Christiansen, D.H., Andersen, M.K., Pedersen-Bjergaard, J. Mutations of AML1 are common in therapy-related myelodysplasia following therapy with alkylating agents and are significantly associated with deletion or loss of chromosome arm 7q and with subsequent leukemic transformation. *Blood* 2004; 104: 1474–1481.
221 Okuda, T., van Deursen, J., Hiebert, S.W., Grosveld, G., Downing, J.R. AML1, the target of multiple chromosomal translocations in human leukemia, is essential for normal fetal liver hematopoiesis. *Cell* 1996; 84: 321–330.
222 Lam, K., Zhang, D.E. RUNX1 and RUNX1-ETO: roles in hematopoiesis and leukemogenesis. *Front Biosci* 2012; 17: 1120–1139.
223 Tsuzuki, S., Hong, D., Gupta, R., Matsuo, K., Seto, M., Enver, T. Isoform-specific potentiation of stem and progenitor cell engraftment by AML1/RUNX1. *PLoS Med* 2007; 4(5): e172.
224 Cameron, S., Taylor, D.S., TePas, E.C., Speck, N.A., Mathey-Prevot, B. Identification of a critical regulatory site in the human interleukin-3 promoter by in vivo footprinting. *Blood* 1994; 83: 2851–2859.
225 Takahashi, A., Satake, M., Yamaguchi-Iwai, Y., et al. Positive and negative regulation of granulocyte–macrophage colony-stimulating factor promoter activity by AML1-related transcription factor, PEBP2. *Blood* 1995; 86: 607–616.
226 Zhang, D.E., Fujioka, K., Hetherington, C.J., et al. Identification of a region which directs the monocytic activity of the colony-stimulating factor 1 (macrophage colony-stimulating factor) receptor promoter and binds PEBP2/CBF (AML1). *Mol Cell Biol* 1994; 14: 8085–8095.
227 Satoh, Y., Matsumura, I., Tanaka, H., et al. AML1/RUNX1 works as a negative regulator of c-Mpl in hematopoietic stem cells. *J Biol Chem* 2008; 283: 30045–30056.
228 Nuchprayoon, I., Meyers, S., Scott, L.M., Suzow, J., Hiebert, S., Friedman, A.D. PEBP2/CBF, the murine homolog of the human myeloid AML1 and PEBP2 beta/CBF beta proto-oncoproteins, regulates the murine myeloperoxidase and

neutrophil elastase genes in immature myeloid cells. *Mol Cell Biol* 1994; 14: 5558–5568.

229 Ogihara, H., Kanno, T., Morii, E., et al. Synergy of PEBP2/CBF with mi transcription factor (MITF) for transactivation of mouse mast cell protease 6 gene. *Oncogene* 1999; 18: 4632–4639.

230 Chan, E.M., Comer, E.M., Brown, F.C., et al. AML1-FOG2 fusion protein in myelodysplasia. *Blood* 2005; 105: 4523–4526.

231 Imai, Y., Kurokawa, M., Izutsu, K., et al. Mutations of the AML1 gene in myelodysplastic syndrome and their functional implications in leukemogenesis. *Blood* 2000; 96: 3154–3160.

232 Harada, H., Harada, Y., Tanaka, H., Kimura, A., Inaba, T. Implications of somatic mutations in the AML1 gene in radiation-associated and therapy-related myelodysplastic syndrome/acute myeloid leukemia. *Blood* 2003; 101: 673–680.

233 Steensma, D.P., Gibbons, R.J., Mesa, R.A., Tefferi, A., Higgs, D.R. Somatic point mutations in RUNX1/ CBFA2/AML1 are common in high-risk myelodysplastic syndrome, but not in myelofibrosis with myeloid metaplasia. *Eur J Haematol* 2005; 74: 47–53.

234 Harada, Y., Harada, H. Molecular pathways mediating MDS/AML with focus on AML1/RUNX1 point mutations. *J Cell Physiol* 2009; 220: 16–20.

235 Antony-Debré, I., Manchev, V.T., Balayn, N., et al. Level of RUNX1 activity is critical for leukemic predisposition but not for thrombocytopenia. *Blood* 2015; 125: 930–940.

236 Speck, N.A., Gilliland, D.G. Core-binding factors in haematopoiesis and leukaemia. *Nat Rev Cancer* 2002; 2: 502–513.

237 Chen, C.Y., Lin, L.I., Tang, J.L., et al. RUNX1 gene mutation in primary myelodysplastic syndrome – the mutation can be detected early at diagnosis or acquired during disease progression and is associated with poor outcome. *Br J Haematol* 2007; 139: 405–414.

238 Bejar, R., Stevenson, K.E., Caughey, B.A., et al. Validation of a prognostic model and the impact of mutations in patients with lower-risk myelodysplastic syndromes. *J Clin Oncol* 2012; 30: 3376–3382.

239 Zharlyganova, D., Harada, H., Harada, Y., et al. High frequency of AML1/RUNX1 point mutations in radiation-associated myelodysplastic syndrome around Semipalatinsk nuclear test site. *J Radiat Res* 2008; 49: 549–555.

240 Malumbres, M., Barbacid, M. RAS oncogenes: the first 30 years. *Nat Rev Cancer* 2003; 3: 459–465.

241 Schubbert, S., Shannon, K., Bollag, G. Hyperactive Ras in developmental disorders and cancer. *Nat Rev Cancer* 2007; 7: 295–308.

242 Kraulis, P.J., Domaille, P.J., Campbell-Burk, S.L., Van Aken, T., Laue, E.D. Solution structure and dynamics of ras p21.GDP determined by heteronuclear three- and four-dimensional NMR spectroscopy. *Biochemistry* 1994; 33: 3515–3531.

243 Janssen, J., Steenvoorden, A., Lyons, J., et al. Ras gene mutations in acute and chronic myelocytic leukemias, chronic myeloproliferative disorders, and myelodysplastic syndromes. *Proc Natl Acad Sci U S A* 1987; 84: 9228–9232.
244 Padua, R.A., Carter, G., Hughes, D., et al. Ras mutations in myelodysplasia detected by amplification, oligonucteotide hybridization and transformation. *Leukemia* 1988; 2: 503–510.
245 Parker, J., Mufti, G.J. Ras and myelodysplasia: lessons from the last decade. *Semin Hematol* 1996; 33: 206–224.
246 Reuter, C.W., Morgan, M.A., Bergmann, L. Targeting the Ras signaling pathway: a rational, mechanism-based treatment for hematologic malignancies? *Blood* 2000; 96: 1655–1669.
247 Wang, L.C., Kuo, F., Fujiwara, Y., Gilliland, D.G., Golub, T.R., Orkin, S.H. Yolk sac angiogenic defect and intra-embryonic apoptosis in mice lacking the Ets-related factor TEL. *EMBO J* 1997; 16: 4374–4383.
248 Wang, L.C., Swat, W., Fujiwara, Y., et al. The TEL/ETV6 gene is required specifically for hematopoiesis in the bone marrow. *Genes Dev* 1998; 12: 2392–2402.
249 Waga, K., Nakamura, Y., Maki, K., et al. Leukemia-related transcription factor TEL accelerates differentiation of Friend erythroleukemia cells. *Oncogene* 2003; 22: 59–68.
250 Takahashi, W., Sasaki, K., Komatu, N., Mitani, K. TEL/ETV6 accelerates erythroid differentiation and inhibits megakaryocytic maturation in a human leukemia cell line UT-7/GM, *Cancer Sci* 2005; 96: 340–308.
251 Odero, M.D., Carlson, K., Calasanz, M.J., Lahortiga, I., Chinwalla, V., Rowley, J.D. Identification of new translocations involving ETV6 in hematologic malignancies by fluorescence in situ hybridization and spectral karyotyping. *Genes Chromosomes Cancer* 2001; 31: 134–142.
252 Zhang, M.Y., Churpek, J.E., Keel, S.B., et al. Germline ETV6 mutations in familial thrombocytopenia and hematologic malignancy. *Nat Genet* 2015; 47: 180–185.
253 Lane, D.P., Crawford, L.V. T antigen is bound to a host protein in SV40-transformed cells. *Nature* 1979; 278: 261–263.
254 Marcel, V., Perrier, S., Aoubala, M., et al. Δ160p53 is a novel N-terminal p53 isoform encoded by Δ133p53 transcript. *FEBS Lett* 2010; 584: 4463–4468.
255 Linzer, D.I.H., Levine, A.J. Characterization of a 54K dalton cellular SV40 tumor antigen present in SV40-transformed cells and uninfected embryonal carcinoma cells. *Cell* 1979; 17: 43–52.
256 Baker, S.J., Fearon, E.R., Nigro, J.M., et al. Chromosome 17 deletions and p53 gene mutations in colorectal carcinomas. *Science* 1989; 244: 217–221.
257 Donehower, L.A., Harvey, M., Slagle, B.L., et al. Mice deficient for p53 are developmentally normal but susceptible to spontaneous tumours. *Nature* 1992; 356: 215–221.
258 Bargonetti, J., Friedman, P.N., Kern, S.E., Vogelstein, B., Prives, C. Wild-type but not mutant p53 immunopurified proteins bind to sequences adjacent to the SV40 origin of replication. *Cell* 1991; 65: 1083–1091.
259 el-Deiry, W.S., Kern, S.E., Pietenpol, J.A., Kinzler, K.W., Vogelstein, B. Definition of a consensus binding site for p53. *Nat Genet* 1992; 1: 45–49.

260 Bourdon, J.C., Deguin-Chambon, V., Lelong, J.C., et al. Further characterisation of the p53 responsive element – identification of new candidate genes for trans-activation by p53. *Oncogene* 1997; 14: 85–94.

261 el-Deiry, W.S., Tokino, T., Velculescu, V.E., et al. WAF1, a potential mediator of p53 tumor suppression. *Cell* 1993; 75: 817–825.

262 Nakano, K., Vousden, K.H. PUMA, a novel proapoptotic gene, is induced by p53. *Mol Cell* 2001; 7: 683–694.

263 Bourdon, J.C., Renzing, J., Robertson, P.L., Fernandes, K.N., Lane, D.P. Scotin, a novel p53-inducible proapoptotic protein located in the ER and the nuclear membrane. *J Cell Biol* 2002; 158: 235–246.

264 Mercer, W.E., Shields, M.T., Amin, M., et al. Negative growth regulation in a glioblastoma tumor cell line that conditionally expresses human wild-type p53. *Proc Natl Acad Sci U S A* 1990; 87: 6166–6170.

265 Serrano, M., Lin, A.W., McCurrach, M.E., Beach, D., Lowe, S.W. Oncogenic ras provokes premature cell senescence associated with accumulation of p53 and p16INK4a. *Cell* 1997; 88: 593–602.

266 Shaw, P., Bovey, R., Tardy, S., Sahli, R., Sordat, B., Costa, J. Induction of apoptosis by wild-type p53 in a human colon tumor-derived cell line. *Proc Natl Acad Sci U S A* 1992; 89: 4495–4499.

267 Yonish-Rouach, E., Resnitzky, D., Lotem, J., Sachs, L., Kimchi, A., Oren, M. Wild-type p53 induces apoptosis of myeloid leukaemic cells that is inhibited by interleukin-6. *Nature* 1991; 352: 345–347.

268 Liu, G., Parant, J.M., Lang, G., et al. Chromosome stability, in the absence of apoptosis, is critical for suppression of tumorigenesis in Trp53 mutant mice. *Nat Genet* 2004; 36: 63–68.

269 Levine, A.J., H.u., W., Feng, Z. The p53 pathway: what questions remain to be explored? *Cell Death Differ* 2006; 13: 1027–1036.

270 Harris, S.L., Levine, A.J. The p53 pathway: positive and negative feedback loops. *Oncogene* 2005; 24: 2899–2908.

271 Oren, M. Decision making by p53: life, death and cancer. *Cell Death Differ* 2003; 10: 431–442.

272 Scheffner, M., Werness, B.A., Huibregtse, J.M., Levine, A.J., Howley, P.M. The E6 oncoprotein encoded by human papillomavirus types 16 and 18 promotes the degradation of p53. *Cell* 1990; 63: 1129–1136.

273 Olivier, M., Eeles, R., Hollstein, M., Khan M.A., Harris, C.C., Hainaut, P. The IARC TP53 database: new online mutation analysis and recommendations to users. *Hum Mutat* 2002; 19: 607–614.

274 Varley, M. Germline TP53 mutations and Li–Fraumeni syndrome. *Hum Mutat* 2003; 21: 313–320.

275 Bullock, A.N., Fersht, A.R. Rescuing the function of mutant p53. *Nat Rev Cancer* 2001; 1: 68–76.

276 Rücker, F.G., Schlenk, R.F., Bullinger, L., et al. TP53 alterations in acute myeloid leukemia with complex karyotype correlate with specific copy number alterations, monosomal karyotype, and dismal outcome. *Blood* 2012; 119: 2114–2121.

277 Sebaa, A., Ades, L., Baran-Marzack, F., et al. Incidence of 17p deletions and TP53 mutation in myelodysplastic syndrome and acute myeloid leukemia with 5q deletion. *Genes Chromosomes Cancer* 2012; 51: 1086–1092.

278 Kulasekararaj, A.G., Smith, A.E., Mian, S.A., et al. TP53 mutations in myelodysplastic syndrome are strongly correlated with aberrations of chromosome 5, and correlate with adverse prognosis. *Br J Haematol* 2013; 160: 660–672.

279 Delhommeau, F., Dupont, S., Della-Valle, V., et al. Mutation in TET2 in myeloid cancers. *N Engl J Med* 2009; 360: 2289–2301.

280 Tahiliani, M., Koh, K.P., Shen, Y., et al. Conversion of 5-methylcytosine to 5-hydroxymethylcytosine in mammalian DNA by MLL partner TET1. *Science* 2009; 324: 930–935.

281 Iyer, L.M., Tahiliani, M., Rao, A., Aravind, L. Prediction of novel families of enzymes involved in oxidative and other complex modifications of bases in nucleic acids. *Cell Cycle* 2009; 8: 1698–1710.

282 Valinluck, V., Tsai, H.H., Rogstad, D.K., Burdzy, A., Bird, A., Sowers, L.C. Oxidative damage to methyl-CpG sequences inhibits the binding of the methyl-CpG binding domain (MBD) of methyl-CpG binding protein 2 (MeCP2). *Nucleic Acids Res* 2004; 32: 4100–4108.

283 Valinluck, V., Sowers, L.C. Endogenous cytosine damage products alter the site selectivity of human DNA maintenance methyltransferase DNMT1. *Cancer Res* 2007; 67: 946–950.

284 Langemeijer S.M., Kuiper, R.P., Berends, M., et al. Acquired mutations in TET2 are common in myelodysplastic syndromes. *Nat Genet* 2009; 41: 838–842.

285 Abdel-Wahab, O., Manshouri, T., Patel, J., et al. Genetic analysis of transforming events that convert chronic myeloproliferative neoplasms to leukemias. *Cancer Res* 2010; 70: 447–452.

286 Schaub, F.X., Looser, R., Li, S., et al. Clonal analysis of TET2 and JAK2 mutations suggests that TET2 can be a late event in the progression of myeloproliferative neoplasms. *Blood* 2010; 115: 2003–2007.

287 Smith, A.E., Mohamedali, A.M., Kulasekararaj, A., et al. Next-generation sequencing of the TET2 gene in 355 MDS and CMML patients reveals low-abundance mutant clones with early origins, but indicates no definite prognostic value. *Blood* 2010; 116: 3923–3932.

288 Pastor, W.A., Pape, U.J., Huang, Y., et al. Genome-wide mapping of 5-hydroxymethylcytosine in embryonic stem cells. *Nature* 2011; 473: 394–397.

289 La Starza, R., Crescenzi, B., Nofrini, V., et al. FISH analysis reveals frequent co-occurrence of 4q24/TET2 and 5q and/or 7q deletions. *Leuk Res* 2012; 36: 37–41.

290 Ko, M., Huang, Y., Jankowska, A.M., et al. Impaired hydroxylation of 5-methylcytosine in myeloid cancers with mutant TET2. *Nature* 2010; 468: 839–843.

291 Li, Z., Cai, X., Cai, C.L., et al. Deletion of Tet2 in mice leads to dysregulated hematopoietic stem cells and subsequent development of myeloid malignancies. *Blood* 2011; 118: 4509–4518.

292 Lin, T.L., Nagata, Y., Kao, H.W., et al. Clonal leukemic evolution in myelodysplastic syndromes with TET2 and IDH1/2 mutations. *Haematologica* 2014; 99: 28–36.

293 Kosmider, O., Gelsi-Boyer, V., Cheok, M., et al. TET2 mutation is an independent favorable prognostic factor in myelodysplastic syndromes (MDSs). *Blood* 2009; 114: 3285–3291.
294 Jones, P.A., Baylin, S.B. The epigenomics of cancer. *Cell* 2007; 128: 683–692.
295 Chédin, F. The DNMT3 family of mammalian de novo DNA methyltransferases. *Prog Mol Biol Transl Sci* 2011; 101: 255–285.
296 Yan, X.J., Xu, J., Gu, Z.H., et al. Exome sequencing identifies somatic mutations of DNA methyltransferase gene DNMT3A in acute monocytic leukemia. *Nat Genet* 2011; 43: 309–315.
297 Fan, H., Liu, D., Qiu, X., et al. A functional polymorphism in the DNA methyltransferase-3A promoter modifies the susceptibility in gastric cancer but not in esophageal carcinoma. *BMC Med* 2010; 8: 12.
298 Ley, T.J., Ding, L., Walter, M.J., et al. DNMT3A mutations in acute myeloid leukemia. *N Engl J Med* 2010; 363: 2424–2433.
299 Ewalt, M., Galili, N.G., Mumtaz, M., et al. DNMT3a mutations in high-risk myelodysplastic syndrome parallel those found in acute myeloid leukemia. *Blood Cancer J* 2011; 1(3): e9.
300 Walter, M.J., Ding, L., Shen, D., et al. Recurrent DNMT3A mutations in patients with myelodysplastic syndromes. *Leukemia* 2011; 25: 1153–1158.
301 Yamashita, Y., Yuan, J., Suetake, I., et al. Array-based genomic resequencing of human leukemia. *Oncogene* 2010; 29: 3723–3731.
302 Gowher, H., Loutchanwoot, P., Vorobjeva, O., et al. Mutational analysis of the catalytic domain of the murine Dnmt3a DNA-(cytosine C5)-methyltransferase. *J Mol Biol* 2006; 357: 928–941.
303 Challen, G.A., Sun, D., Jeong, M., et al. Dnmt3a is essential for hematopoietic stem cell differentiation. *Nat Genet* 2011; 44: 23–31.
304 Sjöblom, T., Jones, S., Wood, L.D., et al. The consensus coding sequences of human breast and colorectal cancers. *Science* 2006; 314: 268–274.
305 Yan, H., Parsons, D.W., Jin, G., et al. IDH1 and IDH2 mutations in gliomas. *N Engl J Med* 2009; 360: 765–773.
306 Marcucci, G., Maharry, K., Wu, Y.Z., et al. IDH1 and IDH2 gene mutations identify novel molecular subsets within de novo cytogenetically normal acute myeloid leukemia: a Cancer and Leukemia Group B study. *J Clin Oncol* 2010; 28: 2348–2355.
307 Mardis, E.R., Ding, L., Dooling, D.J., et al. Recurring mutations found by sequencing an acute myeloid leukemia genome. *N Engl J Med* 2009; 361: 1058–1066.
308 Paschka, P., Schlenk, R.F., Gaidzik, V.I., et al. IDH1 and IDH2 mutations are frequent genetic alterations in acute myeloid leukemia and confer adverse prognosis in cytogenetically normal acute myeloid leukemia with NPM1 mutation without FLT3 internal tandem duplication. *J Clin Oncol* 2010; 28: 3636–3643.
309 Tefferi, A., Lasho, T.L., Abdel-Wahab, O., et al. IDH1 and IDH2 mutation studies in 1473 patients with chronic-, fibrotic- or blast-phase essential thrombocythemia, polycythemia vera or myelofibrosis. *Leukemia* 2010; 24: 1302–1309.
310 Wagner, K., Damm, F., Gohring, G., et al. Impact of IDH1 R132 mutations and an IDH1 single nucleotide polymorphism in cytogenetically normal acute

myeloid leukemia: SNP rs11554137 is an adverse prognostic factor. *J Clin Oncol* 2010; 28: 2356–2364.

311 Ward, P.S., Patel, J., Wise, D.R., et al. The common feature of leukemia-associated IDH1 and IDH2 mutations is a neomorphic enzyme activity converting α-ketoglutarate to 2-hydroxyglutarate. *Cancer Cell* 2010; 17: 225–234.

312 Kosmider, O., Gelsi-Boyer, V., Slama, L., et al. Mutations of IDH1 and IDH2 genes in early and accelerated phases of myelodysplastic syndromes and MDS/myeloproliferative neoplasms. *Leukemia* 2010; 24: 1094–1096.

313 Xu, X., Zhao, J., Xu, Z., et al. Structures of human cytosolic NADP-dependent isocitrate dehydrogenase reveal a novel self-regulatory mechanism of activity. *J Biol Chem* 2004; 279: 33946–33957.

314 Dang, L., Whit, D.W., Gross, S., et al. Cancer-associated IDH1 mutations produce 2-hydroxyglutarate. *Nature* 2009; 462: 739–744.

315 Gross, S., Cairns, R.A., Minden, M.D., et al. Cancer-associated metabolite 2-hydroxyglutarate accumulates in acute myelogenous leukemia with isocitrate dehydrogenase 1 and 2 mutations. *J Exp Med* 2010; 207: 339–344.

316 Klose, R.J., Kallin, E.M., Zhang, Y. JmjC-domain-containing proteins and histone demethylation. *Nat Rev Genet* 2006; 7: 715–727.

317 Xu, W., Yang, H., Liu, Y., et al. Oncometabolite 2-hydroxyglutarate is a competitive inhibitor of α-ketoglutarate-dependent dioxygenases. *Cancer Cell* 2011: 19: 17–30.

318 Figueroa, M.E., Abdel-Wahab, O., Lu, C., et al. Leukemic IDH1 and IDH2 mutations result in a hypermethylation phenotype, disrupt TET2 function, and impair hematopoietic differentiation. *Cancer Cell* 2010; 18: 553–567.

319 Loenarz, C., Schofield, C.J. Expanding chemical biology of 2-oxoglutarate oxygenases. *Nat Chem Biol* 2008; 4: 152–156.

320 Thol, F., Weissinger, E.M., Krauter, J., et al. IDH1 mutations in patients with myelodysplastic syndromes are associated with an unfavorable prognosis. *Haematologica* 2010; 95: 1668–1674.

321 Cho, Y.S., Kim, E.J., Park, U.H., Sin, H.S., Um, S.J. Additional sex comb-like 1 (ASXL1), in cooperation with SRC-1, acts as a ligand-dependent coactivator for retinoic acid receptor. *J Biol Chem* 2006; 281: 17588–17598.

322 Abdel-Wahab, O., Adli, M., LaFave, L.M., et al. ASXL1 mutations promote myeloid transformation through loss of PRC2-mediated gene repression. *Cancer Cell* 2012; 22: 180–193.

323 Davies, C., Yip, B.H., Fernandez-Mercado, M., et al. Silencing of ASXL1 impairs the granulomonocytic lineage potential of human $CD34^+$ progenitor cells. *Br J Haematol* 2013; 160: 842–850.

324 Hoischen, A., van Bon, B.W., Rodríguez-Santiago, B., et al. De novo nonsense mutations in ASXL1 cause Bohring–Opitz syndrome. *Nat Genet* 2011; 43: 729–731.

325 Magini, P., Della Monica, M., Uzielli, M.L., et al. Two novel patients with Bohring–Opitz syndrome caused by de novo ASXL1 mutations. *Am J Med Genet A* 2012; 158A: 917–921.

326 An, Q., Wright, S.L., Moorman, A.V., et al. Heterogeneous breakpoints in patients with acute lymphoblastic leukemia and the dic(9;20)(p11~13;q11) show recurrent involment of genes at 20q11.21. *Haematologica* 2009; 94: 1164–1169.

327 Schnittger, S., Eder, C., Jeromin, S., et al. ASXL1 exon 12 mutations are frequent in AML with intermediate risk karyotype and are independently associated with an adverse outcome. *Leukemia* 2013; 27: 82–91.
328 Gelsi-Boyer, V., Trouplin, V., Adélaïde, J., et al. Mutations of polycomb-associated gene ASXL1 in myelodysplastic syndromes and chronic myelomonocytic leukaemia. *Br J Haematol* 2009; 145: 788–800.
329 Rocquain, J., Carbuccia, N., Trouplin, V., et al. Combined mutations of ASXL1, CBL, FLT3, IDH1, IDH2, JAK2, KRAS, NPM1, NRAS, RUNX1, TET2 and WT1 genes in myelodysplastic syndromes and acute myeloid leukemias. *BMC Cancer* 2010; 10: 401.
330 Gelsi-Boyer, V., Brecqueville, M., Devillier, R., Murati, A., Mozziconacci, M.J., Birnbaum, D. Mutations in ASXL1 are associated with poor prognosis across the spectrum of malignant myeloid diseases. *J Hematol Oncol* 2012; 5: 12.
331 Tiu, R.V., Visconte, V., Traina, F., Schwandt, A., Maciejewski, J.P. Updates in cytogenetics and molecular markers in MDS. *Curr Hematol Malig Rep* 2011; 6: 126–135.
332 Chang, C.-J., Hung, M.-C. The role of EZH2 in tumour progression. *Br J Cancer* 2012; 106: 243–247.
333 Dillon, S.C., Zhang, X., Trievel, R.C., Cheng, X. *The SET-domain protein superfamily: protein lysine methyltransferases. Genome Biol* 2005; 6: 227.
334 Simon, J.A., Lange, C.A. Roles of the EZH2 histone methyltransferase in cancer epigenetics. *Mutat Res* 2008; 647: 21–29.
335 Xu, F., Liu, X. The role of histone methyltransferase EZH2 in myelodysplastic syndromes. *Expert Rev Hematol* 2012; 5: 177–185.
336 Nikoloski, G., Langemeijer, S.M., Kuiper, R.P., et al. Somatic mutations of the histone methyltransferase gene EZH2 in myelodysplastic syndromes. *Nat Genet* 2010; 42: 665–667.
337 Makishima, H., Jankowska, A.M., Tiu, R.V., et al. Novel homo- and hemizygous mutations in EZH2 in myeloid malignancies. *Leukemia* 2010; 24: 1799–1804.
338 Jaiswal, S., Fontanillas, P., Flannick, J., et al. Age-related clonal hematopoiesis associated with adverse outcomes. *N Engl J Med* 2014; 371: 2488–2498.
339 Wahl, M.C., Will, C.L., Lührmann, R.The spliceosome: design principles of a dynamic RNP machine. *Cell* 2009; 136: 701–718.
340 Thol, F., Kade, S., Schlarmann, C., et al. Frequency and prognostic impact of mutations in SRSF2, U2AF1, and ZRSR2 in patients with myelodysplastic syndromes. *Blood* 2012; 119: 3578–3584.
341 Yoshida, K., Sanada, M., Shiraishi, Y., et al. Frequent pathway mutations of splicing machinery in myelodysplasia. *Nature* 2011; 478: 64–69.
342 Graubert, T.A., Shen, D., Ding, L., et al. Recurrent mutations in the U2AF1 splicing factor in myelodysplastic syndromes. *Nat Genet* 2011; 44: 53–57.
343 Qian, J., Yao, D.-m., Lin, J., et al. U2AF1 mutations in Chinese patients with acute myeloid leukemia and myelodysplastic syndrome. *PLoS One* 2012; 7(9): e45760.
344 Damm, F., Kosmider, O., Gelsi-Boyer, V., et al. Mutations affecting mRNA splicing define distinct clinical phenotypes and correlate with patient outcome in myelodysplastic syndromes. *Blood* 2012; 119: 3211–3218.

345 Patnaik, M.M., Lasho, T.L., Finke, C.M., et al. Spliceosome mutations involving SRSF2, SF3B1, and U2AF35 in chronic myelomonocytic leukemia: prevalence, clinical correlates, and prognostic relevance. *Am J Hematol* 2013; 88: 201–206.
346 Wu, S.J., Kuo, Y.Y., Hou, H.A., et al. The clinical implication of SRSF2 mutation in patients with myelodysplastic syndrome and its stability during disease evolution. *Blood* 2012; 120: 3106–3111.
347 Ogawa, S. Splicing factor mutations in myelodysplasia. *Int J Hematol* 2012; 96: 438–442.
348 NCBI. SF3B1 splicing factor 3b, subunit 1, 155 kDa [Homo sapiens (human)], 2015, http://www.ncbi.nlm.nih.gov/gene/23451 (accessed May 2015).
349 Malcovati, L., Papaemmanuil, E., Bowen, D.T., et al. Clinical significance of SF3B1 mutations in myelodysplastic syndromes and myelodysplastic/myeloproliferative neoplasms. *Blood* 2011; 118: 6239–6246.
350 Damm, F., Thol, F., Kosmider, O., et al. SF3B1 mutations in myelodysplastic syndromes: clinical associations and prognostic implications. *Leukemia* 2012; 26: 1137–1140.
351 Visconte V, Avishai N, Mahfouz R, et al. Distinct iron architecture in SF3B1-mutant myelodysplastic syndrome patients is linked to an SLC25A37 splice variant with a retained intron. *Leukemia*, 2015; 29: 188–195
352 Papaemmanuil, E., Cazzola, M., Boultwood, J., et al. Somatic SF3B1 mutation in myelodysplasia with ring sideroblasts. *N Engl J Med* 2011; 365: 1384–1395.
353 Cazzola, M., Della Porta, M.G., Malcovati, L. The genetic basis of myelodysplasia and its clinical relevance. *Blood* 2013; 122: 4021–4034.
354 Dolatshad, H., Pellagatti, A., Fernandez-Mercado, M., et al. Disruption of SF3B1 results in deregulated expression and splicing of key genes and pathways in myelodysplastic syndrome hematopoietic stem and progenitor cells. *Leukemia* 2015; 29: 1092–1103.
355 Parganas, E., Wang, D., Stravopodis, D., et al. Jak2 is essential for signaling through a variety of cytokine receptors. *Cell* 1998; 93: 385–395.
356 Neubauer, H., Cumano, A., Müller, M., Wu, H., Huffstadt, U., Pfeffer, K. Jak2 deficiency defines an essential developmental checkpoint in definitive hematopoiesis. *Cell* 1998; 93: 397–409.
357 Kralovics, R., Passamonti, F., Buser, A.S., et al. A gain-of-function mutation of JAK2 in myeloproliferative disorders. *N Engl J Med* 2005; 352: 1779–1790.
358 Hellström-Lindberg, E., Cazzola, M. The role of JAK2 mutations in RARS and other MDS. *Hematology Am Soc Hematol Educ Program* 2008; 52–59.
359 Musto, P., Simeon, V., Guariglia, R., et al. Myelodysplastic disorders carrying both isolated del(5q) and JAK2(V617F) mutation: concise review, with focus on lenalidomide therapy. *Onco Targets Ther* 2014; 7: 1043–1050.
360 Swaminathan, G., Tsygankov, A.Y. The Cbl family proteins: ring leaders in regulation of cell signaling. *J Cell Physiol* 2006; 209: 21–43.
361 Levkowitz, G., Waterman, H., Zamir, E., et al. c-Cbl/Sli-1 regulates endocytic sorting and ubiquitination of the epidermal growth factor receptor. *Genes Dev* 1998; 12: 3663–3674.
362 Panigada, M., Porcellini, S., Barbier, E., et al. Constitutive endocytosis and degradation of the pre-T cell receptor. *J Exp Med* 2002; 195: 1585–1597.

363 Kaabeche, K., Guenou, H., Bouvard, D., Didelot, N., Listrat, A., Marie, P.J. Cbl-mediated ubiquitination of alpha5 integrin subunit mediates fibronectin-dependent osteoblast detachment and apoptosis induced by FGFR2 activation. *J Cell Sci* 2005; 118: 1223–1232.
364 Kales, S.C., Ryan, P.E., Nau, M.M., Lipkowitz, S. Cbl and human myeloid neoplasms: the Cbl oncogene comes of age. *Cancer Res* 2010; 70: 4789–4794.
365 Tan, Y.H., Krishnaswamy, S., Nandi, S., et al. CBL is frequently altered in lung cancers: its relationship to mutations in MET and EGFR tyrosine kinases. *PLoS One* 2010; 5(1): e8972.
366 Ryan, P.E., Davies, G.C., Nau, M.M., Lipkowitz, S. Regulating the regulator: negative regulation of Cbl ubiquitin ligases. *Trends Biochem Sci* 2006; 31: 79–88.
367 Makishima, H., Cazzolli, H., Szpurka, H., et al. Mutations of e3 ubiquitin ligase cbl family members constitute a novel common pathogenic lesion in myeloid malignancies. *J Clin Oncol* 2009; 27: 6109–6116.
368 Sanada, M., Suzuki, T., Shih, L.Y., et al. Gain-of-function of mutated C-CBL tumour suppressor in myeloid neoplasms. *Nature* 2009; 460: 904–908.
369 Kao, H.W., Sanada, M., Liang, D.C., et al. A high occurrence of acquisition and/or expansion of C-CBL mutant clones in the progression of high-risk myelodysplastic syndrome to acute myeloid leukemia. *Neoplasia* 2011; 13: 1035–1042.
370 Dumbar, A.J., Gondek, L.P., O'Keefe, C.L., et al. 250K single nucleotide polymorphism array karyotyping identifies acquireduniparental disomy and homozygous mutations, including novel missense substitutions of c-Cbl, in myeloid malignancies. *Cancer Res* 2008; 68: 10349–10357.
371 Slape, C., Liu, L.Y., Beachy, S., Aplan, P.D. Leukemic transformation in mice expressing a NUP98-HOXD13 transgene is accompanied by spontaneous mutations in Nras, Kras, and Cbl. *Blood* 2008; 112: 2017–2019.
372 Jaenisch, R., Bird, A. Epigenetic regulation of gene expression: how the genome integrates intrinsic and environmental signals. *Nat Genet*, 2003; 33(Suppl): 245–254.
373 Álvarez-Errico, D., Vento-Tormo, R., Sieweke, M., Ballestar, E. Epigenetic control of myeloid cell differentiation, identity and function. *Nat Rev Immunol* 2015; 15(1): 7–17.
374 Griffiths, E.A., Gore, S.D. DNA methyltransferase and histone deacetylase inhibitors in the treatment of myelodysplastic syndromes. *Semin Hematol* 2008; 45: 23–30.
375 Fenaux, P., Mufti, G.J., Hellstrom-Lindberg, E., et al. Efficacy of azacitidine compared with that of conventional care regimens in the treatment of higher-risk myelodysplastic syndromes: a randomised, open-label, phase III study. *Lancet Oncol* 2009; 10: 223–232.
376 Figueroa, M.E., Skrabanek, L., Li, Y., et al. MDS and secondary AML display unique patterns and abundance of aberrant DNA methylation. *Blood* 2009; 114: 3448–3458.
377 Voso, M.T., D'Alò, F., Greco, M., et al. Epigenetic changes in therapy-related MDS/AML. *Chem Biol Interact* 2010; 184: 46–49.

378 Potapova, A., Hasemeier, B., Römermann, D., et al. Epigenetic inactivation of tumour suppressor gene KLF11 in myelodysplastic syndromes. *Eur J Haematol* 2010; 84: 298–303.
379 Shen, L., Kantarjian, H., Guo, Y., et al. DNA methylation predicts survival and response to therapy in patients with myelodysplastic syndromes. *J Clin Oncol* 2010; 28: 605–613.
380 Bies, J., Sramko, M., Fares, J., et al. Myeloid-specific inactivation of p15Ink4b results in monocytosis and predisposition to myeloid leukemia. *Blood* 2010; 116: 979–987.
381 Aggerholm, A., Holm, M.S., Guldberg, P., Olesen, L.H., Hokland, P. Promoter hypermethylation of p15INK4B, HIC1, CDH1, and ER is frequent in myelodysplastic syndrome and predicts poor prognosis in early-stage patients. *Eur J Haematol* 2006; 76: 23–32.
382 Kim, M., Oh, B., Kim, S.Y., et al. p15INK4b methylation correlates with thrombocytopenia, blast percentage, and survival in myelodysplastic syndromes in a dose dependent manner: quantitation using pyrosequencing study. *Leuk Res* 2010; 34: 718–722.
383 Zhao, Y., Fei, C., Zhang, X., et al. Methylation of the p73 gene in patients with myelodysplastic syndromes: correlations with apoptosis and prognosis. *Tumour Biol* 2013; 34: 165–172.
384 Jiang, Y., Dunbar, A., Gondek, L.P., et al. Aberrant DNA methylation is a dominant mechanism in MDS progression to AML. *Blood* 2009; 113: 1315–1325.
385 Lee, S., Kwon, H.C., Kim, S.H., et al. Identification of genes underlying different methylation profiles in refractory anemia with excess blast and refractory cytopenia with multilineage dysplasia in myelodysplastic syndrome. *Korean J Hematol* 2012; 47: 186–193.
386 Wang, H., Wang, X.Q., Xu, X.P., Lin, G.W. ID4 methylation predicts high risk of leukemic transformation in patients with myelodysplastic syndrome. *Leuk Res* 2010; 34: 598–604.
387 Lin, J., Yao, D.M., Qian, J., et al. Methylation status of fragile histidine triad (FHIT) gene and its clinical impact on prognosis of patients with myelodysplastic syndrome. *Leuk Res* 2008; 32: 1541–1545.
388 Del Rey, M., O'Hagan, K., Dellett, M., et al. Genome-wide profiling of methylation identifies novel targets with aberrant hypermethylation and reduced expression in low-risk myelodysplastic syndromes. *Leukemia* 2013; 27: 610–618.
389 Hopfer, O., Nolte, F., Mossner, M., et al. Epigenetic dysregulation of GATA1 is involved in myelodysplastic syndromes dyserythropoiesis. *Eur J Haematol* 2012; 88: 144–153.
390 Jones, P.S. Functions of DNA methylation: islands, start sites, gene bodies and beyond. *Nat Rev Genet* 2012; 13: 484–492.
391 Elliott, G., Hong, C., Xing, X., et al. Intermediate DNA methylation is a conserved signature of genome regulation. *Nat Commun* 2015; 6: 6363.
392 Lister, R., Pelizzola, M., Dowen, R.H., et al. Human DNA methylomes at base resolution show widespread epigenomic differences. *Nature* 2009; 462: 315–322.

393 Kouzarides, T. Chromatin modifications and their functions. *Cell* 2007; 128: 693–705.
394 Jenuwein, T., Allis, CD. Translating the histone code. *Science* 2001; 293: 1074–1080.
395 Lachner, M., Jenuwein, T. The many faces of histone lysine methylation. *Curr Opin Cell Biol* 2002; 14: 286–298.
396 Wei, Y., Gañán-Gómez, I., Salazar-Dimicoli, S., McCay, S.L., Garcia-Manero, G. Histone methylation in myelodysplastic syndromes. *Epigenomics* 2011; 3: 193–205.
397 Ryan, B.M., Robles, A.I., Harris, CC. Genetic variation in microRNA networks: the implications for cancer research. *Nat Rev Cancer* 2010; 10: 389–402.
398 Fang, J., Varney, M., Starczynowski, DT. Implication of microRNAs in the pathogenesis of MDS. *Curr Pharm Des* 2012; 18: 3170–3179.
399 Erdogan, B., Facey, C., Qualtieri, J., et al. Diagnostic microRNAs in myelodysplastic syndrome. *Exp Hematol* 2011; 39: 915–926.
400 Rhyasen, G.W., Starczynowski, DT. Deregulation of microRNAs in myelodysplastic syndrome. *Leukemia* 2012; 26: 13–22.
401 Hussein, K., Theophile, K., Büsche, G., et al. Significant inverse correlation of microRNA-150/MYB and microRNA-222/p27 in myelodysplastic syndrome. *Leuk Res* 2010; 34: 328–334.
402 Hussein, K., Theophile, K., Büsche, G., et al. Aberrant microRNA expression pattern in myelodysplastic bone marrow cells. *Leuk Res* 2010; 34: 1169–1174.
403 Erdogan, B., Bosompem, A., Peng, D., et al. Methylation of promoters of microRNAs and their host genes in myelodysplastic syndromes. *Leuk Lymphoma* 2013; 54: 2720–2727.
404 Sokol, L., Caceres, G., Volinia, S., et al. Identification of a risk dependent microRNA expression signature in myelodysplastic syndromes. *Br J Haematol* 2011; 153: 24–32.
405 Lee, D.W., Futami, M., Carroll, M., et al. Loss of SHIP-1 protein expression in high-risk myelodysplastic syndromes is associated with miR-210 and miR-155. *Oncogene* 2012; 31: 4085–4094.
406 Bhagat, T.D., Zhou, L., Sokol, L., et al. miR-21 mediates hematopoietic suppression in MDS by activating TGF-β signaling. *Blood* 2013; 121: 2875–2881.
407 Vasilatou, D., Papageorgiou, S.G., Kontsioti, F., et al. Expression analysis of mir-17-5p, mir-20a and let-7a microRNAs and their target proteins in $CD34^+$ bone marrow cells of patients with myelodysplastic syndromes. *Leuk Res* 2013; 37: 251–258.
408 Beck, D., Ayers, S., Wen, J., et al. Integrative analysis of next generation sequencing for small non-coding RNAs and transcriptional regulation in myelodysplastic syndromes. *BMC Med Genomics* 2011; 4: 19.
409 Raab, J.R., Chiu, J., Zhu, J., et al. Human tRNA genes function as chromatin insulators. *EMBO J* 2012; 31: 330–350.
410 Van Raam, B.J., Salvesen, GS. Transferring death: a role for tRNA in apoptosis regulation. *Mol Cell* 2010; 37: 591–592.
411 Mei, Y., Yong, J., Liu, H., et al. tRNA binds to cytochrome c and inhibits caspase activation. *Mol Cell* 2010; 37: 668–678.

CHAPTER 2

Molecular genetics of the myeloproliferative neoplasms

Philip A. Beer

Introduction

The human myeloproliferative neoplasms (MPNs) are a group of clonal stem-cell malignancies characterized by the overproduction of mature blood cells. The predominant cell type in excess defines the specific disease entity and is closely correlated with clinical complications, although overlap exists in the phenotypic and clinical manifestations of the individual conditions (Table 2.1). In addition, MPNs share a variable tendency to phenotypic transformation, which may manifest as uncontrolled cellular proliferation, bone marrow failure, bone marrow fibrosis or development of acute leukaemia.

Overview of the different types of mutation found in MPN patients

The recent acceleration in sequencing and related technologies has revealed the MPNs as genetically complex disorders. Even at initial diagnosis, MPN patients often harbour several distinct genetic mutations in disparate pathways, with further mutations acquired with disease progression. Constitutional alleles have also been identified that alter the risk of acquiring a clonal MPN or modulate its clinical phenotype. In addition, a group of rare inherited conditions lead to polyclonal overproduction of mature blood cells and represent important differential diagnoses for the clonal MPNs. The first section of this chapter gives an overview of these genetic events divided into five broad categories. The first three categories outline the genetic events that are central to MPN pathogenesis, comprising mutations in cytokine signalling pathways,

The Genetic Basis of Haematological Cancers, First Edition. Edited by Sabrina Tosi and Alistair G. Reid.

Table 2.1 MPN phenotypes and major disease complications.

MPN	Predominant phenotype	Clinical complications	Disease progression
Essential thrombocythaemia (ET)	Thrombocytosis	Thrombosis Haemorrhage	Myelofibrosis (10–30%) AML (2–5%)
Polycythaemia vera (PV)	Erythrocytosis ± thrombocytosis ± granulocytosis	Thrombosis Haemorrhage	Myelofibrosis (10–30%) AML (2–5%)
Primary myelofibrosis (PMF)	Bone marrow fibrosis Splenomegaly Anaemia ± cytopaenias ± thrombocytosis ± granulocytosis	Bone marrow failure Thrombosis Splenic discomfort Cachexia	AML (10–30%) Intractable wasting syndrome
Chronic eosinophilic leukaemia (CEL)	Eosinophilia ± BM mastocytosis	Endomyocardial fibrosis Neuropathy Lung damage	Acute leukaemia: myeloid or lymphoid (rare)
Systemic mastocytosis (SM)	BM mastocytosis Tissue mastocytosis (skin, lymph nodes, spleen)	Syncope, diarrhoea, anaphylaxis Bone marrow failure Hepatosplenic infiltration	Myelofibrosis AML (mast cell or blastic)

BM, bone marrow; ±: variable feature; AML, acute myeloid leukaemia.

mutations in pathways controlling transcriptional regulation and events associated with transformation to advanced-phase disease (including acute leukaemia). The fourth category comprises constitutional alleles that are associated with an increased risk of developing a clonal MPN. The fifth category comprises inherited conditions characterized by polyclonal blood cell overproduction that phenocopy the clonal MPN.

Following a general overview of the different mutation categories, the specific mutations in these five categories are discussed for the individual MPNs. Owing to extensive phenotypic and molecular overlap, polycythaemia vera (PV), essential thrombocythaemia (ET) and primary myelofibrosis (PMF) are considered in a single section. PV, ET and PMF also represent the most extensively studied and characterized of the MPNs and serve as a paradigm for understanding MPN biology in general.

Acquired mutations in cytokine signalling pathways

Genetic alterations that result in constitutive signalling through cytokine pathways are cardinal events in MPN pathogenesis and are directly implicated as drivers of the specific disease phenotypes (Tables 2.2 and 2.3). Signalling pathway mutations target either cytokine receptors or downstream signalling intermediaries (Fig. 2.1). Mutations affecting cytokine receptors result in increased signalling that is limited to a specific cytokine pathway and is therefore usually associated with overproduction of a single blood lineage. Examples include mast cell proliferation with *KIT* mutations and eosinophilia with alterations of *PDGFRA*. Mutations affecting molecules that lie downstream of cytokine receptors result in increased signalling across a group of related cytokine pathways and may, therefore, be associated with overproduction of several cell types. For example, mutations in *JAK2* result in constitutive signalling through the thrombopoietin, erythropoietin and G-CSF receptors, resulting in thrombocytosis, erythrocytosis and granulocytosis. Signalling pathway mutations are generally mutually exclusive, with each individual patient harbouring only one genetic alteration in this

Table 2.2 The prevalence of genetic alterations in specific genes in different myeloid neoplasms.

	PV	ET	PMF	CEL	SM	MPN-BC	CML	CML-BC	AML	MDS	CMML
JAK2	>50%	>50%	>50%	Not reported	Not reported	10-50%	Not reported	Not reported	<10%	<10%	<10%
CALR	Not reported	10-50%	10-50%	Not reported	Not reported	10-50%	Not reported	Not reported	Not reported	<10%	<10%
MPL	Not reported	<10%	10-50%	Not reported	Not reported	10-50%	Not reported	Not reported	<10%	Not reported	Not reported
LNK	<10%	<10%	<10%	Not reported	Not reported	<10%	Not reported	Not reported	<10%	Not reported	Not reported
PDGFRA/B	Not reported	Not reported	Not reported	10-50%	<10%	<10%	Not reported	Not reported	<10%	Not reported	Not reported
KIT	Not reported	Not reported	Not reported	Not reported	>50%	<10%	Not reported	Not reported	<10%	Not reported	Not reported
EZH2	<10%	<10%	<10%	<10%	Not reported	<10%	Not reported	Not reported	<10%	<10%	<10%
SUZ12	<10%	Not reported	<10%	Not reported	Not reported	<10%	Not reported	Not reported	Not reported	Not reported	<10%
ASXL1	<10%	<10%	10-50%	Not reported	10-50%	10-50%	Not reported	<10%	<10%	10-50%	10-50%
TET2	10-50%	<10%	10-50%	Not reported	10-50%	10-50%	Not reported	<10%	10-50%	10-50%	10-50%
IDH1/2	<10%	<10%	<10%	Not reported	Not reported	10-50%	Not reported	<10%	10-50%	<10%	<10%
DNMT3A	<10%	<10%	<10%	Not reported	<10%	<10%	Not reported	Not reported	10-50%	<10%	<10%
Splicing	<10%	<10%	<10%	Not reported	10-50%	<10%	Not reported	Not reported	<10%	>50%	>50%

Key:
- <10% of cases
- 10-50% of cases
- >50% of cases
- Not reported

Table 2.3 Signalling pathway mutations.

Gene	**Normal protein function**	**Mutations**	**Type**	**Functional consequences**	**Phenotypic associations**	**MPN**
JAK2	Mediates signalling via Epo, Tpo and G-CSF receptors	Missense or small deletions	GOF	Constitutive cytokine signalling	↑Hb, ↑granulocytes, ↑Plts	ET, PV, PMF
MPL	Thrombopoietin cytokine receptor	Missense	GOF	Constitutive cytokine signalling	↑Plts	ET, PMF
LNK	Negative regulator of JAK2	Missense or nonsense	LOF	Amplification of JAK2 signalling	↑Hb, ↑granulocytes, ↑Plts	ET, PV, PMF
PDGFRA/ *PDGFRB*	PDGF cytokine receptors	Rearrangement or missense	GOF	Constitutive cytokine signalling	↑Eosinophils ± ↑granulocytes/monocytes	CEL
KIT	SCF (steel factor) cytokine receptor	Missense	GOF	Constitutive cytokine signalling	↑Mast cells	SM

Epo, erythropoietin; Tpo, thrombopoietin; GOF, gain of function; LOF, loss of function; ↑, increased; Hb, haemoglobin; Plts, platelets; ET, essential thrombocythaemia; PV, polycythaemia vera; PMF, primary myelofibrosis; CEL, chronic eosinophilic leukaemia; SM, systemic mastocytosis.

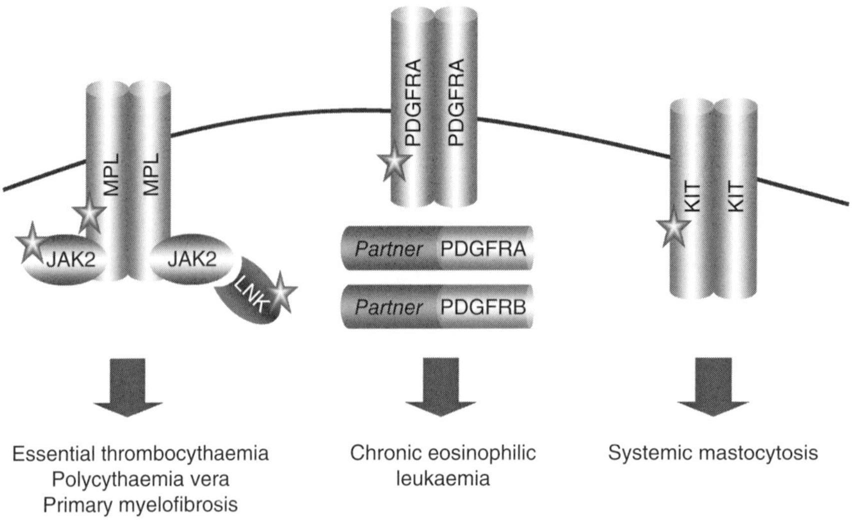

Figure 2.1 Acquired mutations in cytokine signalling pathways. The myeloproliferative neoplasms are characterized by genetic alterations affecting tyrosine kinase molecules and other components of intracellular signalling pathways, comprising point mutations (depicted by stars) and chromosomal rearrangements (where genes are fused with a *partner* gene).

category. The relative specificity for overproduction of one or a few related blood lineages means that mutations in this category are closely associated with specific disorders (Fig. 2.1, Table 2.3).

Acquired mutations in pathways controlling transcriptional regulation

Mutations in genes implicated in the control of transcriptional regulation are common in MPN patients and affect pathways regulating gene expression or RNA splicing (Fig. 2.2). The control of gene expression is regulated at the histone and DNA levels. Nuclear DNA is wound around histone proteins, with the resulting DNA–histone complex referred to as chromatin. The histone protein elements of chromatin can undergo a number of different post-translational modifications, including methylation and acetylation. These histone modifications serve as signals to transcription factors and other mediators of gene expression and play a permissive or repressive role, resulting in either increased or decreased rates of gene transcription. At the DNA level, gene transcription is controlled in part by modification of cytosine residues found within areas of DNA that are rich in cytosine–guanine repeats

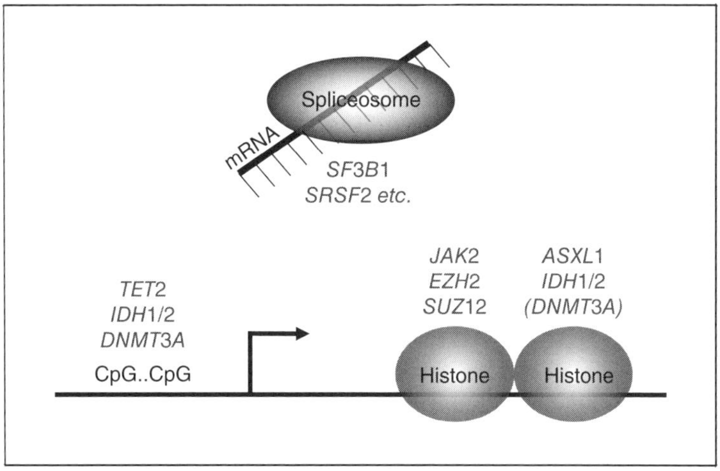

Figure 2.2 Acquired mutations in pathways controlling transcriptional regulation. Mutations in genes implicated in the control of transcriptional regulation are common in the myeloproliferative neoplasms in addition to a wide variety of other myeloid neoplasms and affect pathways regulating histone modification, DNA methylation and RNA splicing. Specific genes targeted by mutation are shown in italics.

(referred to as CpG islands). CpG islands are generally found in gene promoters or close to transcriptional start sites. Methylation of cytosine residues within CpG islands usually has a suppressive effect on gene expression, whereas in actively transcribed genes the cytosine residues are generally unmethylated. More recently, 5-hydroxymethylcytosine has been revealed as an important intermediary in this process.[1] Prior to translation, the nascent RNA molecule undergoes a process of editing known as RNA splicing. This process involves the removal of intronic sequences to generate a mature, protein-coding RNA. In addition, this process can generate diverse RNA species from the same gene, by the inclusion or exclusion of specific exons to generate variant RNAs (referred to as splice variants) encoding divergent proteins. The splicing process is directed by specific DNA sequences within the introns of the gene (referred to as splice donor and splice acceptor sites) and catalysed by a large protein complex known as the spliceosome.

Mutations in genes implicated in histone modification or the control of DNA methylation are found in a wide variety of myeloid malignancies, including the MPN, myelodysplasia and acute myeloid leukaemia (Tables 2.2 and 2.4), and also in lymphoid malignancies, particularly angioimmunoblastic T-cell lymphoma. Mutations in different elements

Table 2.4 Mutations in pathways controlling transcriptional regulation.

Gene	Normal protein function	Mutations	Type	Functional consequences	Phenotypic consequences: mouse models
EZH2	H3K27 methyltransferase	Mainly nonsense, occasional missense	LOF	Reduced H3K27 methylation	↑ HSC competitiveness
SUZ12	H3K27 methyltransferase	Missense	LOF	Reduced H3K27 methylation	↑ HSC competitiveness
ASXL1	H3K4 demethylase	Mainly nonsense, occasional missense	LOF	Altered histone methylation ↑HOX gene expression	Unclear
JAK2	H3Y41 phosphorylation	Missense or small deletions	GOF	Dysregulated gene expression	Consequences of nuclear activity unclear
TET2	DNA 5-hmc production DNA CpG demethylation	Missense or nonsense	LOF	Reduced DNA 5hmC	Progenitor expansion ↑ Myelopoiesis
IDH1/2	Isocitrate metabolism	Missense	NEO	↑ 2-Hydroxyglutarate Reduced DNA 5hmC Altered histone methylation	Progenitor expansion Anaemia
DNMT3A	DNA methyltransferase DNA CpG methylation	Missense or nonsense	LOF	Reduced DNA methylation	HSC accumulation Impaired differentiation
SF3B1	RNA editing	Missense	LOF	Abnormal RNA splicing	Iron deposition in erythroid precursors
Other splicing	RNA editing	Missense	LOF	Abnormal RNA splicing	Unclear

GOF, gain of function; LOF, loss of function; NEO, neomorph mutation alters enzymatic function (see text for details); ↑, increased; 5hmc, 5-hydroxymethylcytosine.

of the spliceosome are distributed even more widely, through myeloid, lymphoid and solid tumours. In contrast to signalling pathway mutations, mutations affecting transcriptional regulation may not lead directly to the overproduction of mature blood cells, are not associated with specific disease phenotypes and often coexist with other transcriptional control mutations in the same patient.

Acquired mutations associated with transformation to advanced-phase disease

Disease progression is a shared feature of the MPNs, although the risk is somewhat variable between the different disorders. Disease progression may involve transformation to acute leukaemia (blastic-phase disease) or the development of aggressive disease in the absence of overt leukaemia (referred to as advanced- or accelerated-phase disease). The phenotypic manifestations of advanced-phase disease are variable and may include uncontrolled cellular proliferation, increasing splenomegaly, bone marrow failure or bone marrow fibrosis. Disease progression is likely driven by the acquisition of additional genetic mutations. Consistent with this, disease progression is often associated with cytogenetic alterations such as chromosomal deletions or duplications, indicating the accumulation of additional genetic damage. Overall, however, the specific molecular lesions that drive disease transformation are not as well characterized as those present in early-phase MPNs.

Inherited predisposition to clonal MPNs

Two distinct patterns of familial MPN predisposition have been identified: (i) rare but strong inherited traits associated with multiple affected individuals within a single family and (ii) common but low-penetrance traits imparting a statistical increase in the risk of acquiring an MPN. In both of these situations, the MPNs that develop are true clonal blood disorders and need to be distinguished from inherited conditions associated with polyclonal blood cell overproduction.

Inherited non-clonal disorders that phenocopy distinct MPNs

Inherited mutations may result in the overproduction of various different blood lineages. These conditions, which are all rare, display phenotypic similarities to the clonal MPNs. The clinical complications and management strategies for inherited disorders, however, are completely different from their clonal counterparts and it is important that these conditions are not confused. Of note, inherited mutations affect some of the same genes that are mutated in acquired clonal MPNs (Table 2.5).

Polycythaemia vera (PV), essential thrombocythaemia (ET) and primary myelofibrosis (PMF)

PV and ET are characterized by overproduction of erythrocytes and platelets, respectively. In both conditions, patients are prone to clinical complications of thrombosis and haemorrhage (Table 2.1). The PMF phenotype is characterized by bone marrow fibrosis secondary to hyperproliferation of clonal megakaryocytes, associated with either thrombocytosis or thrombocytopaenia. The clinical syndrome of PMF is variable and includes complications of thrombocytosis and granulocytosis (thrombosis, haemorrhage), bone marrow failure (anaemia, bleeding, infection), catabolic state (cachexia, lassitude) and increase in spleen size (pancytopaenia, abdominal discomfort). PV, ET and PMF overlap clinically and share a tendency to undergo phenotypic shift, such that patients with ET may develop PV and those with ET or PV may develop a myelofibrotic transformation that is phenotypically indistinguishable from PMF. ET and PV, which represent early-stage disease, generally follow a benign clinical course, with thrombotic complications representing the major source of morbidity. PMF often follows an aggressive course associated with a much-reduced life expectancy and several

Table 2.5 Inherited syndromes with MPN-like phenotypes (mutated genes in parentheses).

Erythrocytosis	High oxygen-affinity haemoglobin (*HBB*, *HBG1*, *HBG2*) 2,3-Biphosphoglycerate mutase deficiency (*BPGM*) Erythropoietin receptor mutation (*EPOR*) Oxygen-sensing pathway mutation (*VHL*, *EGLN1*, *EPAS1*)
Thrombocytosis	Thrombopoietin overproduction (*THPO*) TPO-receptor or JAK2 mutation (*MPL*, *JAK2*)
Myelofibrosis	Grey platelet syndrome (*NBEAL2*)
Eosinophilia	Familial eosinophilia (unknown)
Mastocytosis	Familial mastocytosis (*KIT*)

lines of evidence suggest that PMF represents presentation with an advanced-phase MPN, as discussed in more detail below.

Acquired mutations in cytokine signalling pathways (Table 2.3)

Mutations in *JAK2*

Mutations in *JAK2* are the most common molecular events in the MPN. A *JAK2* V617F mutation is present in around 50% of patients with ET or PMF and up to of 97% of those with PV.[2] *JAK2* V617F mutations are also common in patients with a proliferative variant of chronic myelomonocytic leukaemia[3] and are detected in occasional cases of *de novo* AML, although at least some of the latter represent presentation in the blastic-phase of a previously undiagnosed MPN.[4] The majority of PV patients without a *JAK2* V617F mutation instead harbour a mutation in exon 12 of *JAK2*.[5] A large number of different *JAK2* exon 12 alleles have been reported, comprising either missense mutations or small insertions and deletions leading to in-frame gain or loss of a small number of amino acids.

JAK2 is an intracellular tyrosine kinase that binds to and is essential for signalling through the erythropoietin and thrombopoietin receptors and also contributes to signalling through the receptors for granulocyte colony-stimulating factor (G-CSF), granulocyte–macrophage colony-stimulating factor (GM-CSF), interleukin-3 and interferon-γ. Mutations in *JAK2* reduce the activity of an important autoinhibitory region, resulting in increased basal kinase activity and activation of the JAK2–receptor complex in the absence of ligand binding. This increased signalling activity leads to constitutive activation of downstream effectors, including STAT5, AKT and ERK, resulting in alterations in cellular behaviour that include increased proliferation, resistance to apoptosis and cytokine-independent differentiation.[2,6] In biochemical studies, *JAK2* exon 12 mutations produce similar but stronger activation of downstream targets than the *JAK2* V617F allele.[5] Taken together, these findings imply a direct role for mutant JAK2 in driving the overproduction of the megakaryocyte, erythroid and granulocyte lineages that characterize the clinical phenotypes of ET and PV.

A central role for mutant JAK2 in MPN pathogenesis is corroborated by mouse models of *JAK2* V617F-positive disease. A number of different approaches have been pursued, using expression of either murine or human *JAK2*. These include the use of viruses to induce expression of mutant *JAK2* in transplanted bone marrow[7] or the permanent integration of a mutant *JAK2* gene into the mouse genome, either in a random position (transgenic model)[8] or in place of the normal murine *JAK2* locus (knock-in model).[9] These different approaches have recapitulated most of the major phenotypic features that characterize the *JAK2*-mutant MPN, including thrombocytosis, erythrocytosis, bone marrow fibrosis and extramedullary haematopoiesis. The findings from these studies reinforce a central role for mutant JAK2 in the pathogenesis of these conditions and facilitate further dissection of the disease process. Expression of *JAK2* V617F in mouse cells in which the *Stat5a/b* genes have been deleted abrogates the development of erythrocytosis *in vivo*, but does not prevent the development of bone marrow fibrosis[10,11] This observation, in conjunction with previous *in vitro* studies,[12] cements the role of STAT5 as a key mediator of JAK2 signalling, but implies that other downstream signalling pathways are also important for phenotypic manifestations such as the development of myelofibrosis.

In MPN patients, the *JAK2* V617F mutation is present in progenitors with B-cell, T-cell, NK-cell and myeloid lineage potential, demonstrating that the mutation arises in a multipotent haematopoietic stem cell (HSC).[13] *JAK2* mutations, however, are detected in mature lymphoid cells only in a minority of cases and the proportion of lymphocytes carrying the mutation is usually low. Of relevance to this observation, transformation of cell lines occurs only when mutant JAK2 is co-expressed with a type I homodimeric receptor.[14] Type I homodimeric receptors include the erythropoietin, thrombopoietin and G-CSF receptors. It has been hypothesized that JAK2 requires the presence of one of these receptors as a scaffold in order to activate downstream targets, thus restricting the transforming effects of mutant JAK2 to myeloid lineage cells. In addition, JAK2 is expressed at relatively low levels in haematopoietic stem and progenitor cells and increases at least 10-fold during myeloid differentiation. As such, increasing levels of both JAK2 and its receptors (EpoR, TpoR and G-CSFR) result in an amplification of mutant JAK2 signalling during the later stages of myeloid differentiation. Consistent with this, the mutant allele burden in the stem and progenitor compartment of MPN patients is often low, but increases in terminally differentiated cells; this effect appears particularly pronounced in those harbouring a *JAK2* V617F-homozygous subclone.[15,16]

Evolution of subclones harbouring additional copies of the *JAK2* V617F allele is common in patients with ET and PV. These clones are generated at cell division by mitotic recombination, resulting in duplication of the *JAK2* V617F mutation and loss of the wild-type allele (Fig. 2.3).[17] Homozygosity for *JAK2* exon 12 mutations also arises by the same mechanism. Of interest, fine mapping of the chromosomal cross-over region has indicated that individual patients often harbour several different *JAK2* mutant homozygous clones,[18] consistent with a role for mutant JAK2 in enhancing homologous recombination.[19] An alternative pathway to amplification of the mutant allele involves duplication of the whole of chromosome 9, most likely through chromosomal non-dysjunction at mitosis, producing a trisomy 9 subclone with one wild-type and two mutant *JAK2* alleles (Fig. 2.3).[20,21] Duplication of signalling pathway mutations by mitotic recombination or chromosomal duplication is a common feature of myeloid malignancies, for example *FLT3* and *CBL* mutations in AML. This process is presumed to provide an additional selective advantage to the homozygous clone, either by increased expression of the mutant allele or loss of the wild-type protein.

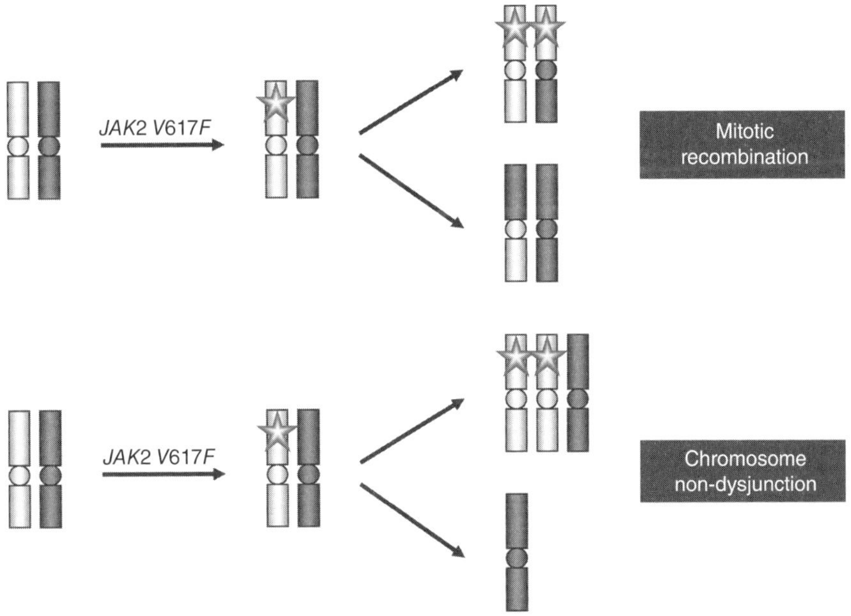

Figure 2.3 Duplication of the *JAK2* V617F allele may arise by mitotic recombination or chromosomal non-dysjunction. Mitotic recombination leads to exchange of genetic material between chromosomes of different parental origin (depicted in white and grey) at cell division, giving rise to daughter clones with either two mutant or two wild-type copies of the *JAK2* allele. Chromosomal non-dysjunction, thought to result from abnormal segregation of a whole chromosome at cell division, gives rise to a daughter clone with two mutant and one wild-type *JAK2* allele. The *JAK2* mutation is depicted by a star.

Progenitor cells from patients with PV or ET harbouring a *JAK2* V617F mutation produce both thrombopoietin-independent megakaryocyte colonies and erythropoietin-independent erythroid colonies, consistent with ligand-independent activation of the thrombopoietin and erythropoietin receptors by mutant JAK2. In the presence of erythropoietin, cultured patient samples typically produce both *JAK2*-mutant and *JAK2*-wild-type erythroid colonies, but only *JAK2*-mutant erythroid progenitors are able to form colonies in the absence of erythropoietin.[15,22] Progenitor cells from *JAK2* V617F-positive MPN patients are competent to engraft immunocompromised mice (NOD/SCID strain), with long-term maintenance of *JAK2*-mutant haematopoiesis in some instances. Of note, whereas normal human progenitors give rise to predominantly lymphoid output in this xenograft system, *JAK2* V617F-positive progenitors produce mainly myeloid cells.[23] This finding provides further evidence that the *JAK2* mutation is acquired in a stem cell and provides a selective advantage for differentiating myeloid but not lymphoid cells.

It is noteworthy that the same *JAK2* V617F mutation is associated with predominant erythrocytosis in patients with PV and thrombocytosis in those with ET. Several lines of evidence implicate the degree of STAT5 activation in determining the disease phenotype, with more pronounced STAT5 activation resulting in erythrocytosis, including (i) the observation that *JAK2* V617F homozygous clones (with stronger STAT5 activation) are over-represented in PV compared with ET,[18,22,24] (ii) the occurrence of *JAK2* exon 12 mutations (which are associated with stronger downstream signalling than *JAK2* V617F) in patients with PV but not ET[5] and (iii) *in vitro* studies indicating that strong, induced STAT5 activation in human progenitor cells results in erythroid differentiation, whereas weaker activation favours megakaryopoiesis.[25] Other contributing factors include the magnitude of STAT1 activation and modulating effects of patient gender.[24,26]

Mutations in *MPL*

Mutations in *MPL*, the thrombopoietin receptor, are present in around 5% of those with ET or PMF.[27,28] Of the *MPL* alleles reported to date, the majority lead to amino acid changes within or close to the intracellular juxtamembrane region of the receptor. This five amino acid motif had previously been identified as an important negative regulator of MPL signalling, with deletion or amino substitution resulting in constitutive MPL signalling due to loss of autoinhibition.[29] The most common

MPN mutations result in amino acid substitutions at position W515 and include changes to leucine, lysine, alanine or arginine. Other rare mutations include amino acid insertions next to or within this motif. Subclones that are homozygous for the *MPL* W515L or W515K allele may arise in patients with ET or PMF. As with mutations in *JAK2*, homozygous clones are generated by mitotic recombination[30,31] (Fig. 2.3).

In vitro expression of *MPL* W515L in cell line models resulted in ligand-independent receptor activation with consequent constitutive activation of key downstream signalling molecules including JAK2, STAT3, STAT5, AKT and ERK. Similar consequences were observed with other alleles encoding mutations or deletions within the juxtamembrane region.[27,29] *In vivo* expression of the *MPL* W515L allele in mouse bone marrow cells recapitulated features of ET and PMF, producing thrombocytosis and progressive bone marrow fibrosis.[27] Overall, the downstream signalling consequences of *MPL* mutations appear similar to mutant JAK2; however, the lineage-restricted expression of the MPL receptor means that the phenotypic consequences of *MPL* mutations are confined to the megakaryocyte lineage.

In patients with an *MPL*-mutant MPN, the mutation may be present in platelets, erythroid cells, granulocytes, B-cells and NK-cells.[32] Progenitors from patients with ET or PMF harbouring an *MPL* W515 mutation produce thrombopoietin-independent megakaryocyte colonies, but not erythropoietin-independent erythroid colonies.[28,32] Progenitors from *MPL*-mutant PMF patients are able to engraft immunocompromised mice.[32] Together, these findings indicate that *MPL* mutations arise in a multipotent stem cell and reinforce the clinical observation that the phenotypic consequences of mutant MPL are limited to the megakaryocyte lineage.

A second, rarer type of alteration targets the transmembrane region of MPL, with a single missense mutation reported to date (*MPL* S505N).[28] *In vitro* expression of this allele in a cell line model resulted in ligand-independent dimerization and consequent activation of the receptor.[33] Substitution of other amino acids at position S505 (such as glutamine, glutamic acid or aspartic acid) also cause constitutive receptor activation *in vitro*,[33] although these mutations have not been reported in MPN patients. Of interest, the same *MPL* S505N allele also occurs as an inherited mutation in kindreds with autosomal dominant familial thrombocytosis (see below).[34] This is an unusual occurrence, as inherited and acquired mutations are generally mutually exclusive, for example, RAS pathway mutations in sporadic and syndromic juvenile myelomonocytic leukaemia[35] and *KIT* mutations in sporadic and familial mastocytosis (see Fig. 2.9 and the systemic mastocytosis section below).

Mutations in *SH2B3* (LNK)

Mutations in *SH2B3* have been detected in rare patients with thrombocytosis, erythrocytosis or myelofibrosis.[36–38] *SH2B3* encodes an adaptor protein, LNK, which binds to and inhibits phosphorylated JAK2, thus exerting a negative effect on signalling through the JAK2–receptor complex. Several different *SH2B3* alleles have been reported, including missense and nonsense mutations, which often coexist with mutations in *JAK2* or *MPL*.

Knock-out of LNK in a mouse model resulted in accumulation of mature myeloid and lymphoid cells. This appears to be driven by hypersensitivity to a number of different myeloid and lymphoid cytokines.[39] In a mouse MPN model driven by activated JAK2 (ETV6–JAK2 fusion), loss of LNK reduced the latency and increased the severity of the disease.[40] In cell-line studies, over-expression of wild-type LNK resulted in marked suppression of signalling through the MPL receptor, in keeping with a role for LNK as a negative regulator of cytokine signalling.[36] In similar experiments, mutant LNK protein showed a partial loss of ability to inhibit both wild-type and mutant (V617F) JAK2 signalling.[36,41,42] Together, these findings imply a role for mutant *SH2B3* alleles in augmenting the signalling effects of mutant JAK2 and MPL.

Mutations in *CALR*

Somatic mutations in the gene encoding calreticulin (*CALR*) have been reported in 15–35% of patients with ET or PMF and occasionally in other myeloid malignancies, but not in PV.[43,44] Of interest, *CALR* mutations are largely mutually exclusive with mutations in *JAK2* and *MPL*, suggesting a role for mutant CALR in signalling pathway activation. Calreticulin is a key endoplasmic reticulum (ER) protein with calcium buffering and protein chaperone activity. It has been implicated in the homeostasis of calcium-dependent signalling and in the ER stress response.[45] A number of different insertion–deletion mutations have been reported, encoding proteins of different sizes.[43,44] Notably, all mutants reported to date harbour an identical and novel C-terminus. The mutant C-terminus is basic (whereas the wild-type is acidic), is lacking the ER retention motif and is predicted to reduce calcium binding activity.[43,44,46] At the time of writing, the cellular consequences of these mutations and their precise impact on patient phenotype were unknown.

Patients with multiple signalling pathway mutations

Mutations in *JAK2* and *MPL* rarely coexist in the same patient. This is consistent with biochemical and cellular studies indicating that these mutations lead to activation of a shared set of downstream pathways and are likely, therefore, to show functional redundancy. Consistent with this redundancy is the observation that in rare patients harbouring two signalling pathway alterations, the mutations are generally present in different clonal proliferations. Two distinct patterns of disease can be observed in such patients: (i) those in whom two clones harbouring different signalling pathway mutations have arisen from a shared ancestral clone and (ii) those harbouring two independent and genetically unrelated MPN clones (Fig. 2.4).[4,20] A related phenomenon is the independent acquisition of the *JAK2* V617F mutation on both parental alleles, a phenomenon that is observed in 5–10% of ET patients and may also indicate the presence of biclonal disease.[47–50]

Acquired mutations in pathways controlling transcriptional regulation (Table 2.4)

Mutations in molecules implicated in the control of transcriptional regulation are present in a proportion of patients with ET, PV or PMF. Mutations in these genes, which affect proteins implicated in histone modification, DNA methylation or RNA splicing, often coexist with a cytokine signalling pathway mutation. In contrast to signalling pathway mutations, patients not uncommonly harbour more than one transcriptional regulation mutation and these mutations do not show strong correlations with specific disease phenotypes.

Mutations in genes implicated in histone modification

Mutations in PRC2

The Polycomb Repressive Complex-2 (PRC2) is a protein complex with histone H3K27 methyltransferase activity involved in the repression of gene expression. Mutations have been reported in components of the core PRC2 complex (*EZH2, EED* and *SUZ12*) and in important PRC2 cofactors (*ASXL1* and *JARID2*). Mutations are usually homozygous or

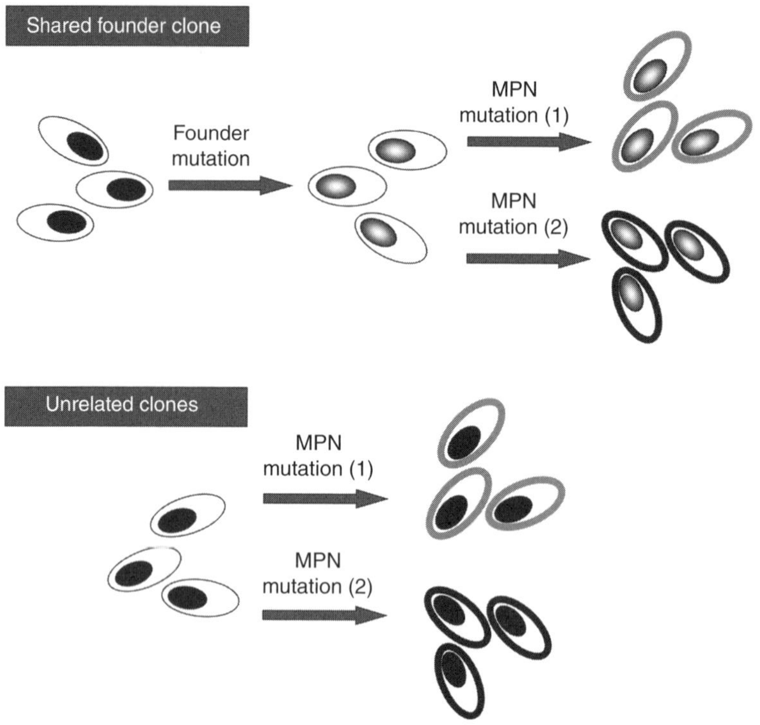

Figure 2.4 Different pathways account for the evolution of oligoclonal disease in patients with a myeloproliferative neoplasm. Two distinct patterns have been observed in patients with biclonal disease. In some cases, two clones are the related progeny of a shared ancestral (founder) clone, for example where a *TET2*-mutant clone gives rise to *TET2/MPL*-mutant and *TET2/JAK2*-mutant daughter clones. In other cases, biclonal disease represents the coexistence of two unrelated MPN clones in the same patient.

hemizygous and result in a loss or decrease in protein function.[49,51–55] The consequent reduction in PRC2 H3K27 methyltransferase activity is predicted to result in increased expression of PRC2-controlled genes. Reduction of ASXL1 expression in human neonatal cells by shRNA resulted in increased expression of *HOX* family genes, which are known to be involved in HSC maintenance and self-renewal. Similar results were obtained in a mouse model, where reduction of ASXL1 was also associated with global reductions in histone H3K27 methylation.[56] Consistent with this, mice with reduced activity of different components of the PRC2 complex show enhanced HSC activity.[57,58] Together, these results imply a role for the PRC2 complex in the negative regulation of HSC activity and suggest that clones with reduced PRC2 activity due to somatic mutation may acquire a selective advantage.

A dual role for mutant JAK2

JAK2 plays a key role in cytokine receptor signalling, as outlined above. In addition, JAK2 acts as a regulator of gene transcription, mediated by its nuclear translocation and direct modification of histone proteins (Fig. 2.5). JAK2 phosphorylates a specific histone tyrosine residue (H3Y41), resulting in displacement of the heterochromatin protein HP1α and consequent de-repression of gene expression.[59] In normal haematopoietic cells, nuclear localization of JAK2 appears contingent upon activation of JAK2 by cytokine signalling, whereas in the presence of a *JAK2* V617F mutation, nuclear localization occurs in the absence of receptor activation.[59,60] JAK2 therefore plays a dual role as a key mediator of cytokine signalling and as a modulator of histone structure and gene expression. At present, the precise transcriptional consequences of JAK2 mutations, and the relative contribution of cytoplasmic versus nuclear JAK2 to cellular transformation, remain to be determined.

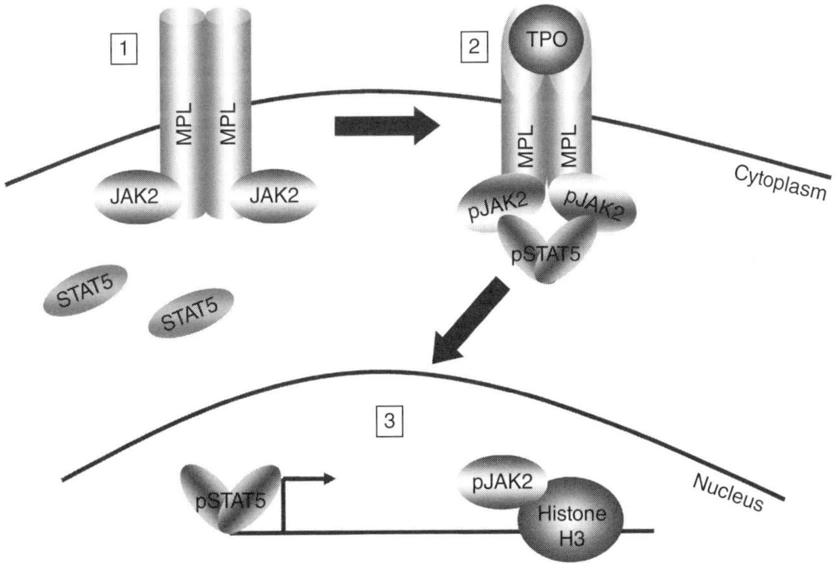

Figure 2.5 Dual roles for JAK2: cytokine signalling and histone modification. JAK2 plays a key role in cytokine receptor signalling, mediated by activation and nuclear translocation of the transcription factor STAT5. In addition, JAK2 acts as a direct regulator of gene transcription, mediated by nuclear translocation of JAK2 and phosphorylation of histone H3Y41, resulting in displacement of the heterochromatin protein HP1α and consequent de-repression of gene expression.

Mutations in genes implicated in DNA methylation

Mutations in *TET2*

Mutations in *TET2* are found in 10–20% of patients with ET, PV or PMF, and also a similar proportion of patients with *de novo* AML, myelodysplasia and chronic myelomonocytic leukaemia (CMML), making *TET2* one of the most frequently mutated genes in myeloid neoplasia (Table 2.2).[61] The TET family of proteins convert 5-methylcytosine to 5-hydroxymethylcytosine, the latter representing a key intermediary in the demethylation of cytosine.[1] It is currently thought that TET proteins are involved in the demethylation of DNA cytosine residues. According to this model, TET proteins play a role in the control of gene transcription by altering the methylation status of promoter CpG islands. Mutations affecting the *TET2* locus, which are generally heterozygous, comprise deletions and missense or nonsense mutations that result in loss of normal TET2 function.[62] In murine cells, reduction of TET2 leads to reduced levels of genomic 5-hydroxymethylcytosine. Human cells from patients harbouring a *TET2* mutation also show reduced levels of 5-hydroxymethylcytosine along with genome-wide changes in CpG island methylation.[62–64] At the cellular level, studies of TET2-mutant MPN patients suggest that loss of TET2 activity results in the clonal expansion of early haematopoietic progenitors.[61] These data are corroborated by mice where the *Tet2* gene has been deleted, which display a gradual increase in the size of the stem cell compartment over time, associated with a competitive advantage of TET2-null stem cells over their wild-type counterparts following transplantation. Of note, these phenotypes are also observed following deletion of only one *Tet2* allele in the mouse, albeit to a lesser degree. This is in keeping with the observation that the majority of *TET2* deletions and mutations in MPN patients are heterozygous and implies that haploinsufficiency for *TET2* may be sufficient to impart a selective advantage to stem and progenitor cells.[65–67]

In addition to effects at the stem cell level, loss of TET2 activity also has consequences for differentiating blood cells. Reduction of TET2 activity in human progenitor cells resulted in skewing towards the monocytic lineage at the expense of lymphoid, erythroid and granulocytic differentiation.[64] Consistent with this, loss of TET2 activity in mouse models is associated with the development a myeloproliferative phenotype comprising monocytosis, splenomegaly and anaemia.[66,67]

These data suggest that loss of TET2 activity exerts effects at multiple levels of the haematopoietic hierarchy, by expanding stem and progenitor cells and by skewing the lineage output of mature cells.

Mutations in *IDH1* and *IDH2*

Mutations in isocitrate dehydrogenase 1 and 2 (*IDH1/2*) are found in a wide variety of myeloid malignancies. The metabolic enzymes IDH1 and IDH2 (cytoplasmic and mitochondrial isoforms, respectively) form part of the citric acid cycle and catalyse the conversion of isocitrate to α-ketoglutarate (α-KG, also known as 2-oxoglutarate). Missense mutations in *IDH1/2*, which are invariably heterozygous, have a dual effect on enzymatic function, resulting in loss of normal function and gain of a novel enzymatic activity. This novel (neomorphic) activity results in the conversion of α-KG to 2-hydroxyglutarate (2-HG),[68] leading to a marked accumulation of 2-HG in IDH-mutant cells. 2-HG is a competitive inhibitor of α-KG and neomorphic IDH mutants inhibit α-KG-dependent enzymes, including TET-family enzymes and Jumonji-family histone demethylases.[63,69] Inhibition of normal TET2 activity results in changes in DNA methylation similar to those seen with mutant TET2 and mutations in *TET2* and *IDH1/2* are mutually exclusive in AML patients. Inhibition of histone demethylase enzymes results in increased methylation at several different histone residues with pleotropic effects on gene expression. These include increased expression of *HOX* family genes, which have been linked to self-renewal of both normal and leukaemic progenitors.[69] Consistent with this, mice harbouring a mutant *IDH1* allele showed expansion of the progenitor compartment, in addition to extramedullary haematopoiesis.[70]

In MPN patients, mutations in *IDH1/2* are most prevalent in blastic-phase disease, although they have also been reported occasionally in patients with ET, PV or PMF.[71,72] As with mutations in *TET2*, it appears that *IDH1/2* mutations can arise as early clonal events, as evidenced by their detection in differentiated haematopoietic colonies in MPN patients and their persistence with normal haematopoiesis in patients successfully treated for AML.[72,73]

Mutations in *DNMT3A*

DNA methyltransferase 3A (DNMT3A) is a *de novo* cytosine methylase implicated in the repression of gene expression via methylation of DNA at CpG islands. Genetic alterations in *DNMT3A* are generally heterozygous and include frameshift and nonsense mutations implying loss of DNMT3A activity;[74,75] in addition, dominant negative effects have been reported for the most common missense mutation.[76] Knock-out of *DNMT3A* in a mouse model resulted in progressive accumulation of haematopoietic stem cells with impaired ability to differentiate.[77] Of note, mutations in *DNMT3A* may coexist with mutations in other DNA methylation pathway genes, including *IDH1/2* and *TET2*, implying non-overlapping functions for these proteins. One possible explanation for this is the potential contribution of DNMT3A to protein complexes with histone-modifying activity.[78] *DNMT3A* mutations appear as early events in *de novo* AML, where they do not appear to impair haematopoietic differentiation.[79]

Mutations in the RNA spliceosome

Mutations affecting the RNA splicing apparatus are found in 5–10% of patients with ET or PMF.[80,81] Alterations in several different genes have been reported to date (*SF3B1, SRSF2, U2AF35, PRPF40B, ZRSR2*), composed almost entirely of heterozygous missense mutations. Mutations in this pathway are also found in other myeloid malignancies, lymphoid neoplasms including chronic lymphocytic leukaemia and multiple myeloma and several different epithelial tumours. Of note, mutations in *SF3B1* show a particularly strong association with the presence of ring sideroblasts (erythroid precursors containing abnormal iron-laden mitochondria). Ring sideroblasts are a morphological feature of a subtype of myelodysplasia (RARS: refractory anaemia with ring sideroblasts) and are also found in a subset of patients with ET or PMF. Mutations in the spliceosome are mutually exclusive in a given patient, implying that different mutations have shared consequences for spliceosome function. In patients with a myeloid neoplasm, spliceosome mutations may coexist with a signalling pathway mutation and/or mutations affecting histone modification or DNA methylation.[81–83]

In model systems, expression of a spliceosome mutant resulted in widespread RNA mis-splicing, including the inclusion of introns in mature RNA species.[81] Analysis of sliceosome mutation-bearing samples from patients with myeloid or lymphoid neoplasms suggests more subtle effects that recurrently target specific genes.[84–86] Although a consensus is emerging of alterations in RNA splicing as a result of these mutations, their cellular consequences are currently obscure.[87]

Acquired mutations associated with progression to advanced and blastic-phase disease

The clinical features of advanced-phase disease, which include uncontrolled cellular proliferation, increasing splenomegaly, bone marrow failure and bone marrow fibrosis, arise in a proportion of ET and PV patients during the course of their disease. Of note, these features are also characteristic of PMF at initial diagnosis. PMF is clinically indistinguishable from myelofibrotic transformation of ET or PV and is associated with several other features of advanced-phase disease, including a high prevalence of karyotypic abnormalities, evidence of ineffective and dysplastic haematopoiesis, high rates of progression to AML and shortened overall survival. Hence PMF may be best considered as presentation with an advanced-phase MPN.

Advanced-phase disease likely arises through a process of clonal evolution characterized by the acquisition of additional genetic events. Constitutive activation of JAK2 has been implicated as a driver of this process, as expression of mutant JAK2 leads to the accumulation of reactive oxygen species, increased DNA damage and aberrant DNA repair.[19,88–90] The degree to which other MPN-associated signalling mutations phenocopy these effects is currently unclear.

Overall, mutations in pathways controlling transcriptional regulation (including *ASXL1*, *IDH1/2* and *EZH2*) appear more common in patients with myelofibrosis (either secondary or PMF) compared with ET or PV, although a causative role for these mutations in disease progression has not been established. Mutations leading to increased signalling via the RAS pathway (affecting *KRAS*, *NRAS* or *CBL*) also appear more common in myelofibrosis compared with ET or PV.[91–94] Cellular studies have suggested that RAS signalling downstream of the erythropoietin receptor exerts a negative effect on erythroid differentiation, whereas RAS signalling downstream of MPL (the thrombopoietin receptor) enhances megakaryocyte differentiation.[95–97] Thus activation of the RAS pathway mimics several phenotypic features of myelofibrosis, including suppression of erythropoiesis and enhanced megakaryopoiesis; however, RAS pathway mutations have only been identified in a minority of patients with primary or post-ET/PV myelofibrosis. At the time of writing, no mutations have been identified that distinguish PMF or post-ET/PV myelofibrosis from early-stage ET and PV.

Transformation to acute myeloid leukaemia (AML) affects 10–30% of those with PMF and a small minority of patients with ET or PV. The mutational profile of post-MPN AML shares some similarities with *de novo*

AML, including mutations in transcriptional control pathways (e.g. *TET2, ASXL1, EZH2* and *IDH1*), which may also be present in earlier stages of disease, and alterations in DNA repair and cellular differentiation pathways (e.g. *TP53, RUNX1* and *IKZF1*), which are rare in early-stage disease. In contrast to *de novo* AML, balanced chromosomal translocations are rare in post-MPN AML.[4,31,98–100]

Surprisingly, leukaemic transformation of a *JAK2* V617F-positive MPN is negative for the *JAK2* mutation in around half of all cases.[4,101,102] Three potential mechanisms may explain this occurrence. One potential mechanism involves leukaemic transformation arising in a wild-type daughter cell resulting from mitotic recombination in a *JAK2* V617F-heterozygous cell (Fig. 2.3). This mechanism has been excluded in all cases examined to date, leaving two possible alternatives. In the first model, the two phases of disease arise from a shared founder clone. In the second model, the MPN and AML are clonally unrelated and reflect transformation of independent stem cells. As noted above, direct evidence exists for both of these models in patients with an early-stage MPN (Fig. 2.4).[4,20] It seems likely, therefore, that either mechanism can account for *JAK2* wild-type AML following a *JAK2* mutant MPN.

Synthesis: genetic complexity, clonal progression and functional overlap of mutations in ET, PV and PMF

The expanding lexicon of genetic mutations has uncovered a hitherto unappreciated complexity in the clonal architecture of the MPNs. Individual patients, including those with early-stage disease, often harbour mutations in several different genes implicated in divergent cellular pathways (Table 2.2). Mutations in signalling pathways are usually mutually exclusive, implying similar functional consequences for mutations in genes such as *JAK2, MPL* and possibly *CALR* (Table 2.3). Mutations in *TET2* and *IDH1/2* are also mutually exclusive, as are spliceosome mutations. The same is not true, however, for other pathways and patients with MPNs or other myeloid neoplasm often harbour more than one mutated gene in the histone modification and/or DNA methylation pathways (Table 2.4), with some mutations showing a positive correlation.[44]

Biochemical studies and mouse models imply a central role for signalling pathway mutations in driving the overproduction of specific mature blood lineages. Although the consequences of mutations affecting histone- or DNA-modifying enzymes are less well characterized, mutations in several enzymes appear to act by imparting haematopoietic stem cells with a competitive advantage (*EZH2/SUZ12, TET2, DNMT3A*) (Table 2.4).

The presence of multiple different mutations in a single patient has allowed studies to track the clonal development of individual MPNs. *TET2* mutations are often acquired as early events during clonal evolution and in some patients precede acquisition of a mutation in *JAK2* or *MPL*.[61] This has led to the suggestion that reduced TET2 activity may be a disease-initiating event, at least in a proportion of patients. In some patients, however, mutations in *JAK2* or *MPL* are acquired before a mutation in *TET2*, implying that several different genetic changes have the ability to initiate a clonal haematopoietic proliferation.[103,104] Overall, many MPN-associated mutations (other than those specifically associated with leukaemic transformation such as alterations of *RUNX1*) may be seen in early-stage (ET/PV) and also advanced-phase disease (PMF or post-ET/PV myelofibrosis), and it appears that the order in which an MPN clone acquires these different mutations is not fixed.

The complex genetic landscape of these disorders suggests that even in the early stages of disease, individual MPN are composed of a collage of mutations with overlapping cellular effects acting at multiple levels of the haematopoietic hierarchy. Signalling pathway mutations appear to play a central role in driving the overproduction of mature blood cells; however, the contributions and interactions of mutations in other pathways are not fully understood.

Inherited predisposition to clonal MPNs

Several families have been reported that contain multiple members affected by an MPN, with inheritance patterns suggestive of an autosomal dominant trait with incomplete penetrance.[105–109] Kindreds with different phenotypes including ET, PV and PMF may be found within the same family and X-chromosome inactivation studies have demonstrated clonal haematopoiesis indicative of a true MPN. Of note, in one study a proportion of apparently unaffected relatives showed erythropoietin-independent erythroid colony growth, suggesting an early-phase or *forme fruste* MPN.[105] Clinical phenotypes and complication rates are reportedly similar in familial and sporadic cases, although in some kindreds the MPN manifests progressively earlier in subsequent generations (genetic anticipation).[108] In addition, the spectrum of somatic mutations reported in these families is similar to those with a sporadic MPN.[110] At the time of writing, the alleles underlying these familial MPNs have yet to be identified.

An increased risk of developing an MPN in first-degree relatives of those with ET or PV (outside families such as those described above) was first identified from cancer registry data.[111] This was subsequently shown to be attributable in part to a constitutional haplotype containing the *JAK2* locus (the 46/1 haplotype). Inheritance of this haplotype is associated with a 3–4-fold increased risk of developing ET or PV.[47,48,112] The presence of the *JAK2* locus within the 46/1 haplotype block suggests that genetic variant(s) affecting JAK2 protein expression or function may account for this effect. This hypothesis is corroborated by the observation that carriers of the 46/1 haplotype are at increased risk of developing a *JAK2*-positive, *MPL*-positive or *JAK2*/*MPL*-wild-type MPN, implying cooperation between the 46/1 variant and different acquired mutations resulting in activation of the JAK2-receptor complex. To date, the 46/1 haplotype has not been shown to exert any consistent effect on disease phenotype or clinical complications. At present, the mechanisms underlying the increased risk in 46/1 carriers is unknown and further studies are under way to determine whether the 46/1 haplotype contains an element imparting a small risk to all those who carry the allele or conceals a less prevalent but more potent allele imparting a greater risk to a minority of carriers.

Inherited non-clonal disorders that phenocopy distinct MPNs

Familial thrombocytosis

Familial thrombocytosis is a rare disorder caused by mutations in thrombopoietin, the thrombopoietin receptor complex or other unknown genes (Fig. 2.6, Table 2.5). Mutations in the 5′-untranslated region or targeting splice sites in the thrombopoietin gene are associated with increased mRNA stability, increased translation of thrombopoietin and consequent thrombocytosis.[113] These mutations are dominantly inherited and have not been reported in acquired, clonal MPN.[114] A mutation in the thrombopoietin receptor (*MPL* S505N) has been reported in Japanese and Italian kindreds.[34,115] This mutation is also dominantly inherited and results in ligand-independent activation of the receptor.[33] Of interest, the same mutation may be acquired in patients with sporadic ET or PMF.[28] Mutations in *JAK2* have been reported in families with autosomal dominant thrombocytosis. These germline mutations

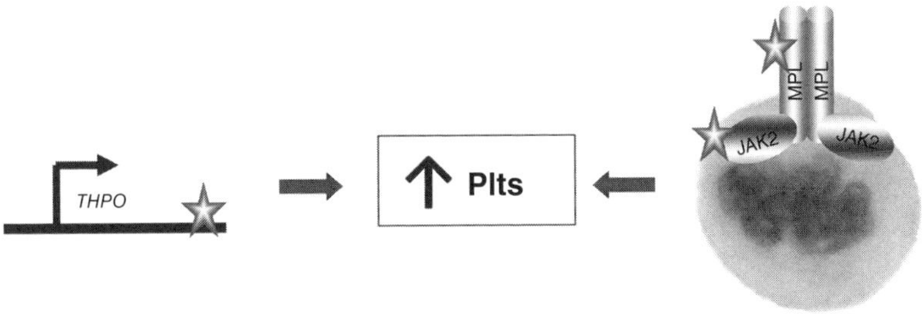

Figure 2.6 Germline mutations associated with hereditary thrombocytosis. Inherited mutations in several genes (marked with red stars) result in thrombocytosis, due to increased translation of thrombopoietin or increased MPL signalling in megakaryocytes.

result in constitutive JAK2 activation, but to a lesser degree than the somatic *JAK2* V617F allele.[116–119] Overall, the clinical course of familial thrombocytosis is generally mild: although complicated by occasional thrombotic or bleeding episodes, development of myelofibrosis or acute leukaemia is not a prominent feature.

Familial polycythaemia

Familial polycythaemia is a rare and genetically heterogeneous disorder, associated with mutations in the haemoglobin molecule, the erythropoietin receptor or elements of the intracellular oxygen-sensing pathway (Fig. 2.7, Table 2.5; http://www.erythrocytosis.org/).[120,121] A number of different mutations in the haemoglobin beta or, rarely, alpha chain are associated with the production of a haemoglobin molecule with increased oxygen affinity. The consequent reduction in tissue oxygen delivery triggers increased erythropoietin production, resulting in a compensatory erythrocytosis. A similar mechanism is responsible for the erythrocytosis seen in a subset of those with hereditary persistence of foetal haemoglobin (foetal haemoglobin has a naturally higher oxygen affinity that adult haemoglobin) and also in those with inherited deficiency of 2,3-biphosphoglycerate mutase, an enzyme that facilitates unloading of oxygen from haemoglobin in the peripheral circulation. High-affinity haemoglobins are usually discovered by chance and are generally of no clinical consequence.

Mutations in the erythropoietin receptor (EpoR) are seen in rare cases of autosomal dominant erythrocytosis. Most reported mutations encode a truncated EpoR that displays hypersensitivity to erythropoietin, with possible mechanisms including loss of docking sites for negative regulators

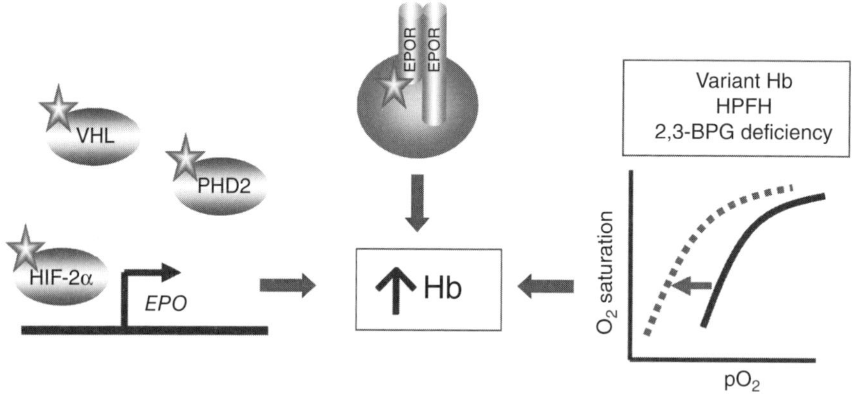

Figure 2.7 Germline mutations associated with hereditary erythrocytosis. Inherited mutations in several genes (marked with stars) result in erythrocytosis, due to production of a haemoglobin molecule with increased oxygen affinity, increased intracellular erythropoietin signalling or inappropriate activation of the hypoxia sensing pathway.

of receptor activation and/or failure to internalize and deactivate the receptor–cytokine complex following ligand binding. Clinical symptoms are variable and an excess mortality from vascular complications has been reported.

Inherited mutations in several genes result in alterations in the intracellular oxygen sensing pathway. Mutations in *VHL* (autosomal recessive), *EGLN1* (PHD2, autosomal dominant) or *EPAS1* (HIF-2α, autosomal dominant) result in increased transcription of hypoxia-inducible genes, including erythropoietin, with consequent erythrocytosis.[122–124] In contrast to other inherited mutations causing erythrocytosis, mutations in this pathway may lead to more profound physiological consequences, including pulmonary hypertension, altered responses to exercise and metabolic changes. The extent of these physiological changes has not been fully characterized but appears to vary according to the specific genetic mutation.[125]

Familial myelofibrosis

Grey platelet syndrome is characterized by variable thrombocytopaenia, abnormal bleeding, large platelets that stain poorly on standard blood films and development of myelofibrosis. The clinical features of this rare autosomal recessive disorder, which is associated with mutations in *NBEAL2*, are due to failure of platelet alpha-granule production. It is thought the associated myelofibrosis is the result of leakage or hyper-secretion of pro-fibrotic factors from bone marrow megakaryocytes. Patients usually present with bleeding or easy bruising, whereas others are discovered incidentally; however, presentation with established myelofibrosis may occur.[126]

Principles and clinical utility of laboratory testing

Molecular testing for genetic mutations has become the first-line investigation for patients with an unexplained increase in platelet count and/or haemoglobin level. A reasonable approach is to screen all such referrals for the *JAK2* V617F mutation. Further testing in V617F-negative cases may comprise screening for *CALR* exon 9 and *MPL* exon 10 mutations in those with thrombocytosis or mutations in *JAK2* exon 12 in those with a raised haemoglobin. The techniques used to detect these mutations should take into account the low mutant allele burden found in some patients (especially those with ET) and the presence of multiple different *CALR* exon 9, *JAK2* exon 12 and *MPL* exon 10 alleles. Suitable techniques include allele-specific or real-time polymerase chain reaction (PCR) for *JAK2* V617F, pyrosequencing or high-resolution melt curve analysis for mutations in *JAK2* exon 12 and *MPL* exon 10 and fragment length analysis for mutations in *CALR* exon 9.[43,127,128] Peripheral blood is a suitable source of material for testing and either DNA or cDNA can be used.

The clinical impact of the identification of mutations in *JAK2*, *MPL* and *CALR* has been significant, resulting in streamlining of the diagnostic process, revisions to diagnostic criteria, reduced cost of diagnosis and reduced need for invasive investigations such as bone marrow aspiration.[129,130] Regarding prognostic value, an increased risk of both venous and arterial thrombosis in *JAK2*-mutant compared to *JAK2*-wild-type ET has been identified in a number of primary studies and confirmed by meta-analysis.[131,132] In addition, a higher mutant allele burden in patients with *JAK2* V617F-positive ET and PV may correlate with an increased risk of thrombotic complications.[133,134] The predictive value of other mutations is currently less clear and different mutations in *JAK2*, *CALR* and *MPL* do not appear to predict for differences in myelofibrotic transformation, acute leukaemia or overall survival.

The clinical utility of testing for mutations in transcriptional regulation pathways is currently not well defined. The widespread occurrence of these mutations limits their diagnostic utility. In addition, mutations in these genes have been detected in patients with apparently normal haematopoiesis.[135] Early data suggest a potential role in patient stratification, with specific mutations (e.g. *ASXL1*) and a higher total mutational load predicting for poor patient outcomes.[104,136–138] At present, however, it is not clear how these data should be used in routine clinical practice.

The identification of the specific molecular events driving MPNs has led to the rapid development of targeted therapies. JAK1/2 inhibitors are effective at reducing spleen size and improving symptoms in patients with myelofibrosis and may increase overall survival, effects which are not limited to patients harbouring a *JAK2* mutation.[139,140] Also of potential interest are small-molecule inhibitors with specificity for mutant forms of IHD1/2.[141,142]

Key points are summarized in Box 2.1.

Box 2.1 Summary of ET, PV and PMF.

Essential thrombocythaemia (ET):
- ET is characterized by overproduction of platelets; clinical complications include thrombosis and haemorrhage.
- Signalling pathway activation is due to mutations in *JAK2* (50%), the thrombopoietin receptor, *MPL* (5%) or rarely *SH2B3* (also known as LNK).
- Mutations in *CALR* (15–35%) are largely mutually exclusive with mutations in *JAK2* and *MPL*, although their functional consequences are currently unknown.
- Inherited thrombocytosis is associated with autosomal dominant mutations in *MPL*, *THPO* or *JAK2* and generally follows a benign clinical course.

Polycythaemia vera (PV):
- PV is characterized by overproduction of erythrocytes with variable excess of platelets and granulocytes; clinical complications include thrombosis and haemorrhage.
- Signalling pathway activation is due to mutations in *JAK2* in most cases (V617F mutation in 97%, exon 12 mutations in 3%).
- Inherited erythrocytosis is associated with mutations in the haemoglobin molecule, the erythropoietin receptor or the intracellular oxygen-sensing pathway (*VHL*, *EGLN1* or *EPAS1*),
- Most inherited erythrocytosis syndromes follow a benign course; however, mutations in the oxygen-sensing pathway may result in more generalized physiological perturbations with consequent morbidity and mortality.

Primary myelofibrosis (PMF):
- PMF is characterized by megakaryocyte proliferation and bone marrow fibrosis, with variable splenomegaly and overproduction of platelets and granulocytes.
- Clinical complications of PMF include thrombosis, haemorrhage, bone marrow failure and a general wasting syndrome, and PMF is associated with a significantly worse prognosis than other MPN.
- PMF patients harbour similar patterns of signalling pathway mutations to those with ET.
- PMF may be best considered as advanced or accelerated-phase disease, although the specific underlying genetic events are currently unknown.
- Inherited myelofibrosis seen in rare kindreds harbouring *NBEAL2* mutations as part of a platelet dysfunction syndrome.

Chronic eosinophilic leukaemia

Chronic eosinophilic leukaemia (CEL) is characterized by an excess of mature eosinophils in the peripheral blood that may be mild to marked in its degree. Eosinophilia is also a feature of other MPNs, including CML, systemic mastocytosis and occasionally PV or PMF. In addition, reactive, polyclonal eosinophilia is seen with haematological malignancies including Hodgkin lymphoma, T-cell lymphoma and B-ALL [especially in association with t(5;14)], along with a number of different epithelial tumours and a broad collection of non-malignant disorders. A diagnosis of CEL is restricted to clonal proliferations where eosinophilia is the predominant or only blood abnormality and other MPNs such as CML have been excluded. Clinical complications of CEL are largely due to eosinophil degranulation, resulting in damage to skin (urticaria, rashes), heart (endomyocardial fibrosis and thrombosis), nervous system (various manifestations: peripheral or central, focal or generalized) and lung (pulmonary fibrosis, infiltrates or pleural effusion). Of note, the presence of end-organ damage is not specific for CEL and may occur as a consequence of any cause of persistent eosinophilia, be it clonal or reactive.

Acquired mutations in cytokine signalling pathways

CEL is characterized by rearrangements leading to constitutive activation of tyrosine kinase molecules, including PDGFRA, PDGFRB and FGFR1 (Fig. 2.1, Table 2.3).[143] In addition, a number of rare fusion genes involving signalling molecules arise in patients with an MPN characterized by eosinophilia, some of which have only been reported in single cases; examples include *PCM1-JAK2* and *ETV6-FLT3*.[144] Of note, a significant number of those diagnosed with CEL do not harbour a rearrangement in one of these genes. The exact proportion of CEL cases lacking a recognized molecular event is uncertain, but may be in excess of 50%. This uncertainty is due to the lack of specific clinical, morphological or laboratory tests for CEL in the absence of a clonal genetic marker and is compounded by myriad causes of reactive eosinophilia.

The platelet-derived growth factor receptors, PDGFRA and PDGFRB, are related transmembrane tyrosine kinase receptors with 46% identity and 73% similarity at the amino acid level. These molecules, which form homodimeric receptors or a heterodimeric complex with each other, are important for the transduction of mitogenic signals in various mesenchymal tissues and are not normally highly expressed in blood cells. Deregulation of PDGFRA/B activity usually arises through genetic rearrangement resulting in the generation of a fusion gene in the orientation *5′ partner–3′ PRDGFRA/B*. The consequences of these fusions include (i) deregulated expression of the *PDGFRA/B*-containing fusion gene under the control of the fusion partner's promoter, (ii) generation of a chimeric oncoprotein with constitutive signalling activity and (iii) expression of a cytoplasmic (rather than transmembrane) protein that escapes normal activation-induced degradation.[145]

Rearrangement of *PDGFRA* occurs in 20–50% of patients diagnosed with CEL. By far the most common *PDGFRA* rearrangement is the *FIP1L1-PDGFRA* fusion, which is due to an intrachromosomal deletion.[146,147] Rare interchromosomal translocations resulting in *PDGFRA* rearrangement have also been reported, all in the orientation *5′ partner–3′ PRDGFRA*, with partners including *ETV6*, *STRN* and *BCR*. The *FIP1L1-PDGFRA* fusion gene encodes a chimeric protein that lacks the transmembrane and extracellular regions of PDGFRA and is therefore entirely intracellular (Fig. 2.8). The fusion is also missing an important PDGFRA autoinhibitory region (the WW-like domain located in the intracellular juxtamembrane region). Loss of this region results in constitutive kinase activity of the chimeric oncoprotein. Other rarer PDGFRA fusion proteins are also intracellular and fusion of PDGFRA to BCR or STRN also disrupts the autoinhibitory WW-like domain. Although the ETV6-PDGFRA fusion retains the WW-like autoinhibitory region, the ETV6 portion contains a strong dimerization motif and biochemical

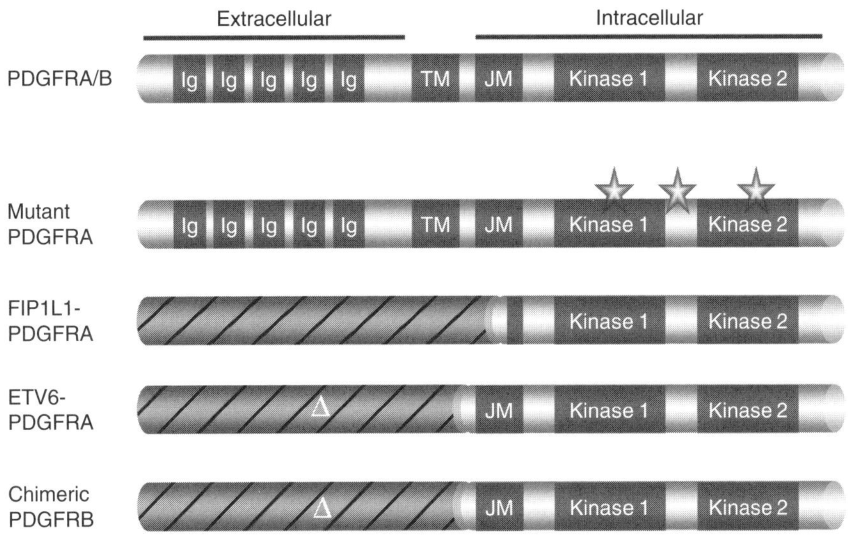

Figure 2.8 Genetic alterations of the platelet-derived growth factor receptors A and B (*PDGFRA/B*) are associated with chronic eosinophilic leukaemia. The PDGFRA/B receptors can be altered by point mutation (depicted by stars) or chromosomal translocation (depicted by fusion to a hatched partner). Some, but not all, fusion partners contain a dimerization motif (depicted by Δ).

studies indicate that this motif is able to overcome autoinhibition, resulting in constitutive tyrosine kinase activity.[148,149]

A minority of CEL patients harbour missense mutations in *PDGFRA*. Several different alleles have been reported, resulting in alterations within or between the kinase domains (Fig. 2.8). Cell-line studies indicate that point mutations in *PDGFRA* result in ligand-independent activation of kinase activity. In contrast to FIP1L1-PDGFRA, which is insensitive to its natural ligands by virtue of being entirely cytoplasmic, PDGFRA point mutants are able to increase further the activation of downstream pathways in response to ligand binding.[150]

Expression of FIP1L1-PDGFRA in human progenitor cells resulted in cytokine-independent differentiation of eosinophil, neutrophil and erythroid lineages.[151,152] This effect was inhibited partially by expression of a dominant negative STAT5 and to a greater extent by co-inhibition of ERK and PI3K,[151] implicating these molecules as downstream mediators of FIP1L1-PDGFRA-induced transformation. Constitutive activation of STAT5 and AKT is also a consequence of missense mutations in the PDGFRA receptor.[150]

In vivo, expression of FIP1L1-PDGFRA in mouse bone marrow cells (by retroviral gene transfer) induced a myeloproliferative disorder similar to CML, associated with a mild eosinophilia and predominant neutrophil proliferation.[153] In mice engineered to express both FIP1L1-PDGFRA and increased levels of IL-5 (a major driver of eosinophil proliferation), the resulting MPN was characterized by predominant eosinophilia, whereas eosinophilia was not as striking in mice expressing *BCR-ABL1* and increased levels of IL-5.[154] Although the relevance of these findings for human CEL is not entirely clear, it is interesting that an inherited polymorphism in the gene encoding the IL-5 receptor has been associated with the degree of eosinophilia and tissue infiltration in *FIP1L1-PDGFRA*-positive CEL patients.[155] Taken together, these findings suggest that other features in addition to the specific oncogene may modulate the CEL disease phenotype, including the genetic background of the individual patient.

In patients with CEL, rearrangements and mutations of *PDGFRA* are associated with an isolated eosinophilia which may be mild or marked. An increase in bone marrow mast cells is also a common feature and occasional cases meet diagnostic criteria for systemic mastocytosis.

Rearrangements of *PDGFRB* arise following interchromosomal translocation, with over 20 different fusion partners reported, all in the *5′ partner–3′ PDGFRB* orientation (Fig. 2.8).[144] The translated chimeric proteins are cytosolic rather than transmembrane, contain a protein–protein dimerization domain contributed by the fusion partner and show constitutive activation of tyrosine kinase activity.

Expression of a PDGFRB fusion in human progenitor cells resulted in cytokine-independent differentiation of eosinophil, neutrophil and erythroid lineages.[152] *In vivo*, expression of *ETV6-PDGFRB* in mouse cells produced an MPN-like phenotype, although as with *FIP1L1-PDGFRA*, eosinophilia was not a prominent feature.[156] *ETV6-PDGFRB* expression in different engineered mouse strains suggested that whereas STAT5 is essential for *ETV6-PDGFRB*-induced disease, STAT1 and SRC are dispensable.[157]

In human CEL, *PDGFRB* rearrangement is associated with phenotypically heterogeneous disease, although eosinophilia is a constant feature. Different MPNs show variable involvement and/or dysplastic maturation of the neutrophil and/or monocyte lineages, with the precise phenotype showing some correlation with the *PDGFRB* fusion partner.

FGFR1-rearranged disease is extremely rare and is associated with a wide range of haematological phenotypes, including CEL, acute myeloid leukaemia and lymphoblastic lymphoma, which often arise sequentially within an individual patient.[158] Rearrangements are in the orientation *5′ partner–3′ FGFR1*. At least 12 different translocation partners have been reported to date, all of which contribute a dimerization domain leading to constitutive activation of FGFR1 signalling.[144]

Acquired mutations in pathways controlling transcriptional regulation

At the time of writing, mutations in genes involved in transcriptional regulation have been reported only rarely in patients with CEL; however, the number of patients studied thus far is small and the occurrence of such mutations remains a possibility.

Acquired mutations associated with progression to advanced and blastic-phase disease

Progression to acute leukaemia is a rare complication of *PDGFRA*- and *PDGFRB*-rearranged disease. Of note, blastic-phase disease may be characterized by a myeloid or lymphoid phenotype and may present with leukaemia or lymphoma. Occasional patients harbour a *PDGFRA* or *PDGFRB* rearrangement at the time of diagnosis with AML, T-cell lymphoma/leukaemia or very rarely B-cell leukaemia. Eosinophilia has been an invariant characteristic of such cases, suggesting they may represent presentation with blastic-phase disease following a previously undiagnosed MPN.[159,160] Blastic-phase disease of myeloid or lymphoid lineage is also a feature of *FGFR1*-rearranged disease, where it is an almost invariable part of the natural history.

Progressive disease in patients with *PDGFRA* or *PDGFRB* rearrangements is likely associated with additional genetic alterations, although the nature of these events is currently unknown. The natural history of *FGFR1*-rearranged disease, with rapid and invariable development of high-grade neoplasia, raises the possibility that disease progression may be driven by the FGFR1 oncoprotein alone. This has not been formally proven, however, and other mechanisms remain possible, such as the induction of marked genetic instability by chimeric FGFR1 oncoproteins.

Inherited predisposition to clonal MPNs

To date, no inherited predisposition to develop CEL has been reported, although studies in this area have been limited.

Inherited non-clonal disorders that phenocopy distinct MPNs

Inherited isolated eosinophilia is a very rare entity. In one large pedigree, inheritance was autosomal dominant and linked to a region on chromosome 5q that is home to a number of cytokines involved in the regulation of eosinophil proliferation and differentiation, including IL-3, IL-5 and GM-CSF. Causative genetic changes, however, have not been identified.[161] Eosinophilia is a feature of a number of rare inherited immunodeficiency syndromes (including Wiskott–Aldrich, IgA deficiency, hyper-IgM and hyper-IgE syndromes); however, other clinical features are generally sufficient to suggest the correct diagnosis.

Principles and clinical utility of laboratory testing

Genetic testing is a key part of the diagnostic process for patients with suspected CEL, both in distinguishing clonal from reactive causes of eosinophilia (which may be difficult on clinical and morphological grounds alone) and in determining response to therapy. *FIP1L1-PDGFRA*, the most common genetic lesion in CEL, is due to a small intrachromosomal deletion that is not visualized by conventional cytogenetic analysis. Approaches to detection include analysis of metaphase or interphase bone marrow cells by fluorescent *in situ* hybridization (FISH) using *FIP1L1-PDGFRA*-specific probes or molecular analysis of peripheral blood or bone marrow cDNA for *FIP1L1-PDGFRA* transcripts. Rearrangements of *PDGFRB* are usually apparent by conventional cytogenetic analysis of metaphase spreads. Alternatively, FISH may be used to detect rearrangement of the *PDGFRB* locus, an approach that is less labour intensive than conventional G-banding analysis and can be performed on interphase cells. In patients with suspected CEL who are negative for rearrangements of *PDGFRA/B*, further testing may include full karyotype analysis and sequencing of *PDGFRA* for point mutations. Alternatively, physicians may opt for an empirical trial of imatinib therapy.

PDGFRA and PDGFRB are highly sensitive to inhibition by tyrosine kinase inhibitors (e.g. imatinib), and patients harbouring fusions or point mutations affecting these proteins can expect to achieve a complete haematological remission with imatinib or similar agents. Molecular response to therapy can be monitored by FISH, although this requires serial bone marrow aspiration. In patients with a *FIL1L1-PDGFRA* fusion, molecular response can be monitored by quantifying the level of *FIL1L1-PDGFRA* transcripts in the peripheral blood by real-time PCR. Occasional patients with *FIL1L1-PDGFRA*-positive disease achieve an initial haematological and molecular response followed by disease relapse associated with increasing *FIL1L1-PDGFRA* transcript levels. These patients often harbour acquired *FIL1L1-PDGFRA* mutations that render the oncoprotein resistant to inhibition by imatinib,[146] a situation analogous to acquired tyrosine kinase resistance secondary to *BCR-ABL1* mutations in patients with chronic myeloid leukaemia. Patients with *PDGFRA/B*-rearranged blastic-phase disease may retain sensitivity to tyrosine kinase inhibitors, although subsequent acquisition of resistance-inducing mutations has been reported. A proportion of patients diagnosed with CEL in the absence of a recognized molecular lesion also show a clinical response to imatinib therapy and an empirical therapeutic trial may be considered.

Patients with *FGFR1* rearrangements are resistant to therapy with first- and second-line tyrosine kinase inhibitors, although agents with potential activity are under development.

Key points are summarized in Box 2.2.

Box 2.2 Summary of CEL.

Chronic eosinophilic leukaemia (CEL):

- CEL is characterized by overproduction of eosinophils; clinical complications result from eosinophil degranulation and include cardiac fibrosis/thrombosis, neuropathy and lung damage.
- In the absence of a clonal genetic alteration, it can be difficult to distinguish CEL from the myriad causes of reactive eosinophilia.
- Signalling pathway activation is commonly due to rearrangements of *PDGFRA* or *PDGFRB*, with rare cases harbouring other constitutive kinases, including FGFR1, JAK2 or FLT3.
- Cases with alterations affecting *PDGFRA/B* are highly responsive to therapy with tyrosine kinase inhibitors (e.g. imatinib), as are a subgroup of cases without an identifiable mutation.
- Familial eosinophilia is very rare and the underlying alleles are currently unknown.

Neoplastic mast cell disease

Systemic mastocytosis (SM) is a rare disorder characterized by proliferation of mast cells in the bone marrow and other tissues, with or without expansion of other myeloid lineages. Clinical manifestations are heterogeneous and include skin rashes, symptoms related to the release from mast cells of vaso-active mediators (hypotension/syncope, diarrhoea, flushing and anaphylaxis) and end-organ damage due to tissue infiltration (bone marrow failure, bone destruction, liver failure, splenomegaly with hypersplenism and gastrointestinal malabsorption). Organ failure is the main source of morbidity and mortality, although a minority of patients develop acute leukaemia.

Acquired mutations in cytokine signalling pathways

SM is characterized by mutations in *KIT*, which encodes the cytokine receptor for stem cell factor (SCF; also known as steel factor or Kit ligand). *KIT* mutations, which are almost invariably heterozygous, result in amino acid changes within the second kinase domain of the receptor. The most common mutation, *KIT* D816V, is present in approximately 90–95% of SM patients, with other variants in adjacent amino acids present in a further 5%. In addition, rare cases harbour mutations in the juxtamembrane domain (Fig. 2.9).[162]

KIT is a transmembrane receptor tyrosine kinase that is expressed on haematopoietic progenitors, mast cells and their progenitors, melanocytes, germ cells and specialist pacemaker cells in the intestine (interstitial cells of Cajal). In the blood system, *KIT* mutations are also prevalent in acute myeloid leukaemia, where they show a strong correlation with rearrangements affecting the RUNX1 core binding factor complex [including inv(16), t(8;21) and mutations in *RUNX1*]. AML-associated *KIT* mutations overlap with SM-associated alleles and additionally affect the extracellular and transmembrane regions of the receptor (Fig. 2.9). Oncogenic *KIT* mutations are also a feature of solid tumours arising in KIT-expressing tissues, including seminoma, melanoma and gastrointestinal stromal tumour (GIST; thought to arise from the interstitial cells of Cajal). *KIT* mutations are a particular feature of GIST, mainly comprising amino acid changes and small deletions affecting the juxtamembrane domain of the receptor.

Mutations in *KIT* result in phosphorylation of the receptor and activation of downstream mediators in the absence of cognate ligand binding.[163] Important downstream targets of KIT include STAT5 and PI3K/AKT. Inhibition of STAT5 impaired the growth of a KIT mutant mast cell leukaemia line and neoplastic mast cells from SM patients show high levels of phosphorylated (activated) STAT5.[164,165] In addition,

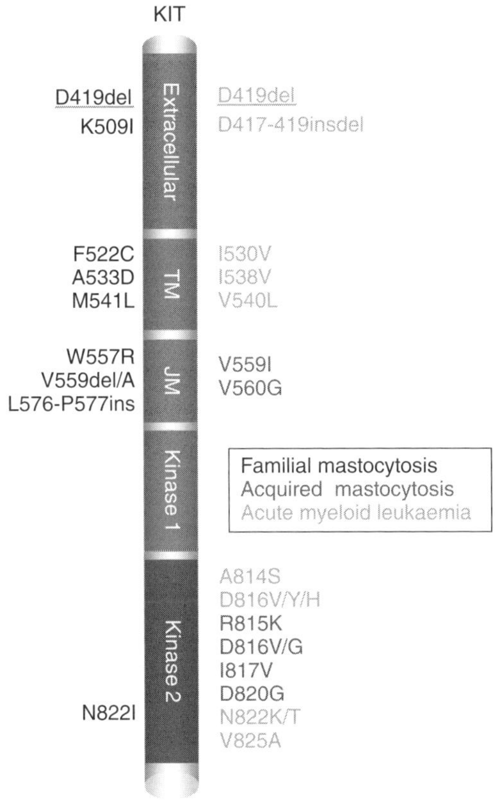

Figure 2.9 An overview of *KIT* mutations in acquired and inherited disease. Schematic representation of the KIT receptor depicting the location of different mutations found in acquired and inherited disorders shown on the right and left hand-side of the diagram respectively, demonstrating that acquired and inherited mutations are almost entirely mutually exclusive, such that with rare exceptions (underlined), inherited mutations are absent from clonal disorders and acquired mutations are not inherited. TM, transmembrane region; JM, intracellular juxtamembrane region.

studies using induced expression of mutant KIT in mouse cells identified an essential role for activation of the PI3K/AKT pathway.[166,167]

In a transgenic mouse model, expression of the human *KIT* D816V allele resulted in mast cell proliferation and infiltration into various tissues, including skin, bone marrow and lymph nodes, along with cytokine-independent expansion of bone marrow cells.[168] In patients with SM, the *KIT* mutation is usually present in peripheral blood B-cells and monocytes in addition to mast cells,[169] consistent with a stem cell disorder.

Approximately 30% of patients with SM have an associated non-mast cell myeloproliferative or myelodysplastic disorder (WHO classification: systemic mastocytosis with associated clonal, haematological non-mast cell lineage disease: SM-AHNMD). The accompanying myeloid proliferation, most CMML or atypical chronic myeloid leukaemia, may be associated with additional genetic mutations, for example *JAK2* V617F with coexisting CMML. Of note, in a patient with coexisting SM and ET, mutations in *KIT* and *JAK2* were found in separate clonal proliferations.[20] Hence in some cases, SM-AHNMD represents morphologically apparent biclonal disease.

Acquired mutations in pathways controlling transcriptional regulation

Mutations in several genes involved in transcriptional control have been reported in SM, including *TET2, ASXL1, DNMT3A* and components of the RNA splicing complex.[170–172] Owing to the rarity of SM and the relatively small numbers of patients enrolled in these studies, it is currently unclear if this mutation pattern differs from other MPNs such as PV or ET. Given that patients with SM-AHNMD may harbour biclonal disease, it is possible that additional mutations are present in a separate neoplastic expansion, rather than representing a cooperating event in the KIT-mutant clone.

Acquired mutations associated with progression to advanced and blastic-phase disease

SM is clinically heterogeneous, ranging from a stable benign disorder to a rapidly progressive condition with an extremely poor prognosis. In any given patient, however, the disease phenotype is usually consistent over time. The presence of mutations affecting transcriptional control pathways has been reported to predict increased disease severity and reduced overall survival;[171,172] however, all studies to date have included small numbers of patients and the clinical utility of these findings remains uncertain. Of interest, constitutional features may also play a role in modulating the SM phenotype, as a specific germline polymorphism in the IL-4 receptor is associated with less aggressive disease.[173]

A small minority of patients progress to acute leukaemia, which may be manifest morphologically by a proliferation of poorly differentiated mast cells (acute mast cell leukaemia) or primitive myeloid blasts (acute myeloid leukaemia). Leukaemic transformation is likely driven by the acquisition of additional genetic events, although the details of these events are currently unknown. For patients with SM-AHNMD, the risk of disease progression is higher, due to transformation of the non-mast cell component of the neoplasm.

Inherited predisposition to clonal MPNs

To date, no inherited predisposition to develop SM has been reported, although studies in this disease have been limited.

Inherited non-clonal disorders that phenocopy distinct MPNs

Inherited *KIT* mutations are associated with an autosomal dominant syndrome characterized by mast cell proliferation, skin hyperpigmentation, mesenteric plexus hypertrophy, dysphagia and GIST tumours. The genetic mutations result in increased KIT signalling and the diverse clinical phenotype reflects hyperproliferation of tissues in which the KIT receptor is normally expressed. The penetrance of the different clinical features is variable, with some kindreds showing mast cell disease or GIST tumours in isolation whereas others manifest the full clinical syndrome. These differences in disease phenotype do not correlate well with specific mutations or mutational hotspots, suggesting that other factors such as the genetic background of the different families may play a role in modulating the clinical phenotype.

The mutations underlying this inherited syndrome are mainly clustered in juxtamembrane region of KIT, although mutations are also seen in the extracellular, transmembrane and kinase domain of the receptor (Fig. 2.9). The same domains of the receptor are also targeted by acquired mutations associated with SM or AML and sometimes affect the same amino acid residues. It is noteworthy, however, that with few exceptions the inherited and acquired alleles are mutually exclusive, such that inherited mutations are unusual in acquired, clonal disorders and acquired mutations are rarely seen as inherited alleles (Fig. 2.9).

The mast cell proliferation associated with familial *KIT* mutations may result in typical skin lesions (e.g. urticaria pigmentosa), symptoms related to mediator release and bone marrow mast cell infiltration. The pattern of infiltrating mast cells in the skin, bone marrow and other tissues may be difficult to distinguish morphologically from acquired neoplastic mast cell disease. Organ failure and other serious manifestations of acquired SM, however, are rarely seen in kindreds with familial mastocytosis.

Principles and clinical utility of laboratory testing

Molecular testing plays a central role in the diagnosis of SM, with the presence of an appropriate *KIT* mutation showing high sensitivity and specificity for this disorder. Testing for *KIT* mutations can be performed on any tissue where an abnormal mast cell infiltrate has been identified. The diagnostic technique used should be of suitable sensitivity as the burden of clonal disease is often low, particularly in the bone marrow, which is the usual source of DNA for molecular testing. For this reason, testing of peripheral blood samples for *KIT* mutations is not usually appropriate. Testing for the common *KIT* D816V mutation is recommended in all cases, with suitable techniques including allele-specific PCR, real-time PCR or melting curve analysis (but not direct sequencing, which lacks sensitivity). Melting curve analysis has the advantage of potential sensitivity to rarer mutations affecting D816 or adjacent amino acids. Additional genetic testing may be considered in SM patients with a second myeloid neoplasm, with the choice of tests based on the clinical diagnosis (for example, testing for a *JAK2* V617F mutation in patients with accompanying ET or CMML).

In patients with clinical features of SM but testing negative for a *KIT* D816V mutation, several further avenues may be explored. Screening for rarer *KIT* alleles may be considered. A family history of mast cell disease and/or gastrointestinal malignancy (suggestive of GIST) should be sought and, if present, testing for inherited *KIT* mutations in the juxtamembrane and other receptor domains may be useful. Bone marrow morphological features of SM may overlap with *FIP1L1-PDGFRA*-positive CEL and testing for the *FIP1L1-PDGFRA* fusion is advised in cases of *KIT*-negative SM (given that *FIP1L1-PDGFRA*-positive disease responds well to therapy with tyrosine kinase inhibitors).

The presence and type of *KIT* mutation play an important role in predicting response to therapy. Although KIT is inhibited by tyrosine kinase inhibitors such as imatinib, mutations at D816 engender resistance to this agent. Second-generation tyrosine kinase inhibitors such as dasatinib show activity against mutant KIT *in vitro*. Results from clinical trials, however, have been disappointing and newer molecules with improved activity are under development. In rare cases where SM is diagnosed in the absence of a *KIT* mutation or in those with familial disease, clinical responses to imatinib or similar agents have been reported, as both wild-type KIT and KIT harbouring mutations in the juxtamembrane domain are sensitive to imatinib and related compounds. The introduction of small-molecule inhibitors with activity against KIT D816 mutants would encourage the development of assays to allow molecular response to therapy to be monitored and quantified.

Key points are summarized in Box 2.3.

Box 2.3 Summary of SM.

Systemic mastocytosis (SM):

- SM is characterized by overproduction of mast cells, occasionally in conjunction with a second myeloid neoplasm.
- Clinical complications result from either mast cell degranulation (hypotension/syncope, diarrhoea, flushing and anaphylaxis) or tissue infiltration (bone marrow failure, bone destruction, liver failure, splenomegaly with hypersplenism and gastrointestinal malabsorption).
- Signalling pathway activation is due to *KIT* mutations in most cases.
- Familial mast cell proliferations are associated with inherited *KIT* mutations and are seen as part of a clinical syndrome which includes gastrointestinal tumours and skin hyperpigmentation.

References

1 Kohli, R.M., Zhang, Y. TET enzymes, TDG and the dynamics of DNA demethylation. *Nature* 2013; 502: 472–479.

2 Levine, R.L., Pardanani, A, Tefferi, A, Gilliland, DG. Role of JAK2 in the pathogenesis and therapy of myeloproliferative disorders. *Nat Rev Cancer* 2007; 7: 673–683.

3 Pich, A., Riera, L., Sismondi, F., et al. JAK2V617F activating mutation is associated with the myeloproliferative type of chronic myelomonocytic leukaemia. *J Clin Pathol* 2009; 62: 798–801.

4 Beer, P.A., Delhommeau, F., Lecouedic, J.P., et al. Two routes to leukemic transformation following a JAK2 mutation-positive myeloproliferative neoplasm. *Blood* 2010; 115(14): 2891–2900.

5 Scott, L.M., Tong, W., Levine, R.L., et al. JAK2 exon 12 mutations in polycythemia vera and idiopathic erythrocytosis. *N Engl J Med* 2007; 356: 459–468.

6 Campbell, P.J., Green, A.R. The myeloproliferative disorders. *N Engl J Med* 2006; 355: 2452–2466.

7 Lacout, C., Pisani, D.F., Tulliez, M., Gachelin, F.M., Vainchenker, W., Villeval, J.L. JAK2V617F expression in murine hematopoietic cells leads to MPD mimicking human PV with secondary myelofibrosis. *Blood* 2006; 108: 1652–1660.

8 Tiedt, R, Hao-Shen, H., Sobas, M.A., et al. Ratio of mutant JAK2-V617F to wild-type Jak2 determines the MPD phenotypes in transgenic mice. *Blood* 2008; 111: 3931–3940.

9 Li, J., Kent, D.G., Chen, E., Green, A.R. Mouse models of myeloproliferative neoplasms: JAK of all grades. *Dis Models Mech* 2011; 4: 311–317.

10 Walz, C., Ahmed, W., Lazarides, K., et al. Essential role for Stat5a/b in myeloproliferative neoplasms induced by BCR-ABL1 and JAK2(V617F) in mice. *Blood* 2012; 119: 3550–3560.
11 Yan, D., Hutchison, R.E., Mohi, G. Critical requirement for Stat5 in a mouse model of polycythemia vera. *Blood* 2012; 119: 3539–3549.
12 Garcon, L., Rivat, C., James, C., et al. Constitutive activation of STAT5 and Bcl-xL overexpression can induce endogenous erythroid colony formation in human primary cells. *Blood* 2006; 108: 1551–1554.
13 Delhommeau, F., Dupont, S., Tonetti, C., et al. Evidence that the JAK2 G1849T (V617F) mutation occurs in a lymphomyeloid progenitor in polycythemia vera and idiopathic myelofibrosis. *Blood* 2007; 109: 71–77.
14 Lu, X., Levine, R., Tong, W., et al. Expression of a homodimeric type I cytokine receptor is required for JAK2V617F-mediated transformation. *Proc Natl Acad Sci U S A* 2005; 102: 18962–18967.
15 Dupont, S., Masse, A., James, C., et al. The JAK2 V617F mutation triggers erythropoietin hypersensitivity and terminal erythroid amplification in primary cells from patients with polycythemia vera. *Blood* 2007; 110: 1013–1021.
16 Anand, S., Stedham, F., Beer, P., et al. Effects of the JAK2 mutation on the hematopoietic stem and progenitor compartment in human myeloproliferative neoplasms. *Blood* 2011; 118: 177–181.
17 Baxter, E.J., Scott, L.M., Campbell, P.J., et al. Acquired mutation of the tyrosine kinase JAK2 in human myeloproliferative disorders. *Lancet* 2005; 365: 1054–1061.
18 Godfrey, A.L., Chen, E., Pagano, F., et al. JAK2V617F homozygosity arises commonly and recurrently in PV and ET, but PV is characterized by expansion of a dominant homozygous subclone. *Blood* 2012; 120: 2704–2707.
19 Plo, I., Nakatake, M., Malivert, L., et al. JAK2 stimulates homologous recombination and genetic instability: potential implication in the heterogeneity of myeloproliferative disorders. *Blood* 2008; 112: 1402–1412.
20 Beer, P.A., Jones, A.V., Bench, A.J., et al. Clonal diversity in the myeloproliferative neoplasms: independent origins of genetically distinct clones. *Br J Haematol* 2009; 144: 904–908.
21 James, C., Ugo, V., Le Couedic, J.P., et al. A unique clonal JAK2 mutation leading to constitutive signalling causes polycythaemia vera. *Nature* 2005; 434: 1144–1148.
22 Scott, L.M., Scott, M.A., Campbell, P.J., Green, A.R. Progenitors homozygous for the V617F mutation occur in most patients with polycythemia vera, but not essential thrombocythemia. *Blood* 2006; 108: 2435–2437.
23 James, C., Mazurier, F., Dupont, S., et al. The hematopoietic stem cell compartment of JAK2V617F-positive myeloproliferative disorders is a reflection of disease heterogeneity. *Blood* 2008; 112: 2429–2438.
24 Godfrey, A.L., Chen, E., Pagano, F., Silber, Y., Campbell, P.J., Green, A.R. Clonal analyses reveal associations of JAK2V617F homozygosity with hematologic features, age and gender in polycythemia vera and essential thrombocythemia. *Haematologica* 2013; 98: 718–721.

25 Olthof, S.G., Fatrai, S., Drayer, A.L., Tyl, M.R., Vellenga, E., Schuringa, J.J. Down-regulation of signal transducer and activator of transcription 5 (STAT5) in CD34$^+$ cells promotes megakaryocytic development, whereas activation of STAT5 drives erythropoiesis. *Stem Cells* 2008; 26: 1732–1742.

26 Chen, E., Beer, P.A., Godfrey, A.L., et al. Distinct clinical phenotypes associated with JAK2V617F reflect differential STAT1 signaling. *Cancer Cell* 2010; 18: 524–535.

27 Pikman, Y., Lee, B.H., Mercher, T., et al. MPLW515L is a novel somatic activating mutation in myelofibrosis with myeloid metaplasia. *PLoS Med* 2006; 3: e270.

28 Beer, P.A., Campbell, P.J., Scott, L.M., et al. MPL mutations in myeloproliferative disorders: analysis of the PT-1 cohort. *Blood* 2008; 112: 141–149.

29 Staerk, J., Lacout, C., Sato, T., Smith, S.O., Vainchenker, W., Constantinescu, S.N. An amphipathic motif at the transmembrane–cytoplasmic junction prevents autonomous activation of the thrombopoietin receptor. *Blood* 2006; 107: 1864–1871.

30 Kawamata, N., Ogawa, S., Yamamoto, G., et al. Genetic profiling of myeloproliferative disorders by single-nucleotide polymorphism oligonucleotide microarray. *Exp Hematol* 2008; 36: 1471–1479.

31 Beer, P.A., Ortmann, C.A., Stegelmann, F., et al. Molecular mechanisms associated with leukemic transformation of MPL-mutant myeloproliferative neoplasms. *Haematologica* 2010; 95: 2153–2156.

32 Chaligne, R., James, C., Tonetti, C., et al. Evidence for MPL W515L/K mutations in hematopoietic stem cells in primitive myelofibrosis. *Blood* 2007; 110: 3735–3743.

33 Ding, J., Komatsu, H., Iida, S., et al. The Asn505 mutation of the c-MPL gene, which causes familial essential thrombocythemia, induces autonomous homodimerization of the c-Mpl protein due to strong amino acid polarity. *Blood* 2009; 114: 3325–3328.

34 Ding, J., Komatsu, H., Wakita, A., et al. Familial essential thrombocythemia associated with a dominant-positive activating mutation of the c-MPL gene, which encodes for the receptor for thrombopoietin. *Blood* 2004; 103: 4198–4200.

35 Schubbert, S., Shannon, K., Bollag, G. Hyperactive Ras in developmental disorders and cancer. *Nat Rev Cancer* 2007; 7: 295–308.

36 Oh, S.T., Simonds, E.F., Jones, C., et al. Novel mutations in the inhibitory adaptor protein LNK drive JAK-STAT signaling in patients with myeloproliferative neoplasms. *Blood* 2010; 116: 988–992.

37 Pardanani, A., Lasho, T., Finke, C., Oh, S.T., Gotlib, J., Tefferi, A. LNK mutation studies in blast-phase myeloproliferative neoplasms and in chronic-phase disease with TET2, IDH, JAK2 or MPL mutations. *Leukemia* 2010; 24: 1713–1718.

38 Lasho, T.L., Pardanani, A., Tefferi, A. LNK mutations in JAK2 mutation-negative erythrocytosis. *N Engl J Med* 2010; 363: 1189–1190.

39 Velazquez, L., Cheng, A.M., Fleming, H.E., et al. Cytokine signaling and hematopoietic homeostasis are disrupted in Lnk-deficient mice. *J Exp Med* 2002; 195: 1599–611.

40 Bersenev, A., Wu, C., Balcerek, J., et al. Lnk constrains myeloproliferative diseases in mice. *J Clin Invest* 2010; 120: 2058–2069.

41 Baran-Marszak, F., Magdoud, H., Desterke, C., et al. Expression level and differential JAK2-V617F-binding of the adaptor protein Lnk regulates JAK2-mediated signals in myeloproliferative neoplasms. *Blood* 2010; 116: 5961–5971.
42 Koren-Michowitz, M., Gery, S., Tabayashi, T., et al. SH2B3 (LNK) mutations from myeloproliferative neoplasms patients have mild loss of function against wild type JAK2 and JAK2 V617F. *Br J Haematol* 2013; 161: 811–820.
43 Klampfl, T., Gisslinger, H., Harutyunyan, A.S., et al. Somatic mutations of calreticulin in myeloproliferative neoplasms. *N Engl J Med* 2013; 369: 2379–2390.
44 Nangalia, J., Massie, C.E., Baxter, E.J., et al. Somatic CALR mutations in myeloproliferative neoplasms with nonmutated JAK2. *N Engl J Med* 2013; 369: 2391–2405.
45 Wang, W.A., Groenendyk, J., Michalak, M. Calreticulin signaling in health and disease. *Int J Biochem Cell Biol* 2012; 44: 842–846.
46 Shivarov, V., Ivanova, M., Tiu, R.V. Mutated calreticulin retains structurally disordered C terminus that cannot bind Ca^{2+}: some mechanistic and therapeutic implications. *Blood Cancer J* 2014; 4: e185.
47 Olcaydu, D., Harutyunyan, A., Jager, R., et al. A common JAK2 haplotype confers susceptibility to myeloproliferative neoplasms. *Nat Genet* 2009; 41: 450–454.
48 Jones, A.V., Campbell, P.J., Beer, P.A., et al. The JAK2 46/1 haplotype predisposes to MPL-mutated myeloproliferative neoplasms. *Blood* 2010; 115: 4517–4523.
49 Beer, P.A., Ortmann, C.A., Campbell, P.J., Green, A.R. Independently acquired biallelic JAK2 mutations are present in a minority of patients with essential thrombocythemia. *Blood* 2010; 116: 1013–1014.
50 Jones, A.V., Chase, A., Silver, R.T., et al. JAK2 haplotype is a major risk factor for the development of myeloproliferative neoplasms. *Nat Genet* 2009; 41: 446–449.
51 Ernst, T., Chase, A.J., Score, J., et al. Inactivating mutations of the histone methyltransferase gene EZH2 in myeloid disorders. *Nat Genet* 2010; 42: 722–726.
52 Score, J., Hidalgo-Curtis, C., Jones, A.V., et al. Inactivation of polycomb repressive complex 2 components in myeloproliferative and myelodysplastic/myeloproliferative neoplasms. *Blood* 2012; 119: 1208–1213.
53 Carbuccia, N., Murati, A., Trouplin, V, et al. Mutations of ASXL1 gene in myeloproliferative neoplasms. *Leukemia* 2009; 23: 2183–2186.
54 Abdel-Wahab, O., Pardanani, A., Patel, J., et al. Concomitant analysis of EZH2 and ASXL1 mutations in myelofibrosis, chronic myelomonocytic leukemia and blast-phase myeloproliferative neoplasms. *Leukemia* 2011; 25: 1200–1202.
55 Stein, B.L., Williams, D.M., O'Keefe, C., et al. Disruption of the ASXL1 gene is frequent in primary, post-essential thrombocytosis and post-polycythemia vera myelofibrosis, but not essential thrombocytosis or polycythemia vera: analysis of molecular genetics and clinical phenotypes. *Haematologica* 2011; 96: 1462–1469.
56 Abdel-Wahab, O., Adli, M., LaFave, L.M., et al. ASXL1 mutations promote myeloid transformation through loss of PRC2-mediated gene repression. *Cancer Cell* 2012; 22: 180–193.
57 Majewski, I.J., Blewitt, M.E., de Graaf, C.A., et al. Polycomb repressive complex 2 (PRC2) restricts hematopoietic stem cell activity. *PLoS Biol* 2008; 6: e93.

58 Majewski, I.J., Ritchie, M.E., Phipson, B., et al. Opposing roles of polycomb repressive complexes in hematopoietic stem and progenitor cells. *Blood* 2010; 116: 731–739.

59 Dawson, M.A., Bannister, A.J., Gottgens, B., et al. JAK2 phosphorylates histone H3Y41 and excludes HP1alpha from chromatin. *Nature* 2009; 461: 819–822.

60 Rinaldi, C.R., Rinaldi, P., Alagia, A., et al. Preferential nuclear accumulation of JAK2V617F in $CD34^+$ but not in granulocytic, megakaryocytic or erythroid cells of patients with Philadelphia-negative myeloproliferative neoplasia. *Blood* 2010; 116: 6023–6026.

61 Delhommeau, F., Dupont, S., Della Valle, V., et al. Mutation in TET2 in myeloid cancers. *N Engl J Med* 2009; 360: 2289–2301.

62 Ko, M., Huang, Y., Jankowska, A.M., et al. Impaired hydroxylation of 5-methylcytosine in myeloid cancers with mutant TET2. *Nature* 2010; 468: 839–843.

63 Figueroa, M.E., Abdel-Wahab, O., Lu, C., et al. Leukemic IDH1 and IDH2 mutations result in a hypermethylation phenotype, disrupt TET2 function and impair hematopoietic differentiation. *Cancer Cell* 2010; 18: 553–567.

64 Pronier, E., Almire, C., Mokrani, H., et al. Inhibition of TET2-mediated conversion of 5-methylcytosine to 5-hydroxymethylcytosine disturbs erythroid and granulomonocytic differentiation of human hematopoietic progenitors. *Blood* 2011; 118: 2551–2555.

65 Ko, M., Bandukwala, H.S., An, J., et al. Ten-Eleven-Translocation 2 (TET2) negatively regulates homeostasis and differentiation of hematopoietic stem cells in mice. *Proc Natl Acad Sci U S A* 2011; 108: 14566–14571.

66 Moran-Crusio, K., Reavie, L., Shih, A., et al. Tet2 loss leads to increased hematopoietic stem cell self-renewal and myeloid transformation. *Cancer Cell* 2011; 20: 11–24.

67 Quivoron, C., Couronne, L., Della Valle, V., et al. TET2 inactivation results in pleiotropic hematopoietic abnormalities in mouse and is a recurrent event during human lymphomagenesis. *Cancer Cell* 2011; 20: 25–38.

68 Dang, L., White, D.W., Gross, S., et al. Cancer-associated IDH1 mutations produce 2-hydroxyglutarate. *Nature* 2009; 462: 739–744.

69 Xu, W., Yang, H., Liu, Y., et al. Oncometabolite 2-hydroxyglutarate is a competitive inhibitor of alpha-ketoglutarate-dependent dioxygenases. *Cancer Cell* 2010; 19: 17–30.

70 Sasaki, M., Knobbe, C.B., Munger, J.C., et al. IDH1(R132H) mutation increases murine haematopoietic progenitors and alters epigenetics. *Nature* 2012; 488: 656–659.

71 Tefferi, A., Lasho, T.L., Abdel-Wahab, O., et al. IDH1 and IDH2 mutation studies in 1473 patients with chronic-, fibrotic- or blast-phase essential thrombocythemia, polycythemia vera or myelofibrosis. *Leukemia* 2010; 24: 1302–1309.

72 Green, A., Beer, P. Somatic mutations of IDH1 and IDH2 in the leukemic transformation of myeloproliferative neoplasms. *N Engl J Med* 2010; 362: 369–370.

73 Chou, W.C., Peng, K.Y., Lei, W.C., et al. Persistence of mutant isocitrate dehydrogenase in patients with acute myeloid leukemia in remission. *Leukemia* 2012; 26: 527–529.

74 Abdel-Wahab, O., Pardanani, A., Rampal, R., Lasho, T.L., Levine, R.L., Tefferi, A. DNMT3A mutational analysis in primary myelofibrosis, chronic myelomonocytic leukemia and advanced phases of myeloproliferative neoplasms. *Leukemia* 2011; 25: 1219–1220.
75 Stegelmann, F., Bullinger, L., Schlenk R.F., et al. DNMT3A mutations in myeloproliferative neoplasms. *Leukemia* 2011; 25: 1217–1219.
76 Kim, S.J., Zhao, H., Hardikar, S., Singh, A.K., Goodell, M.A., Chen, T. A DNMT3A mutation common in AML exhibits dominant-negative effects in murine ES cells. *Blood* 2013; 122: 4086–4089.
77 Challen, G.A., Sun, D., Jeong, M., et al. Dnmt3a is essential for hematopoietic stem cell differentiation. *Nat Genet* 2012; 44: 23–31.
78 Datta, J., Ghoshal, K., Sharma, S.M., Tajima, S., Jacob, S.T. Biochemical fractionation reveals association of DNA methyltransferase (Dnmt) 3b with Dnmt1 and that of Dnmt 3a with a histone H3 methyltransferase and Hdac1. *J Cell Biochem* 2003; 88: 855–864.
79 Shlush, L.I., Zandi, S., Mitchell, A., et al. Identification of pre-leukaemic haematopoietic stem cells in acute leukaemia. *Nature* 2014; 506: 328–333.
80 Papaemmanuil, E., Cazzola, M., Boultwood, J., et al. Somatic SF3B1 mutation in myelodysplasia with ring sideroblasts. *N Engl J Med* 2011; 365: 1384–1395.
81 Yoshida, K', Sanada, M', Shiraishi, Y', et al. Frequent pathway mutations of splicing machinery in myelodysplasia. *Nature* 2011; 478: 64–69.
82 Lasho, T.L., Finke, C.M., Hanson, C.A., et al. SF3B1 mutations in primary myelofibrosis: clinical, histopathology and genetic correlates among 155 patients. *Leukemia* 2012; 26: 1135–1137.
83 Visconte, V., Makishima, H., Jankowska, A., et al. SF3B1, a splicing factor is frequently mutated in refractory anemia with ring sideroblasts. *Leukemia* 2012; 26: 542–545.
84 Makishima, H., Visconte, V., Sakaguchi, H., et al. Mutations in the spliceosome machinery, a novel and ubiquitous pathway in leukemogenesis. *Blood* 2012; 119: 3203–3210.
85 Quesada, V., Conde, L., Villamor, N., et al. Exome sequencing identifies recurrent mutations of the splicing factor SF3B1 gene in chronic lymphocytic leukemia. *Nat Genet* 2012; 44: 47–52.
86 Wang, L., Lawrence, M.S., Wan, Y., et al. SF3B1 and other novel cancer genes in chronic lymphocytic leukemia. *N Engl J Med* 2011; 365: 2497–2506.
87 Scott, L.M., Rebel, V.I. Acquired mutations that affect pre-mRNA splicing in hematologic malignancies and solid tumors. *J Natl Cancer Inst* 2013; 105: 1540–1549.
88 Zhao, R., Follows, G.A., Beer, P.A., et al. Inhibition of the Bcl-xL deamidation pathway in myeloproliferative disorders. *N Engl J Med* 2008; 359: 2778–2789.
89 Marty, C., Lacout, C., Droin, N., et al. A role for reactive oxygen species in JAK2 V617F myeloproliferative neoplasm progression. *Leukemia* 2013; 27: 2187–2195.
90 Nieborowska-Skorska, M., Kopinski, P.K., Ray, R., et al. Rac2-MRC-cIII-generated ROS cause genomic instability in chronic myeloid leukemia stem cells and primitive progenitors. *Blood* 2012; 119: 4253–4263.

91 Reilly, J.T., Wilson, G., Barnett, D., Watmore, A., Potter, A. Karyotypic and ras gene mutational analysis in idiopathic myelofibrosis. *Br J Haematol* 1994; 88: 575–581.

92 Janssen, J.W., Steenvoorden, A.C., Lyons, J., et al. RAS gene mutations in acute and chronic myelocytic leukemias, chronic myeloproliferative disorders and myelodysplastic syndromes. *Proc Natl Acad Sci U S A* 1987; 84: 9228–9232.

93 Grand, F.H., Hidalgo-Curtis, C.E., Ernst, T., et al. Frequent CBL mutations associated with 11q acquired uniparental disomy in myeloproliferative neoplasms. *Blood* 2009; 113: 6182–6192.

94 Schnittger, S., Bacher, U., Eder, C., et al. Molecular analyses of 15,542 patients with suspected BCR-ABL1-negative myeloproliferative disorders allow to develop a stepwise diagnostic workflow. *Haematologica* 2012; 97: 1582–1585.

95 Matsumura, I., Nakajima, K., Wakao, H., et al. Involvement of prolonged ras activation in thrombopoietin-induced megakaryocytic differentiation of a human factor-dependent hematopoietic cell line. *Mol Cell Biol* 1998; 18: 4282–4290.

96 Piu, F., Magnani, M., Ader, M.E. Dissection of the cytoplasmic domains of cytokine receptors involved in STAT and Ras dependent proliferation. *Oncogene* 2002; 21: 3579–3591.

97 Ugo, V., Marzac, C., Teyssandier, I., et al. Multiple signaling pathways are involved in erythropoietin-independent differentiation of erythroid progenitors in polycythemia vera. *Exp Hematol* 2004; 32: 179–187.

98 Abdel-Wahab, O., Manshouri, T., Patel, J., et al. Genetic analysis of transforming events that convert chronic myeloproliferative neoplasms to leukemias. *Cancer Res* 2010; 70: 447–452.

99 Jager, R., Gisslinger, H., Passamonti, F., et al. Deletions of the transcription factor Ikaros in myeloproliferative neoplasms. *Leukemia* 2010; 24: 1290–1298.

100 Milosevic, J.D., Puda, A., Malcovati, L., et al. Clinical significance of genetic aberrations in secondary acute myeloid leukemia. *Am J Hematol* 2012; 87: 1010–1016.

101 Campbell, P.J., Baxter, E.J., Beer, P.A., et al. Mutation of JAK2 in the myeloproliferative disorders: timing, clonality studies, cytogenetic associations and role in leukemic transformation. *Blood* 2006; 108: 3548–3555.

102 Theocharides, A., Boissinot, M., Girodon, F., et al. Leukemic blasts in transformed JAK2-V617F-positive myeloproliferative disorders are frequently negative for the JAK2-V617F mutation. *Blood* 2007; 110: 375–379.

103 Schaub, F.X., Looser, R., Li, S., et al. Clonal analysis of TET2 and JAK2 mutations suggests that TET2 can be a late event in the progression of myeloproliferative neoplasms. *Blood* 2010; 115: 2003–2007.

104 Lundberg, P., Karow, A., Nienhold, R., et al. Clonal evolution and clinical correlates of somatic mutations in myeloproliferative neoplasms. *Blood* 2014; 123: 2220–2228.

105 Kralovics, R., Stockton, D.W., Prchal, J.T. Clonal hematopoiesis in familial polycythemia vera suggests the involvement of multiple mutational events in the early pathogenesis of the disease. *Blood* 2003; 102: 3793–3796.

106 Bellanne-Chantelot, C., Chaumarel, I., Labopin, M., et al. Genetic and clinical implications of the Val617Phe JAK2 mutation in 72 families with myeloproliferative disorders. *Blood* 2006; 108: 346–352.
107 Rumi, E., Passamonti, F., Pietra, D., et al. JAK2 (V617F) as an acquired somatic mutation and a secondary genetic event associated with disease progression in familial myeloproliferative disorders. *Cancer* 2006; 107: 2206–2211.
108 Rumi, E., Passamonti, F., Della Porta, M.G., et al. Familial chronic myeloproliferative disorders: clinical phenotype and evidence of disease anticipation. *J Clin Oncol* 2007; 25: 5630–5635.
109 Pietra, D., Li, S., Brisci, A., et al. Somatic mutations of JAK2 exon 12 in patients with JAK2 (V617F)-negative myeloproliferative disorders. *Blood* 2008; 111: 1686–1689.
110 Saint-Martin, C., Leroy, G., Delhommeau, F., et al. Analysis of the ten–eleven translocation 2 (TET2) gene in familial myeloproliferative neoplasms. *Blood* 2009; 114: 1628–1632.
111 Landgren, O., Goldin, L.R., Kristinsson, S.Y., Helgadottir, E.A., Samuelsson, J., Bjorkholm, M. Increased risks of polycythemia vera, essential thrombocythemia and myelofibrosis among 24,577 first-degree relatives of 11,039 patients with myeloproliferative neoplasms in Sweden. *Blood* 2008; 112: 2199–2204.
112 Kilpivaara, O., Mukherjee, S., Schram, A.M., et al. A germline JAK2 SNP is associated with predisposition to the development of JAK2(V617F)-positive myeloproliferative neoplasms. *Nat Genet* 2009; 41: 455–459.
113 Skoda, R. The genetic basis of myeloproliferative disorders. Education Program Book 2007, *Hematology Am Soc Hematol Educ Program* 2007; 1–10.
114 Harrison, C.N., Gale, R.E., Wiestner, A.C., Skoda, R.C., Linch, D.C. The activating splice mutation in intron 3 of the thrombopoietin gene is not found in patients with non-familial essential thrombocythaemia. *Br J Haematol* 1998; 102: 1341–1343.
115 Teofili, L., Giona, F., Martini, M., et al. Markers of myeloproliferative diseases in childhood polycythemia vera and essential thrombocythemia. *J Clin Oncol* 2007; 25: 1048–1053.
116 Mead, A.J., Rugless, M.J., Jacobsen, S.E., Schuh, A. Germline JAK2 mutation in a family with hereditary thrombocytosis. *N Engl J Med* 2012; 366: 967–969.
117 Marty, C., Saint-Martin, C., Pecquet, C., et al. Germ-line JAK2 mutations in the kinase domain are responsible for hereditary thrombocytosis and are resistant to JAK2 and HSP90 inhibitors. *Blood* 2014; 123: 1372–1383.
118 Mead, A.J., Chowdhury, O., Pecquet, C., et al. Impact of isolated germline JAK2V617I mutation on human hematopoiesis. *Blood* 2013; 121: 4156–4165.
119 Etheridge, S.L., Cosgrove, M.E., Sangkhae, V., et al. A novel activating, germline JAK2 mutation, JAK2R564Q, causes familial essential thrombocytosis. *Blood* 2014; 123: 1059–1068.
120 Huang, L.J., Shen, Y.M., Bulut, G.B. Advances in understanding the pathogenesis of primary familial and congenital polycythaemia. *Br J Haematol* 2010; 148: 844–852.
121 McMullin, M.F. Idiopathic erythrocytosis: a disappearing entity. *Hematology Am Soc Hematol Educ Program* 2009; 629–635.

122 Percy, M.J., Furlow, P.W., Lucas, G.S., et al. A gain-of-function mutation in the HIF2A gene in familial erythrocytosis. *N Engl J Med* 2008; 358: 162–168.
123 Percy, M.J., Furlow, P.W., Beer, P.A., Lappin, T.R., McMullin, M.F., Lee, F.S. A novel erythrocytosis-associated PHD2 mutation suggests the location of a HIF binding groove. *Blood* 2007; 110: 2193–2196.
124 Gordeuk, V.R., Sergueeva, A.I., Miasnikova, G.Y., et al. Congenital disorder of oxygen sensing: association of the homozygous Chuvash polycythemia VHL mutation with thrombosis and vascular abnormalities but not tumors. *Blood* 2004; 103: 3924–3932.
125 Formenti, F., Beer, P.A., Croft, Q.P., et al. Cardiopulmonary function in two human disorders of the hypoxia-inducible factor (HIF) pathway: von Hippel–Lindau disease and HIF-2alpha gain-of-function mutation. *FASEB J* 2011; 25: 2001–2011.
126 Gunay-Aygun, M., Falik-Zaccai, T.C., Vilboux, T., et al. NBEAL2 is mutated in gray platelet syndrome and is required for biogenesis of platelet alpha-granules. *Nat Genet* 2011; 43: 732–734.
127 Jones, A.V., Cross, N.C., White, H.E., Green, A.R., Scott, L.M. Rapid identification of JAK2 exon 12 mutations using high resolution melting analysis. *Haematologica* 2008; 93: 1560–1564.
128 Boyd, E.M., Bench, A.J., Goday-Fernandez, A., et al. Clinical utility of routine MPL exon 10 analysis in the diagnosis of essential thrombocythaemia and primary myelofibrosis. *Br J Haematol* 2010; 149: 250–257.
129 McMullin, M.F., Reilly, J.T., Campbell, P., et al. Amendment to the guideline for diagnosis and investigation of polycythaemia/erythrocytosis. *Br J Haematol* 2007; 138: 821–822.
130 Beer, P.A., Erber, W.N., Campbell, P.J., Green, A.R. How I treat essential thrombocythemia. *Blood* 2010; 117: 1472–1482.
131 Dahabreh, I.J., Zoi, K., Giannouli, S., Zoi, C., Loukopoulos, D., Voulgarelis, M. Is JAK2 V617F mutation more than a diagnostic index? A meta-analysis of clinical outcomes in essential thrombocythemia. *Leuk Res* 2009; 33: 67–73.
132 Lussana, F., Caberlon, S., Pagani, C., Kamphuisen, P.W., Buller, H.R., Cattaneo M. Association of V617F Jak2 mutation with the risk of thrombosis among patients with essential thrombocythaemia or idiopathic myelofibrosis: a systematic review. *Thromb Res* 2009; 124: 409–417.
133 Vannucchi, A.M., Antonioli, E., Guglielmelli, P., et al. Prospective identification of high-risk polycythemia vera patients based on JAK2V617F allele burden. *Leukemia* 2007; 21: 1952–1959.
134 Vannucchi, A.M., Antonioli, E., Guglielmelli, P., et al. Clinical profile of homozygous JAK2 617V>F mutation in patients with polycythemia vera or essential thrombocythemia. *Blood* 2007; 110: 840–846.
135 Busque, L., Patel, J.P., Figueroa, M.E., et al. Recurrent somatic TET2 mutations in normal elderly individuals with clonal hematopoiesis. *Nat Genet* 2012; 44: 1179–1181.
136 Brecqueville, M., Rey, J., Devillier, R., et al. Array comparative genomic hybridization and sequencing of 23 genes in 80 patients with myelofibrosis at chronic or acute phase. *Haematologica* 2014; 99: 37–45.

137 Guglielmelli, P., Lasho, T.L., Rotunno G., et al. The number of prognostically detrimental mutations and prognosis in primary myelofibrosis: an international study of 797 patients. *Leukemia* 2014; 28: 1804–1810.
138 Vannucchi, A.M., Lasho, T.L., Guglielmelli, P., et al. Mutations and prognosis in primary myelofibrosis. *Leukemia* 2013; 27: 1861–1869.
139 Verstovsek, S., Kantarjian, H., Mesa, R.A., et al. Safety and efficacy of INCB018424, a JAK1 and JAK2 inhibitor, in myelofibrosis. *N Engl J Med* 2010; 363: 1117–1127.
140 Cervantes, F., Vannucchi, A.M., Kiladjian, J.J., et al. Three-year efficacy, safety and survival findings from COMFORT-II, a phase 3 study comparing ruxolitinib with best available therapy for myelofibrosis. *Blood* 2013; 122: 4047–4053.
141 Wang, F., Travins, J., DeLaBarre, B., et al. Targeted inhibition of mutant IDH2 in leukemia cells induces cellular differentiation. *Science* 2013; 340: 622–626.
142 Rohle, D., Popovici-Muller, J., Palaskas, N., et al. An inhibitor of mutant IDH1 delays growth and promotes differentiation of glioma cells. *Science* 2013; 340: 626–630.
143 Gotlib J., Cools J. Five years since the discovery of FIP1L1–PDGFRA: what we have learned about the fusion and other molecularly defined eosinophilias. *Leukemia* 2008; 22: 1999–2010.
144 *Atlas of Genetics and Cytogenetics in Oncology and Haematology* 2013, http://AtlasGeneticsOncology.org (accessed 1 November 2015).
145 Toffalini, F., Kallin, A., Vandenberghe, P., et al. The fusion proteins TEL–PDGFRbeta and FIP1L1–PDGFRalpha escape ubiquitination and degradation. *Haematologica* 2009; 94: 1085–1093.
146 Cools, J., DeAngelo, D.J., Gotlib, J., et al. A tyrosine kinase created by fusion of the PDGFRA and FIP1L1 genes as a therapeutic target of imatinib in idiopathic hypereosinophilic syndrome. *N Engl J Med* 2003; 348: 1201–1214.
147 Griffin, J.H., Leung, J., Bruner, R.J., Caligiuri, M.A., Briesewitz, R. Discovery of a fusion kinase in EOL-1 cells and idiopathic hypereosinophilic syndrome. *Proc Natl Acad Sci U S A* 2003; 100: 7830–7835.
148 Stover, E.H., Chen, J., Folens, C., et al. Activation of FIP1L1–PDGFRA requires disruption of the juxtamembrane domain of PDGFRA and is FIP1L1-independent. *Proc Natl Acad Sci U S A* 2006; 103: 8078–8083.
149 Curtis, C.E., Grand, F.H., Musto, P., et al. Two novel imatinib-responsive PDGFRA fusion genes in chronic eosinophilic leukaemia. *Br J Haematol* 2007; 138: 77–81.
150 Elling, C., Erben, P., Walz, C., et al. Novel imatinib-sensitive PDGFRA-activating point mutations in hypereosinophilic syndrome induce growth factor independence and leukemia-like disease. *Blood* 2010; 117: 2935–2943.
151 Buitenhuis, M., Verhagen, L.P., Cools, J., Coffer, P.J. Molecular mechanisms underlying FIP1L1--PDGFRA-mediated myeloproliferation. *Cancer Res* 2007; 67: 3759–3766.
152 Montano-Almendras, C.P., Essaghir, A., Schoemans, H., et al. ETV6–PDGFRB and FIP1L1–PDGFRA stimulate human hematopoietic progenitor proliferation and differentiation into eosinophils: role of NF-κB. *Haematologica* 2012; 97: 1064–1072.

153 Cools, J., Stover, E.H., Boulton, C.L., et al. PKC412 overcomes resistance to imatinib in a murine model of FIP1L1–PDGFRalpha-induced myeloproliferative disease. *Cancer Cell* 2003; 3: 459–469.
154 Yamada, Y., Rothenberg, M.E., Lee, A.W., et al. The FIP1L1–PDGFRA fusion gene cooperates with IL-5 to induce murine hypereosinophilic syndrome (HES)/chronic eosinophilic leukemia (CEL)-like disease. *Blood* 2006; 107: 4071–4079.
155 Burgstaller, S., Kreil, S., Waghorn, K., et al. The severity of FIP1L1–PDGFRA-positive chronic eosinophilic leukaemia is associated with polymorphic variation at the IL5RA locus. *Leukemia* 2007; 21: 2428–2432.
156 Tomasson, M.H., Sternberg, D.W., Williams, I.R., et al. Fatal myeloproliferation, induced in mice by TEL/PDGFbetaR expression, depends on PDGFbetaR tyrosines 579/581. *J Clin Invest* 2000; 105: 423–432.
157 Cain, J.A., Xiang, Z., O'Neal, J., et al. Myeloproliferative disease induced by TEL–PDGFRB displays dynamic range sensitivity to Stat5 gene dosage. *Blood* 2007; 109: 3906–3914.
158 Jackson, C.C., Medeiros, L.J., Miranda, R.N. 8p11 myeloproliferative syndrome: a review. *Hum Pathol* 2010; 41: 461–476.
159 Huang, Q., Snyder, D.S., Chu, P., Gaal, K.K., Chang, K.L., Weiss, L.M. PDGFRA rearrangement leading to hyper-eosinophilia, T-lymphoblastic lymphoma, myeloproliferative neoplasm and precursor B-cell acute lymphoblastic leukemia. *Leukemia* 2011; 25: 371–375.
160 Metzgeroth, G., Walz, C., Score, J., et al. Recurrent finding of the FIP1L1–PDGFRA fusion gene in eosinophilia-associated acute myeloid leukemia and lymphoblastic T-cell lymphoma. *Leukemia* 2007; 21: 1183–1188.
161 Klion, A.D., Law, M.A., Riemenschneider, W., et al. Familial eosinophilia: a benign disorder? *Blood* 2004; 103: 4050–4055.
162 Orfao, A., Garcia-Montero, A.C., Sanchez, L., Escribano, L. Recent advances in the understanding of mastocytosis: the role of KIT mutations. *Br J Haematol* 2007; 138: 12–30.
163 Furitsu, T., Tsujimura, T., Tono, T., et al. Identification of mutations in the coding sequence of the proto-oncogene c-kit in a human mast cell leukemia cell line causing ligand-independent activation of c-kit product. *J Clin Invest* 1993; 92: 1736–1744.
164 Harir, N., Boudot, C., Friedbichler, K., et al. Oncogenic Kit controls neoplastic mast cell growth through a Stat5/PI3-kinase signaling cascade. *Blood* 2008; 112: 2463–2473.
165 Baumgartner, C., Cerny-Reiterer, S., Sonneck, K., et al. Expression of activated STAT5 in neoplastic mast cells in systemic mastocytosis: subcellular distribution and role of the transforming oncoprotein KIT D816V. *Am J Pathol* 2009; 175: 2416–2429.
166 Hashimoto, K., Matsumura, I., Tsujimura, T., et al. Necessity of tyrosine 719 and phosphatidylinositol 3′-kinase-mediated signal pathway in constitutive activation and oncogenic potential of c-kit receptor tyrosine kinase with the Asp814Val mutation. *Blood* 2003; 101: 1094–1102.

167 Shivakrupa, R., Bernstein, A., Watring, N., Linnekin, D. Phosphatidylinositol 3′-kinase is required for growth of mast cells expressing the kit catalytic domain mutant. *Cancer Res* 2003; 63: 4412–4419.

168 Zappulla, J.P., Dubreuil, P., Desbois, S., et al. Mastocytosis in mice expressing human Kit receptor with the activating Asp816Val mutation. *J Exp Med* 2005; 202: 1635–1641.

169 Yavuz, A.S., Lipsky, P.E., Yavuz, S., Metcalfe, D.D., Akin, C. Evidence for the involvement of a hematopoietic progenitor cell in systemic mastocytosis from single-cell analysis of mutations in the c-kit gene. *Blood* 2002; 100: 661–665.

170 Tefferi, A., Levine, R.L., Lim, K.H., et al. Frequent TET2 mutations in systemic mastocytosis: clinical, KITD816V and FIP1L1–PDGFRA correlates. *Leukemia* 2009; 23: 900–904.

171 Schwaab, J., Schnittger, S., Sotlar, K., et al. Comprehensive mutational profiling in advanced systemic mastocytosis. *Blood* 2013; 122: 2460–2466.

172 Traina, F, Visconte, V, Jankowska, A.M., et al. Single nucleotide polymorphism array lesions, TET2, DNMT3A, ASXL1 and CBL mutations are present in systemic mastocytosis. *PLoS One* 2012; 7: e43090.

173 Daley, T., Metcalfe, D.D., Akin, C. Association of the Q576R polymorphism in the interleukin-4 receptor alpha chain with indolent mastocytosis limited to the skin. *Blood* 2001; 98: 880–882.

CHAPTER 3

Acute myeloid leukaemia

Matthew L. Smith and Thomas McKerrell

Introduction

Acute myeloid leukaemia (AML) is a clonal haematopoietic malignancy of haematopoietic stem cells (HSCs) characterized by a block in myeloid differentiation together with uncontrolled proliferation of abnormal myeloid precursors that accumulate in the blood and bone marrow. AML can affect children, but this malignancy is most prevalent in adults, with an incidence that increases with age such that the median age of onset is 69 years.[1] Most cases arise *de novo* with no known aetiological factors; other cases develop from a variety of myeloid malignancies, such as myelodysplastic syndrome (MDS), classical myeloproliferative disorders (MPDs) or other types of MPDs such as chronic myelomonocytic leukaemia (CMML).

AML is conventionally classified according to karyotype as 50% have non-random chromosomal aberrations (i.e. balanced translocations, deletions, inversions, monosomies and trisomies), many of which have prognostic relevance.[2] The remaining 50% is comprised of cytogenetically normal AML (CN-AML), a highly heterogeneous subgroup with varied patient outcomes (Fig. 3.1). Stratification of disease risk based on karyotype and a few molecular markers are currently used to guide treatment decisions. Induction chemotherapy leads to good rates of remission; however, the majority of patients relapse and this is the main vehicle for patient mortality. Chemotherapy regimens have not changed in decades, but advances in supportive care and better risk stratification have led improvements in patient prognosis. Despite this, 5-year survival rates remain disappointingly low at 40–45% in young patients and less than 10% in the elderly.[3,4]

In recent years, next-generation sequencing (NGS) technology has revolutionized our understanding of the molecular pathogenesis of cancer, especially in AML, which was the first cancer genome to be

The Genetic Basis of Haematological Cancers, First Edition. Edited by Sabrina Tosi and Alistair G. Reid.

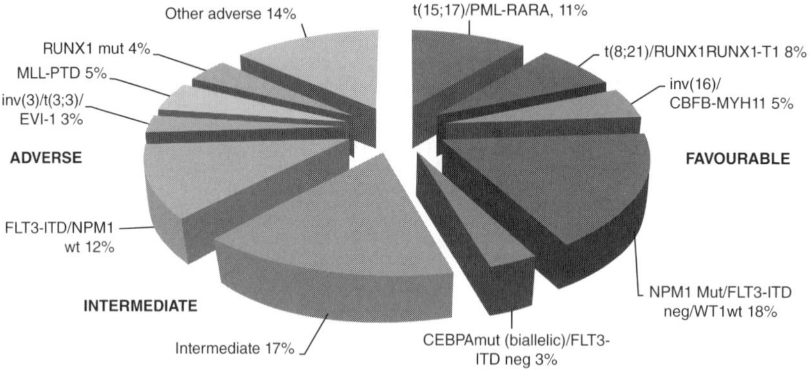

Figure 3.1 Illustration of the distribution frequency of karyotype and molecular aberrations for patients aged under 60 years entered into recent MRC/NCRI clinical studies. From Smith et al.[278] Reproduced with permission from Elsevier.

sequenced.[5] Genomic technology has not only helped us to identify a number of unknown somatic mutations that drive disease, especially in CN-AML, it has also illuminated processes involved in the clonal evolution of this aggressive malignancy. In this chapter, we discuss recent findings in AML and emerging concepts in the genomic era alongside more established principles.

AML classification

The main classification system for AML used for many years was that of the French–American–British (FAB) group, devised in 1976[6] and revised in 1985.[7] The main categories in this classification system are listed in Table 3.1. The older morphology-based FAB classification is nowadays replaced by the World Health Organization (WHO) classification, which also takes genetic abnormalities into account (Table 3.2).[8] Since 2001, AML has been classified using the WHO classification (see Table 3.2). In 2008, the WHO updated this classification, including molecular markers (*NPM1* and *CEBPA*) as subcategories for the first time. It is anticipated that further molecular markers will be incorporated into future updated WHO classifications of AML.

The study of genetic alterations, such as chromosomal abnormalities and gene mutations, has improved our understanding of the mechanisms at the basis of leukaemia initiation and progression. This has enabled clinicians to achieve more precise diagnosis and to provide appropriate therapies. A better understanding of the genetic and epi-genetic events in

Table 3.1 FAB classification system for AML.[281]

Category	Morphology	Incidence (%)[3]
M0	AML with no differentiation	3
M1	AML without maturation	15–20
M2	AML with granulocytic maturation	25–30
M3	Hypergranular APML	5–10
M3 variant	Hypogranular variant APML	
M4	Acute myelomonocytic leukaemia	25–30
M5a	Acute monoblastic leukaemia	2–10
M5b	Acute monocytic leukaemia	
M6	Erythroleukaemia	3–5
M7	Megakaryoblastic leukaemia	3–12

recent years has revealed a number of targets that could have therapeutic potential.

Cytogenetic aberrations

Karyotypic changes, either numerical or structural (or a combination of the two), occur in ~55% of adult AML patients (see Fig. 3.1 for a pie chart of major cytogenetic subgroups). Metaphase G-banding can detect the majority of these chromosomal abnormalities but should be supplemented by fluorescence *in situ* hybridization (FISH) analysis if G-banding is negative, incomplete or not possible owing to a lack of dividing cells.

Fusion genes arising from structural rearrangements

Over 100 balanced, non-random, recurrent, structural chromosomal translocations have been identified and cloned from AML patients, and many of these have been associated with characteristic morphological features and also particular outcomes to standard therapies. The most common translocations in AML fall into three major types: core-binding factor translocations [t(8;21) and inv(16)], t(15;17) and *MLL* rearrangements. These are discussed in detail below:

Core-binding factor fusion genes

Translocations affecting components of the core-binding factor complex (CBF) are seen to occur in ~10% of cases of AML[9] (Fig. 3.2). *RUNX1* (runt transcription factor 1; *AML1*; *CBFA2*; *PEBPA2B*) forms the DNA-binding α

Table 3.2 Previous and Current World Health Organization classification of AML.[37,116,282,]

WHO 2001		WHO 2008	
Category	**Morphology**	**Category**	**Morphology**
AML with recurrent cytogenetic translocations	AML with features of t(8;21)(q22;q22)	AML with recurrent genetic translocations	AML with t(8;21)(q22;q22) *RUNX1-RUNX1T1*
	AML with features of t(15;17)(q24;q12)		AML with t(15;17)(q24;q12); *PML-RARA* or variants
	AML with features of inv(16)(p13q22)		AML with inv(16)(p13q22) or t(16;16)(p13.1;q22); *CBFB-MYH11*
	AML with 11q23 abnormalities		AML with t(9;11)(p22;q23); *MLLT3-MLL* or variants
			AML with t(6;9)(p23;q34); *DEK-NUP214*
			AML with inv(3)(q21q26.2) or t(3;3)(q21;q26.2); *RPN1-EVI1*
			AMKL with t(1;22)(p13;q13); *RBM15-MKL1*
			AML with mutated *NPM1*
			AML with mutated *CEBPA*
AML with multi-lineage dysplasia	AML arising in previous MDS/MPD Without antecedent MDS	AML with MDS-related changes	Prior history of MDS/MPN Multilineage dysplasia MDS related cytogenetic abnormality
AML, therapy related	Alkylating agent related Epipodophyllotoxin related Other types	Therapy-related myeloid neoplasms	Alkylating agents Ionizing radiation Topoisomerase II inhibitors Others
Acute myeloid leukaemia, not otherwise specified	AML minimally differentiated	Acute myeloid leukaemia, not otherwise specified	AML with minimal differentiation
	AML without maturation		AML without maturation
	AML with maturation		AML with maturation

Table 3.2 (*continued*)

WHO 2001		WHO 2008	
Category	**Morphology**	**Category**	**Morphology**
	Acute myelomonocytic leukaemia		Acute myelomonocytic leukaemia
	Acute monocytic leukaemia		Acute monoblastic/monocytic leukaemia
	Acute erythroid leukaemia		Acute erythroid leukaemia
			Pure erythroid leukaemia
			Erythroleukaemia, erythroid/myeloid
	Acute megakaryocytic leukaemia		Acute megakaryoblastic leukaemia
	Acute basophilic leukaemia		Acute basophilic leukaemia
	Acute panmyelosis with myelofibrosis		Acute panmyelosis with myelofibrosis
	Myeloid sarcoma		Myeloid sarcoma

subunit of the heterodimeric transcription factor complex CBF. The runt homology domain of *RUNX1* is named after its sequence homology with the *Drosophila* pair rule gene involved in segmentation, *runt,* and consists of a protein motif of 128 highly conserved amino acids spread over exons 3, 4 and 5 responsible for binding to the consensus sequence TGTGGT and also dimerizing with the β subunit of CBF.[10] *RUNX1* regulates the expression of a variety of genes involved in haematopoiesis such as those for the granulocyte–macrophage colony-stimulating factor (GM-CSF), interleukin 3 (IL-3) or macrophage colony-stimulating factor (M-CSF) receptors via its C-terminal transactivating domain.[11] CBFB forms the DNA-binding β subunit that allosterically binds to Runx1 and stabilizes its conformation, in addition to preventing it from being ubiquitinated and hence degraded (Fig. 3.3a). *RUNX1* is indispensable for definitive haematopoiesis and *RUNX1* null mice die of CNS haemorrhage *in utero* with no evidence of fetal haematopoiesis.[12,13] The phenotype for *CBFB* null mice[9] is the same as for *RUNX1* null mice, suggesting that both elements of the CBF complex are essential for normal haematopoiesis.

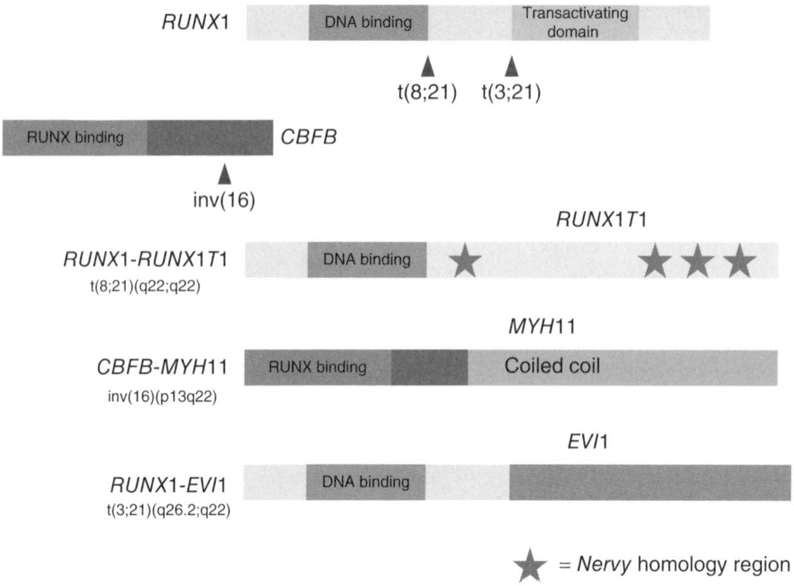

Figure 3.2 Diagram illustrating the structure of the RUNX1 and CBFB components of the heterodimeric core-binding factor (CBF) transcription factor, the chromosomal breakpoint locations and the resultant configurations in the three CBF translocations referred to in the text. Adapted from Speck and Gilliland.[9]

The t(8;21)(q22;q22) translocation results in a fusion between the first five exons of the *RUNX1* gene on chromosome 21 and the second exon of *RUNX1T1* (*ETO* 'eight twenty-one'; *MTG8* 'myeloid translocation gene on chromosome 8') on chromosome 8 (Fig. 3.2).[14] *RUNX1T1* is a transcription factor with four *nervy* homology regions in its C-terminus that recruit co-repressors to *RUNX1* target genes. These include nCor, Sin3 and histone deacetylase enzymes (HDAC) (Fig. 3.3b). Transgenic mice models have shown that this translocation is critical for the development of AML, although additional mutations are also required.[9,15] Knock-in experiments of the t(8;21)(q22;q22) fusion gene show inhibition of normal definitive haematopoiesis and generation of dysplastic progenitors.[16] Retroviral transduction of *RUNX1-RUNX1T1* results in increased self-renewal capacity.[17] The *RUNX1-RUNX1T1* fusion gene is a feature of 5–10% of cases of AML, occurring in 10% of those with FAB type M2.[37] This entity is associated with characteristic Auer rods, pseudo-Pelger–Huet nuclei, marrow eosinophilia and salmon-coloured cytoplasmic granules in early myeloid cells with perinuclear clearing. Additional chromosomal abnormalities are seen in 70% of cases which are typically –X, –Y or del(9q).[37]

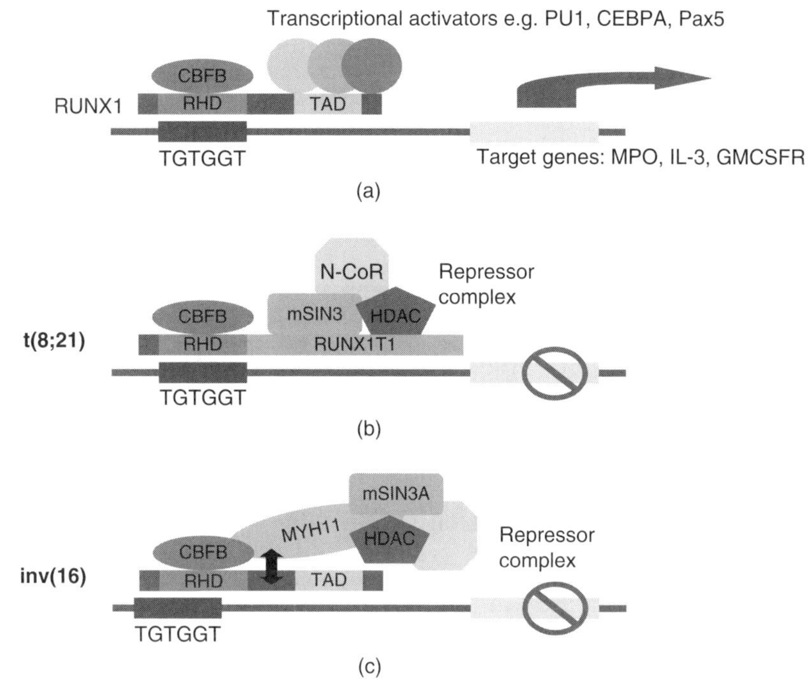

Figure 3.3 (a) Normal RUNX1 and CBFB function. CBFB stabilizes the CBF complex allosterically and prevents *RUNX1* ubiquitination and degradation. (b) RUNX1-RUNX1T1 fusion protein recruits various repressor proteins to the *RUNX1T1* tail that result in *RUNX1* target gene silencing. (c) CBFB-MYH11 fusion protein possesses a high-affinity RUNX1 binding domain in MYH11 while also recruiting various repressor proteins to the MYH11 tail that similarly result in gene silencing.

The β subunit of CBF is disrupted by either the inv(16)(p13q22) inversion or the t(16;16)(q24;q22) translocation (Fig 3.4), where fusion occurs with the myosin heavy-chain gene *MYH11* producing a *CBFB-MYH11* fusion gene[18] (Fig. 3.2). *MYH11* possess a high-affinity *RUNX1* binding domain that increases affinity for *RUNX1*,[19] and also a C-terminal co-repressor that recruits a repressor complex to *RUNX1* target genes[20] (Fig. 3.3c). Such abnormalities occur in 5–8% of cases of AML and 90% of such cases are classified as acute myelomonocytic leukaemia with eosinophilia (AML M4Eo). Clinically, these leukaemias may be associated with myeloid sarcomas ('chloromas') including CNS tumours, and morphologically demonstrate abnormal marrow eosinophil precursors with large, dense, purple–violet granules and nuclear hyposegmentation. Secondary cytogenetic changes are noted in 40% and typically include +22, +8, del(7q) or +21.[21,22]

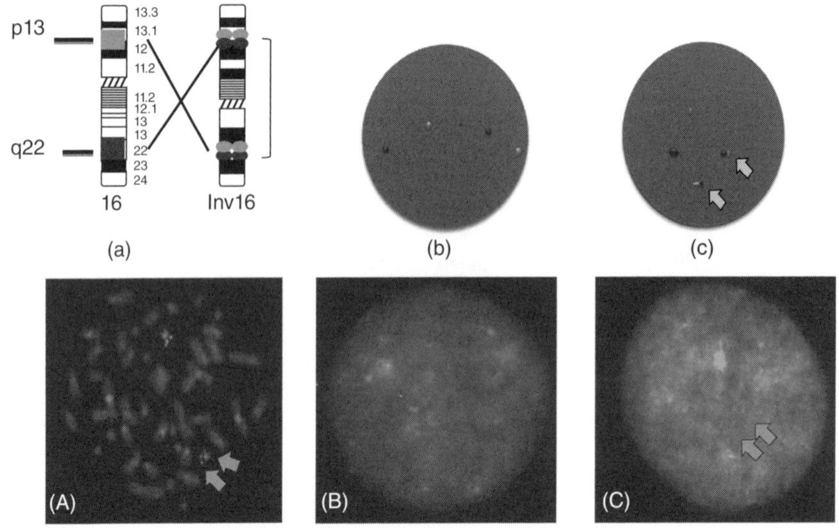

Figure 3.4 Rearrangement inv(16) in AML. Fluorescence *in situ* hybridization performed on the bone marrow metaphase spread of a patient with inv(16) using a dual colour probe set (Metasystems, Altlussheim, Germany). Green and red signals correspond to the *MYH11* and the *CBFB* regions, respectively. A schematic representation of the distribution of FISH signals is shown in the upper row, on both normal and rearranged chromosomes (a), on a normal interphase nucleus (b) and on an interphase nucleus carrying the inv(16) (c). Arrows indicate the fusion signals in (c) and (C). The corresponding photomicrographs are shown in the bottom row (A, B and C). Image courtesy of S. Tosi and A. Naiel, Leukaemia and Chromosome Research Laboratory, Brunel University, London, UK; the patient sample was provided by Professor Jochen Harbott, Department of Paediatric Haematology and Oncology, Justus Liebig University, Giessen, Germany. *(See plate section for color representation of this figure.)*

Multiple additional partner genes have been reported for *RUNX1* that remain to be cloned and identified.[23,24] For example, this complex may be dysregulated by the t(3;21)(q26;q22) *RUNX1-EVI1* fusion gene seen in blastic transformation of CML (Fig. 3.2).

RARA and partner translocations

The t(15;17)(q24;q12) results in fusion of *PML* on 15q24 with *RARA* (retinoic acid receptor alpha) on 17q12 to create a *PML-RARA* fusion on the derivative chromosome 15 and *RARA-PML* on the derivative chromosome 17.[25,26] It is seen in ~12% of cases of AML and is pathognomonic for acute promyelocytic leukaemia (APL), where it is present in >95% of cases. There are three molecular variants of the *PML/RARA* transcript (bcr1, bcr2, bcr3) depending on the breakpoint within the *PML* gene.[27] *PML* has growth suppressor and pro-apoptotic functions and is typically localized in nuclear bodies.[28,29] *RARA* mediates the transcriptional effects of retinoic acid via the heterodimerization of *RARA* with *RXR* (retinoid X receptors) (reviewed in Chambon et al.[30] in 1996). The *PML-RARA* fusion disrupts the localization and probable function of *PML* and represses retinoic acid target genes by recruiting co-repressors and HDAC (reviewed in Grimwade et al.[31] in 1999). ATRA (all-*trans*-retinoic acid), used to treat this condition alongside chemotherapy, results in the release of this co-repressor complex and hence triggers differentiation (Fig. 3.5a). If this fusion gene is expressed in haematopoietic stem cells in mice, a myeloproliferation results in many models with the initiation of APL in a proportion after a latent period. In another mouse model, where *PML-RARA* expression is driven from the *PML* locus in the context

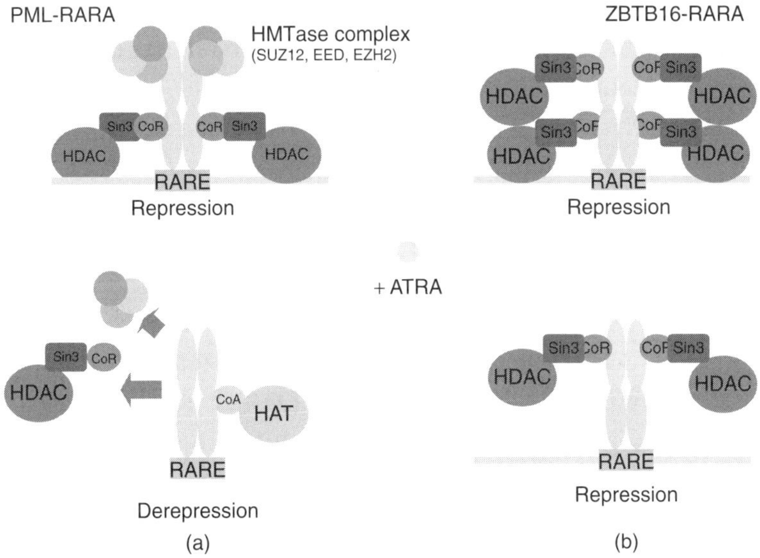

Figure 3.5 Binding of ATRA to repressors in different RARA fusions. (a) The PML-RARA fusion protein binds to retinoic acid repressor elements (RARE) via the RARA moiety. This part of the fusion protein recruits components of a repressor complex containing CoR, Sin3 and histone deacetylases (HDAC). PML similarly recruits components of the repressive hydroxymethyltransferase complex containing SUZ12, EED and EZH2. The addition of ATRA results in loss of these repressor components, gain of histone acetyltransferases (HAT) and gene derepression and transcription. (b) The ZBTB16-RARA fusion protein recruits repressor complexes to RARA and to the ZBTB16 components. Whereas the former are lost on the addition of ATRA, the repressors associated with ZBTB16 remain and lead to ongoing genetic silencing.

of *PML* haploinsufficiency, increased haematopoietic self-renewal with expansion of cell numbers is noted.[32] This expanded stem cell pool is then vulnerable to further mutation-triggering leukaemic change such as *JAK2*[33] or *FLT3*.[34]

In three out of every four cases, these blast cells are typically hypergranular promyelocytes with large numbers of dense azurophilic granules clustered into multiple 'faggots' of Auer rods. One-quarter of cases of APL appear as a microgranular variant where the white count is typically raised and granulation is minimally evident by light microscopy (Fig 3.6).[35,36] The microgranular variant is more frequently associated with a bcr3 breakpoint.[37] APL was formerly associated with a high induction death rate due to the presence of a fulminant coagulopathy at presentation.[36] This has been reduced significantly by the introduction of ATRA. Additional karyotypic abnormalities are seen in 40% of cases, with the most common being +8.[37]

A rapidly available anti-PML fluorescent antibody test can be used to complement metaphase cytogenetics and interphase FISH, confirming a diagnosis of APL and identifying those rare 1–2% of cases with cryptic or interstitial rearrangements beyond resolution of commercially available FISH probes.[38] This reveals a change from the wild-type pattern of PML nuclear bodies (<20/nucleus) to a microspeckled pattern (>30/nucleus) (Fig 3.6). Such a test has been shown to have sensitivity and specificity in excess of 98%, with the additional advantage of providing a rapid result to treating clinicians. Reverse transcriptase polymerase chain reaction (RT-PCR) analysis is the gold standard test for diagnosis as it defines the breakpoint and allows quantitative MRD monitoring. Guidelines have been published on the diagnostic work-up of these cases.[36]

There are also exceptions where *RARA* may fuse with other partner genes, as in t(11;17)(q23;q21) *ZBTB16* (formerly *PLZF*)-*RARA*, t(11;17)(q13;q21) *NuMA-RARA*, t(5;17)(q32;q12) *NPM-RARA*, t(X;17)(p11;q21) *BCOR-RARA* or t(4;17)(q12;q21) *FIP1L1-RARA* translocations,[39] in addition to fusion with *STAT5B* at 17q11.2 via rearrangement within 17q.[36] Those cases with *ZBTB16-RARA* fusions are the most common of these variant APL translocations, occurring in 0.8% of all cases,[39] and typically have characteristic variant morphology and CD56 positivity[40] with a normal PML fluorescence test. They are additionally ATRA resistant owing to the binding of additional co-repressors to *ZBTB16* (Fig. 3.3b).

MLL translocations

Translocations affecting *MLL* (mixed lineage leukaemia) on 11q23 may involve over 60 different partner genes[41] and occur in 5–10% of cases of acute leukaemia. In AML the t(9;11)(p22;q23) with the resultant *MLLT3(AF9)-MLL* fusion gene is the most common 11q23 translocation

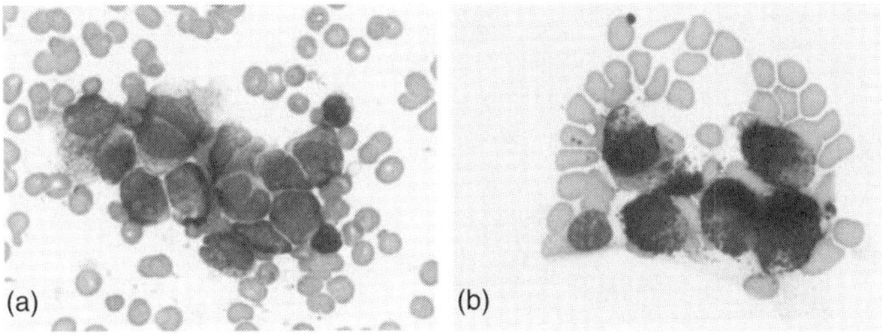

Figure 3.6 A composite of the methodologies for diagnostic evaluation of APL. (a) Bone marrow aspirate smear shows many promyelocytes. Compared with the adjacent normal small lymphocytes, promyelocytes are of medium to large size with irregularly shaped to convoluted to bilobed nuclei, open chromatin, visible to prominent nucleoli and a moderate amount of basophilic cytoplasm with abundant granularity. (b) Cytochemical stain for myeloperoxidase. Compared with the negative small lymphocyte, the promyelocytes are strongly positive, with numerous granules covering the outlines of the nuclei. (c) Negative (macrogranular) immunofluorescent stain for promyelocytic leukaemia (PML). The PML oncogenic domains are observed as several distinct particles in each nucleus. (d) Positive (microgranular) immunofluorescent stain for PML. Numerous (too many to count) fine, dusty granules are present in each nucleus. Panels (a)–(d) reproduced with permission-from Dimov et al.[38] (e, f) Examples of the application of FISH probes specific for and spanning, the PML gene on chromosome 15 (in green) and the RARA gene on chromosome 17 (in red; Kreatech Diagnostics, The Netherlands) to metaphase (e) and interphase (f) cells from a patient with APL. Fusion signals are present on both the der(15) and the der(17), in addition to single green and red signals marking the normal chromosomes 15 and 17, respectively (e). The same hybridization signal pattern is visible in interphase nuclei (f) and is readily distinguishable from normal cells which show two single green and two single red signals (not shown). Panels (e) and (f) courtesy of A Reid and I. Ortiz de Mendibil, Imperial Molecular Pathology, Imperial College Healthcare Trust, London, UK. (g) Example of G-banded karyotype from a bone marrow metaphase of an APL patient. Note that in this case additional abnormalities are present in addition to the typical t(15;17) rearrangement. The full karyotype reads 47,XX,+8,i(8)(q10), t(15;17)(q24;q21). Panel (g) courtesy of John Swansbury, Clinical Cytogenetics; McElwain Laboratories, The Royal Marsden Hospital and the Institute of Cancer Research, Sutton, Surrey, UK. *(See plate section for color representation of this figure.)*

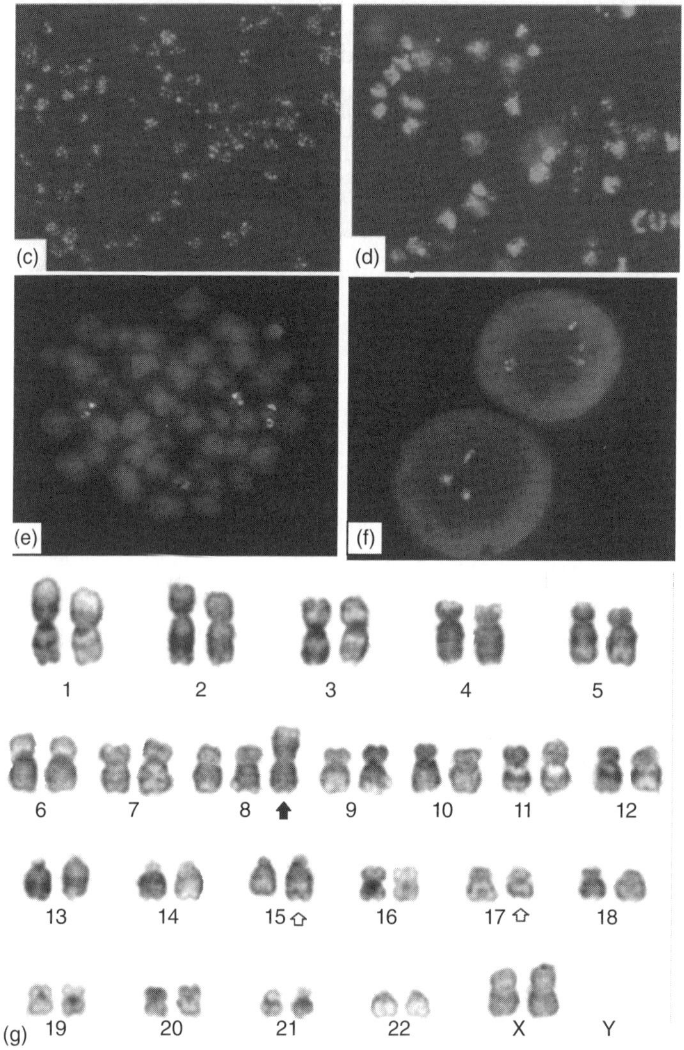

Figure 3.6 (*continued*)

observed, occurring in 2% of all cases of adult AML. Other less common translocations seen in AML include

- t(11;19)(q23;p13.3): *MLL-MLLT1 (ENL)*
- t(11;19)(q23;p13.1): *MLL-ELL*
- t(6;11)(q27;q23): *MLLT4(AF6)-MLL*
- t(10;11)(p12;q23): *MLLT10(AF10)-MLL*
- t(4;11)(q21;q23): *MLLT2 (AF4)-MLL.*

While the majority of such translocations can be seen with leukaemias of either myeloid or lymphoid lineage, *MLLT2-MLL* is predominantly seen in ALL and *MLL-ENL* is mainly seen in AML. They can occur at any age but are more frequent in infant and congenital leukaemias or in podophyllotoxin-related leukaemias.[42,43] Clinically, these leukaemias typically may be associated with disseminated intravascular coagulation (DIC), monoblastic chloromas or gum or skin infiltration and have monoblastic or myelomonocytic morphology. However, not all such translocations have recognizable morphology. One-third are not detected by G-banding, so require FISH analysis if an *MLL* rearrangement is suspected.[44] They do, however, appear to have a recognizable gene expression signature that comprises high levels of *HoxA* and *Meis1* expression.[45]

Chromsome 3q abnormalities affecting *MECOM*

'MDS1 and EVI1 complex locus' (*MECOM*) at 3q26 primarily comprises the 16-exon gene *EVI1* (ecotropic viral integration site-1 locus) situated at 3q26.2, which is a DNA-binding zinc finger transcription factor with a possible role in directing aberrant promoter DNA methylation.[46,47] Within *MECOM* there is also a larger gene, *MDS1-EVI1*, which produces the Evi1 protein with an additional 188αα at the N-terminus due to in-frame splicing. *EVI1* has a role in stem cell self-renewal and maintenance,[46,48] where it has a nuclear localization and interacts with *GATA1* and *GATA2* in addition to *HDAC* and *HMT*. *EVI1*-deficient embryos develop reduced numbers of stem cells and progenitor cells in the para-aortic region with impaired long-term repopulating capacity.[49] Over-expression of *EVI1* in long-term haematopoietic stem cells increases self-renewal and blocks differentiation.[50] It has been proposed that Evi1 acts physically in complex with Dnmt3[51] and results in modulation of miR-1–2 leading to abnormal proliferation.[52] Gene expression profiling experiments demonstrate that *EVI1* over-expressing AML cases have the 'stemness' phenotype similar to that of CD34 stem cells that is associated with a poor prognosis.[53,54]

Chromosome 3q abnormalities dysregulating *EVI1* are seen in 2–4% of cases of AML,[55,56] including cases of myeloid blast crisis of CML and secondary AML in addition to *de novo* AML. The most common 3q abnormalities are inv(3)(q21q26) and t(3;3)(q21;q26) *RPN1-EVI1*, which occur in 1–2% of cases and are seen to increase *EVI1* mRNA expression due to the effect of the relocation of the ribophorin 1 (*RPN1*) enhancer but with no fusion gene formation. However, they can also be seen with translocation partners on 1p36, 5q31 5q35 5q21 and 12p13.[57] The t(3;21)(q26;q22) *MDS1-EVI1-RUNX1* and t(3;12)(q26;p13) *EVI1-MDS1-ETV6* translocations are the most common of these and are primarily confined to therapy-related AML or AML arising from other haematological neoplasms, e.g. CML blast crisis. Abnormalities of chromosome 3q26 usually manifest myeloblastic or myelomonocytic morphology with trilineage dysplasia and abnormal thrombopoiesis such as peripheral thrombocytosis or marrow micromegakaryocytes. Additional cytogenetic changes, such as del(5q), –7 or a complex karyotype, are seen in 50–60% of cases; monosomy 7 accounts for half of these cases.

Uncommon translocations

t(6;9)(p23;q34)

This recurrent translocation, seen in 1% of cases of adult AML, fuses nucleoporin *NUP214* (*CAN*) at 9q34 to the transcription factor *DEK* on 6p23. These rare cases have myeloblastic or myelomonocytic morphology with multilineage marrow dysplasia and marrow and blood basophilia.[58–60] Such cases are often secondary to pre-existing marrow disorders such as MDS or CML.[61]

t(9;22)(q34;q11)

True Philadelphia-positive AML, as opposed to myeloid blast crisis of CML, is rare, accounting for about 0.6% of patients.[62] They usually have heterogeneous morphology with variable evidence of differentiation.[63,64] The translocation may also be found in subclones where it cooperates with other genetic aberrations, e.g. t(8;21).[65] Both the major and minor *BCR* breakpoints, resulting in both *BCR-ABL* p210 and p190, have been reported. Interestingly, the variant *BCR-JAK2* fusion has also been observed.[66]

t(1;22)(p13;q13)

This very rare translocation results in the *RBM15* (RNA-binding motif protein 15; *OTT*)-*MKL1* (megakaryocyte leukaemia 1; *MAL*) fusion. It has been reported in around 40 cases to date, with a propensity for a megakaryoblastic phenotype and a propensity to occur in a paediatric setting.[67]

8p11 translocations

These rare translocations include t(8;16)(p11;q13), t(8;22)(p11;q13) and inv(8)(p11q13) and result in the fusion genes *MOZ-CBP*, *MOZ-p300* and *MOZ-TIF2*, respectively. *MOZ* is a zinc finger protein with histone acetylase function.[68] These cases present with monoblastic features associated with prominent erythrophagocytosis.[69]

t(16;21)(p11;q22)

This translocation results in the *TLS-FUS-ERG* fusion gene and is rare, accounting for less than 1% of all cases of AML.[70] Such cases classically demonstrate CD56 expression, hemophagocytosis and vacuolation of blasts and are associated with a low complete remission (CR) rate and high relapse risk (RR), leading to a poor prognosis overall.[70]

12p13 translocations

These rare cases have myeloblastic or myelomonocytic morphology with marrow basophilia. They are often seen in the setting of therapy-related AML[71]. Although the molecular breakpoints in 12p13 can be heterogeneous, the majority of these translocations involve the *ETV6* gene. More than 40 chromosomal partners for this gene have been described to date.[72]

t(7;12)(q36;p13)

This very rare translocation occurs in approximately one-third of infants with AML, with no particular association with a specific subtype (Fig 3.7). The rearrangement results in the ectopic expression of *HLXB9*, although a fusion transcript *HLXB9-ETV6* has been described in approximately 50% of cases.[73,74]

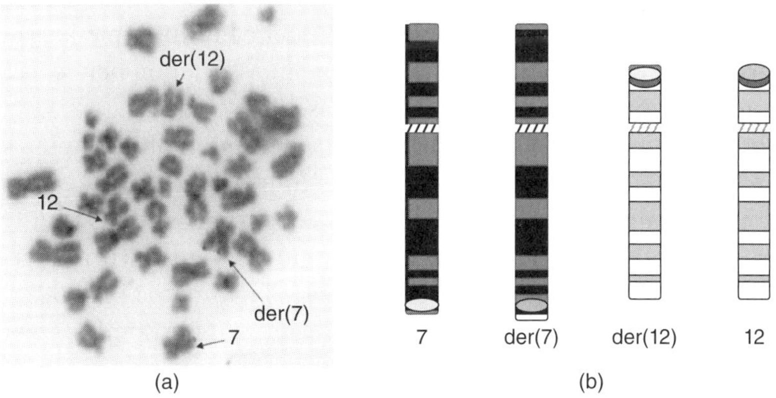

Figure 3.7 Translocation t(7;12) in infant AML. Fluorescence *in situ* hybridization performed on the bone marrow metaphase spread of a patient with t(7;12)(q36;p13) using a three-colour probe set (Metasystems, Altlussheim, Germany). Note localization of FISH signals on chromosome 7 (blue signals), der(7) (green signals), chromosome 12 (green and orange signals) and der(12) (blue and orange signals). The DAPI counterstain used to visualize the chromosomes has been converted into greyscale to simulate a G-like banding pattern (a). The schematic representation of the hybridization pattern is also shown on the ideograms (b). From Naiel et al.[283] *(See plate section for color representation of this figure.)*

1p36 translocations

Translocations affecting *PRDM16* (*MEL*) are seen with *MECOM* in the t(1;3)(p36;q21). Other translocation partners include *RPN1* and also *ETV6* and *IKZF1*.[75] They are often therapy related and are associated with an extremely poor prognosis.

Monosomies

Loss of all or part of a chromosome may be seen, most commonly affecting chromosomes 5, 7 or the Y chromosome. Loss of the Y chromosome is seen in ~3% of cases of AML, but it is also a normal age-related finding in men. Monosomy 7 is the most common sole numerical change affecting an autosome identified in AML, although it more commonly occurs in conjunction with other karyotypic abnormalities.[76] Loss of all or part of chromosomes 5 or 7 is a characteristic finding in MDS (see Chapter 2) and therefore their presence in AML often suggests an MDS prodrome evolved into acute phase disease.

Complex and monosomal karyotypes

Complexity refers to the co-occurrence of multiple karyotypic abnormalities within the same leukaemic cell population (Fig. 3.8). A complex karyotype occurs in 15% of *de novo* MDS/AML and in a larger proportion of treatment-related disease. The frequency of multiple karyotypic abnormalities is shown in Table 3.3, where an association between chromosomal complexity and secondary disease is also shown.

There has been significant debate on the definition of a complex karyotype, with the presence of at least three, four or five unrelated cytogenetic abnormalities proposed by different cooperative groups at different times.[55,77] *TP53* deletion at 17p is also a common event, occurring in 40% of complex karyotype patients, and is seen to be mutated in 60%.[78] Further to this, another group has proposed the term 'monosomal karyotype' defined as an autosomal monosomy in conjunction with at least one other autosomal monosomy, e.g. –5 with –7 or structural abnormality, e.g. inv(3)(q21q26) with –7.[79]

Trisomies

Trisomy 8

Trisomy 8 is the most common numerical cytogenetic aneuploidy seen,[55] present in 12–20% of patients overall, the sole abnormality in 5–8% and interphase FISH identifying a further occult 7%[80] (Fig 3.9). It has

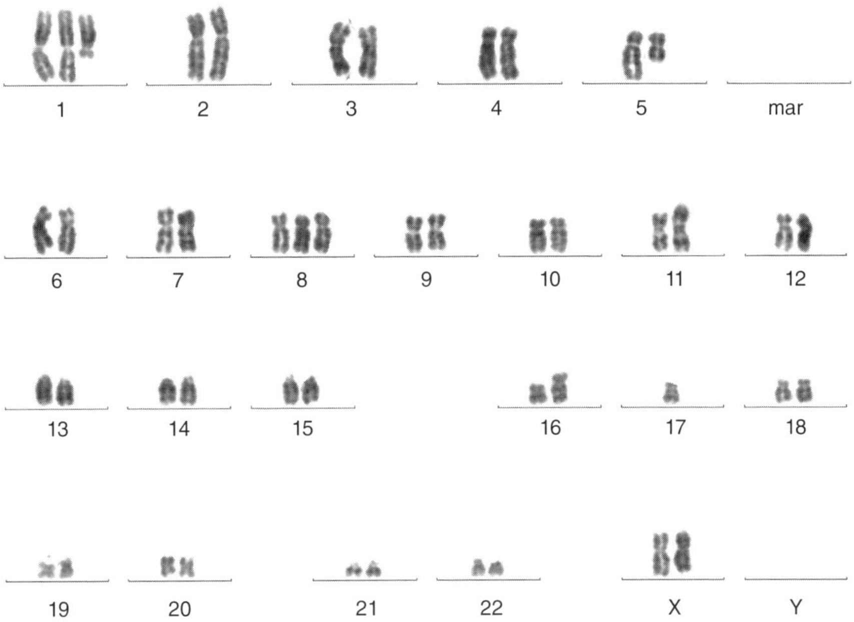

Figure 3.8 A G-banded metaphase spread illustrating a complex karyotype. According to the Medical Research Council Acute Myeloid Leukemia 10 (MRC AML10), the definition of complex karyotype involves the presence of at least five independent chromosomal abnormalities.[129] This is an example of complex karyotype reported in an AML adult patient with. According to the international system of cytogenetics nomenclature (ISCN), the above karyotype reads 47,XX,del(1)(q21),+1,del(5)(q14),i(11)(q11),add(16)(q22),–17. Image courtesy of Giovanni Giudici, Cytogenetics Laboratory, Fondazione Tettamanti, Ospedale San Gerardo, Monza, Italy.

Table 3.3 Frequency of karyotypic complexity and association between karyotypic complexity and secondary disease.[55]

Total number of karyotypic abnormalities	% of patients	Proportion of these that were secondary disease (%)
1	31	7
2	13	9
3	5	6
4	2	11
5	7	16

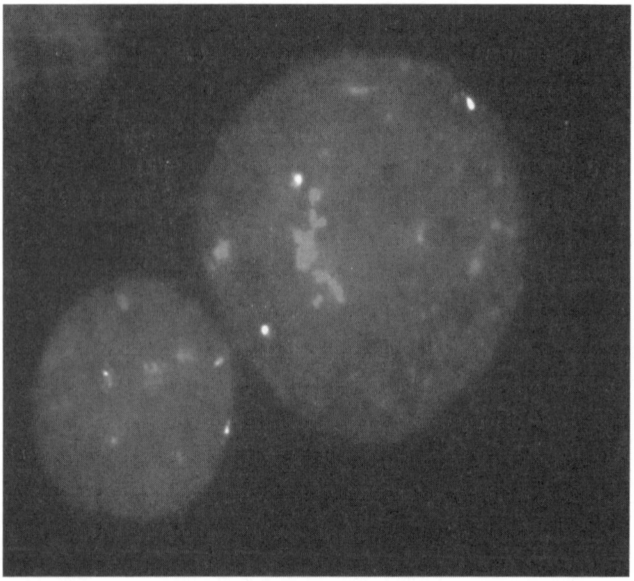

Figure 3.9 Visualization of trisomy 8 in AML. Example of interphase FISH using a probe specific for the centromeric alphoid sequences of chromosome 8. The interphase nuclei were obtained from the bone marrow of a patient with AML. Three hybridization signals (in red) are visible in each of the two nuclei represented here. DAPI was used to counterstain the nuclei in blue. Image capture and analysis were performed using a Zeiss microscope (Axioplan2 Imaging) equipped with a Sensys cooled CCD camera and Smart Capture v2 imaging software (Digital Scientific UK). Image courtesy of S. Tosi, A. Naiel and H. Al-Badri, Leukaemia and Chromosome Research Laboratory, Brunel University, London, UK. *(See plate section for color representation of this figure.)*

been shown to be associated with older patient age, a lower white blood cell (WBC) count and a lower percentage of BM blasts, with an adverse impact on survival also reported.[55] Trisomy 8 demonstrates no specific gene expression profile, suggesting that it may be a disease-modifying secondary event rather than a primary transforming event.[81]

Constitutional trisomy 8 (cT8) is typically seen in a mosaic state (in either blood, skin or both)[82] and is reported to occur in 0.1% of pregnancies and one in 25,000–50,000 live births.[83] Constitutional trisomy 8 mosaicism is termed Warkany syndrome and is typified by abnormal palmar and plantar creases, characteristic facies, mental retardation and other skeletal and ocular abnormalities. CT8 is also associated with myeloid malignancies.[82]

Trisomy 21

Acquired trisomy 21 is one of most common numerical abnormalities, seen in approximately 5% of cases of AML, and also in myeloproliferative neoplasms (MPNs), myelodysplastic syndrome (MDSs), chronic myeloid leukaemia (CML), acute lymphoblastic leukaemia (ALL) and lymphoma.[84] It can occur either as the sole abnormality (in only ~0.4% of cases) or more typically with additional abnormalities, and demonstrates a high variability in phenotype and behaviour.[85]

Trisomy 13

Trisomy 13 or tetrasomy 13 is rare and occurs in a range of subtypes of AML, especially M0. All cases of tetrasomy 13 occur in cases of M0 with small blast morphology and lymphoblast features and a poor prognosis.[86] Nearly all have subsequently been shown to have *RUNX1* mutation and *FLT3* over-expression.[87]

Double minute chromosomes

Double minute chromosomes (dmin) are multiple, small, paired chromatin bodies that do not possess a centromere. There potentially can be more than 80 such dmin within each affected cell and these can act as a method for extrachromosomal amplification of genes, such as *MYC*. They are a rare finding in AML and are associated with a poor prognosis.[88] Associations have been reported with del(9p) and APL morphology[89] and also trisomy 4 in elderly females with AML M2 or M4 and *MYC* amplification.[90]

Normal karyotype – is it really normal?

In many cases, the presentation karyotype will be uninformative. G-banding will fail in some patients owing to the unavailability of sufficient mitotic divisions. Metaphase cytogenetics may also miss translocations owing to poor-quality material, cryptic translocations or small interstitial insertions. Reciprocal translocations can be detected by PCR-based techniques and occasionally molecular techniques highlight discordance between standard cytogenetic analysis, FISH and PCR-based methods. This may reflect a need to screen samples by molecular analysis when morphology is suggestive of a translocation despite normal cytogenetics and FISH.[91] Only approximately 80% of patients with molecular evidence of the *PML-RARA* fusion will have a detectable t(15;17)(q24;q12) by G-banding, therefore FISH should be used to supplement G-banded metaphase analysis and standard probes are available.[92] *CBFB-MYH11* can often be a subtle rearrangement on G-banding and may be missed. Therefore, if the morphology is of AML M4Eo, one should consider FISH or PCR analysis for exclusion of a *CBFB-MYH11* rearrangement. FISH has been used to reveal many cryptic chromosomal abnormalities, and for this reason its use has been routinely implemented in the diagnosis of AML.[93]

Despite intensive evaluation, 40–45% of patients will still be seen to have a normal karyotype, since G-banded metaphase cytogenetics supplemented by FISH has a limit of resolution of approximately 5 Mbp. An increased rate of detection of gross genomic imbalance is obtained using comparative genomic hybridization (CGH) or single nucleotide polymorphism (SNP) array technology (Fig 3.10). Such techniques in one study found acquired copy number aberrations in 24% of normal karyotype AML.[94] It is therefore becoming clearer that 'normal karyotype' AML may be far from normal on a structural genomic level, aside from any mutational changes identified within the DNA sequence itself.

Altered gene expression

Molecular perturbations may occur through mutation; however, up-regulation of gene transcription is increasingly recognized as being of prognostic importance. The molecular basis of such up-regulation remains unclear,[96] but it is possible that the relative expression of these markers reflects mutations yet unidentified, alterations in upstream pathways or epigenetic phenomena.

EVI1

Ecotropic Virus Integration 1 (*EVI1*) translocations have already been discussed with reference to 3q26 abnormalities. *EVI1* is up-regulated as a result of inv(3)(q21q26) and t(3;3)(q21;q26), such that evaluation of AML cases by quantitative real-time PCR allows the identification of cryptic 3q rearrangements.[97,98] *EVI1* is also increased in 5–10% of cases of *de novo* AML without obvious 3q abnormalities.[97,98] Such cases are often associated with coexisting monosomy 7 or *MLL* translocation. *EVI1* over-expression reflects histone hypomethylation of the *EVI1* promoter and a characteristic H3 and H4 acetylation pattern.[99] Elevated *EVI1* expression has been shown to be an independent adverse prognostic

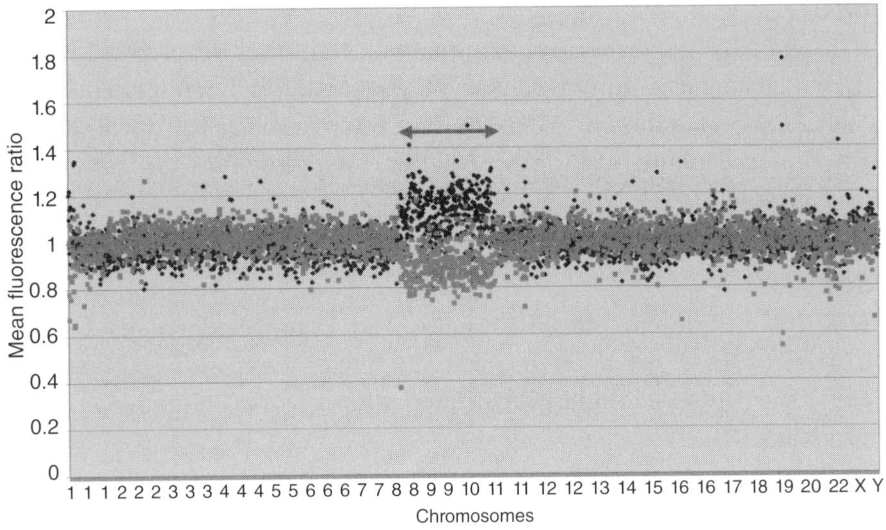

Figure 3.10 Early example of genome-wide array-CGH test data obtained using test DNA from a patient with AML. The earliest genome-wide arrays used for detecting genome imbalance were made up of ~3000 probes spaced at 1 Mb intervals along the genome. Data were obtained by comparing patient versus reference DNAs, each detected with different fluorescent dyes. In this example, dye-swap experiments are shown in blue and pink, respectively, and the mean fluorescent ratios (patient versus reference) reveal genomic gains involving chromosomes 7 and 8 indicated by the red double-ended arrow. Reproduced with permission from Ballabio et al.[95]. *(See plate section for color representation of this figure.)*

factor, being particularly valuable in distinguishing a subgroup of patients who fare poorly with conventional therapy.[97–99]

BAALC

Over-expression of Brain and Acute Leukaemia Cytoplasmic gene (*BAALC*) has been shown to predict an adverse prognosis in normal karyotype patients,[100,101] in addition to predicting refractoriness to treatment.[101] High *BAALC* expression correlates with other adverse molecular signals such as *FLT3*-ITD, *NPM1* wild type and *MLL*-PTD.[101] It is also associated with a specific gene expression profile with expression of stem cell markers, *MDR1* expression and *HOX* gene down-regulation.[102]

MN1

Over-expression of the meningioma-1 gene (*MN1*) is associated with higher *BAALC* expression and wild-type *NPM1*.[103] Over-expression has also been shown to be of independent adverse prognostic significance in normal karyotype AML.[100,104]

ERG

ETS-related gene (*ERG*) over-expression is another reported independent adverse factor in normal karyotype AML.[21,100,102,105] Patients with high *ERG* expression and *FLT3*-ITD had a 5-year overall survival (OS) of 5% compared with 44% for those patients with low *ERG* expression and wild-type *FLT3*.[105]

SET

SET inhibits protein phosphatase 2A and is up-regulated in 28% of unselected cases of *de novo* AML. *SET* over-expression occurs with *EVI1* over-expression and is associated with a worse OS and event-free survival (EFS).[106]

BRE

BRE, a ubiquitination-related gene, is over-expressed in 3% of cases of adult AML. It is seen to be a favourable prognostic marker whose cases typically demonstrate monoblastic morphology and are associated with the *MLL-MLLT3* fusion gene. Indeed, *BRE* expression can stratify this heterogeneous *MLL-MLLT3* group into those with a 5-year OS of 80% for over-expressors compared with 0% for those with normal expression levels.[107]

WT1

WT1 has been shown to be highly expressed in several haematopoietic tumours, including AML.[108] Mechanisms leading to over-expression of *WT1* in AML are poorly understood, but over-expression in AML has been exploited to establish persistence and reappearance of leukaemia. Failure to reduce WT1 transcripts (RT-PCR) below threshold limits defined in normal controls by the end of consolidation has been shown to predict increased relapse risk.[109]

miRNA genes

MicroRNAs are small regulatory RNAs approximately 20 nucleotides in length. They fulfil key roles in post-transcriptional regulation of cellular activities[110] and miRNA patterns reflect the normal regulation of haematopoiesis.[111] In AML there are characteristic miRNA patterns seen with lineage, FAB type, karyotype and mutation status.[112] Unique miRNA signatures are seen with CBF leukaemias, APL, 11q23 translocations and trisomy 8.[113] *NPM1* mutant blasts are seen to over-express miR-10a, miR-10b and miR-196a and down-regulate miR-204 and miR-128a.[111] *CEBPA* mutation is shown to increase miR-181a and miR-335 and down-regulate miR-34a and miR-194.[112] *EVI1* up-regulation results in the increased expression of miR-1–2 via direct binding to its promoter.[52] *FLT3*-ITD is associated with the over-expression of miR-155 and under-expression of miR-144 and miR-451.[113,114] miRNA expression patterns may represent additional markers for adverse outcome· One study has shown that high expression of miR-191 and miR-199a is associated with a worse OS and EFS compared with those cases with low expression.[113] MiR-181a is seen to be over-expressed in normal karyotype AML and those cases with a higher level are seen to have a better CR and a longer OS.[115]

Diagnosis and classification of AML

AML is conventionally classified according to karyotype, as this enables the treating clinician to identify biologically distinct subgroups. In 2001, the World Health Organization (WHO) modified the FAB classification, with the aim of identifying discrete clinical entities within AML as a whole.[116] This work was revised again in 2008 to incorporate an increasing amount of knowledge about the molecular genetics of AML[37] and for the first time incorporated molecular abnormalities, 'AML with mutated *NPM1*' and 'AML with mutated *CEBPA*' as their own distinct sub-categories. The 2001 and 2008 WHO classification systems are summarized above in Table 3.2.

The 2008 WHO classification also provides details of structural and numerical cytogenetic abnormalities that are defined as 'myelodysplasia syndrome (MDS)-related changes'. The diagnosis of AML with MDS-related changes requires ≥20% blasts without prior cytotoxic therapy and any of the following: (i) a prior history of MDS or MPN/MDS; (ii) multi-lineage dysplasia evident in ≥50% of cells in ≥2 lineages or (iii) a, MDS-related cytogenetic abnormality with no prior cytotoxic chemotherapy or diagnostic recurrent balanced translocation. MDS-related cytogenetic changes include a complex karyotype (≥3 abnormalities), unbalanced changes comprising –7/del(7q), –5/del(5q), i(17q)/t(17p), –13/del(13q), del(11q), del(12p)/t(12p), del(9q), idic(X)(q13) and balanced changes comprising t(11;16)(q23;p13.3), t(3;21)(q26.2;q22.1), t(1;3)(p36.3;q21.1), t(2;11)(p21;q23), t(5;12)(q33;p12), t(5;7)(q33;q11.2), t(5;17)(q33;p13), t(5;10)(q33;q21) and t(3;5)(q25;q34). Most patients within this category have frequently observed cytogenetic abnormalities where establishing prognosis is straightforward (e.g. –5, –7, complex karyotype). However, some of the cytogenetic abnormalities that are highlighted in this group are extremely rare, so their significance remains to be firmly established.[55]

Therapy-related AML (t-AML) accounts for 10–20% of cases of AML and often demonstrates multi-lineage dysplasia with an abnormal karyotype in 90%.[37] Therapy-related AML has traditionally been classified according to the agent to which the patient was exposed, which has an important bearing upon disease phenotype, time to onset and outcome. Prior topoisomerase-II inhibitor therapy, such as with etoposide and teniposide, predisposes to leukaemias characterized by balanced reciprocal translocations, particularly involving *MLL* at 11q23, *NUP98* at 11p15, *RUNX1* at 21q22 and *RARA* at 17q21. These typically present following a short latency period of between 1.5 and 3 years from time of first drug exposure with no prior myelodysplastic phase. In contrast, t-AML arising following anti-metabolites, alkylating agents or radiotherapy tends to have a longer latency period of 5–7 years, may have a prior myelodysplastic phase and is characterized by a complex karyotype often featuring loss or deletion of chromosome 5q and/or 7 and a high prevalence of *TP53* mutation.[117] Comparisons of the karyotype of *de novo* AML versus t-MDS/AML show that these therapy-related diseases are more karyotypically complex with more variation in ploidy levels. More classical reciprocal translocations do occur, however, and indeed therapy-related acute promyelocytic leukaemia (t-APL) with the *PML-RARA* fusion gene is the most common second malignancy arising following breast cancer therapy involving epirubicin, mitoxantrone and/or radiotherapy.[118] A major problem in distinguishing between subtypes of t-AML is that the majority of patients who develop this complication have been exposed to combination therapies and it is therefore difficult to identify the causative agent in any particular case. As a result, in the most recent WHO classification of AML this limitation is taken into account and no distinction is made between cases arising following alkylating agents, radiotherapy or drugs targeting topoisomerase-II.[37]

Current risk stratification of AML patients: European LeukemiaNet (ELN) guidelines

European LeukemiaNet (ELN) guidelines, published in 2010,[119] risk stratified AML patients into four groups (see Table 3.4) based upon

Table 3.4 European LeukemiaNet (ELN) molecular risk stratification.[119]

Genetic group	Subsets
Favourable	t(8;21)(q22;q22); *RUNX1-RUNX1T1*
	inv(16)(p13q22) or t(16;16)(p13;q22); *CBFB-MYH11*
	Mutated *NPM1* without *FLT3*-ITD (normal karyotype)
	Mutated *CEBPA* (normal karyotype)
Intermediate 1*	Mutated *NPM1* and *FLT3*-ITD (normal karyotype)
	Wild-type *NPM1* and *FLT3*-ITD (normal karyotype)
	Wild-type *NPM1* without *FLT3*-ITD (normal karyotype)
Intermediate 2	t(9;11)(p22;q23); *MLLT3-MLL*
	Cytogenetic abnormalities not classified as favourable or adverse
Adverse	inv(3)(q21q26.2) or t(3;3)(q21;q26.2); *RPN1-EVI1*
	t(6;9)(p23;q34); *DEK-NUP214*
	t(v;11)(v;q23); *MLL* rearranged
	-5 or del(5q); -7; abn(17p); complex karyotype

*Includes all AMLs with normal karyotype except for those included in the favourable subgroup.

karyotype and the presence or absence of three molecular markers – mutations in *NPM1*, *FLT3*-ITD and *CEBPA*. This stratification is based upon survival data (Fig 3.11) and allows the large group of normal karyotype, intermediate-risk AML, to be divided into a molecularly more favourable and unfavourable groups. These guidelines exclude acute promyelocytic leukaemia, for which recommendations were published separately.[36]

Favourable risk

The 'favourable risk' group based on karyotypic analysis includes those cases of AML with mutant *NPM1* and those with biallelic *CEBPA* mutations in the absence of *FLT3*-ITD.[119–122] Data suggest that only the double and not single mutant CEBPA confers a favourable prognosis.[121,123,124] Patients with single CEBPA mutations generally have a favourable outcome, depending on other concurrent markers such as NPM1/FLT3-ITD status[125].

Intermediate risk

Clinical outcomes vary greatly in the 45–55% of patients with intermediate-risk cytogenetics. This phenotypic heterogeneity is likely to be reflective of the molecular heterogeneity identified in this group via studies using NGS, which is covered in detail later in the section The Genomics of AML.

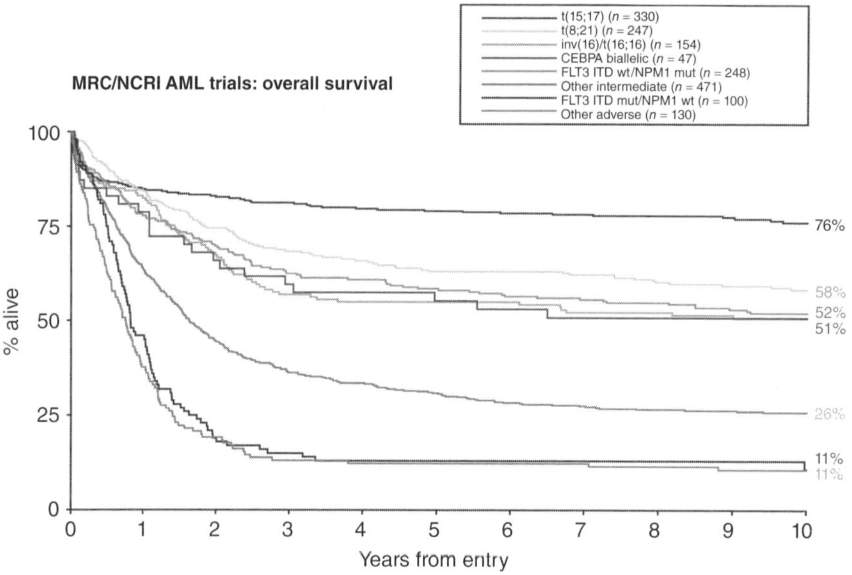

Figure 3.11 Kaplan–Meier plot of overall survival for patients aged under 60 years entered into recent MRC/NCRI clinical studies divided according to karyotype and molecular data. From Smith et al.[278] Reproduced with permission from Elsevier. *(See plate section for color representation of this figure.)*

Adverse risk

About 10–20% of AML patients present with adverse cytogenetics, typically including monosomy 5 or 7, deletions of 5q, abnormalities of 3q or a complex karyotype. Many of these patients tend to be older, often with a prior history of MDS or exposure to chemotherapy. The outcome of patients in this group tends to be very poor with conventional chemotherapy regimens, hence the treatment of choice is allogeneic HSCT (allo-HSCT) in first remission if the patient is fit enough for this aggressive procedure.[126,127]

Therapeutic regimens in AML

AML treatment has remained largely unchanged over the last few decades despite huge advances in our understanding of the molecular basis of this aggressive malignancy. In daily practice, there are two major assessments to be made when deciding patient treatment in non-APL AML: (i) is the patient eligible for intensive standard daunorubicin- and cytarabine-based ('3 + 7') induction therapy?; and (ii) which type of post-remission therapy should be applied? In the context of young, intermediate-risk patients this relates to whether an allo-HSCT is appropriate.[128]

Management of younger adults aged 18–60 years

Induction therapy

Three days of an anthracycline (e.g. daunorubicin at least 60 mg/m^2, idarubicin 10–12 mg/m^2 or mitoxantrone 10–12 mg/m^2) and 7 days of cytarabine (100–200 mg/m^2 continuous i.v.) ('3 + 7') still remains the standard of care in AML. These regimens achieve a CR in 60–80% of younger adults.[119]

Post-remission therapy according to cytogenetic and molecular risk

Favourable risk AML

Repetitive cycles of high-dose cytarabine (HiDAC) (3 g/m^2 per q12 h on days 1, 3 and 5) is considered a reasonable choice for younger adults with CBF AML and also for CN-AML with mutated NPM1 without *FLT3*-ITD and CN-AML with mutated *CEBPA*.[120] In general, patients with favourable-risk AML are not considered candidates for allo-HSCT as the risk of this procedure has been shown to outweigh the benefit in this patient group.[120]

Intermediate-risk AML

For the remaining patients with CN-AML (intermediate I) and those with intermediate II karyotype, repetitive cycles of HiDAC are widely used but the outcome for this subset remains unsatisfactory using just this approach.[119] Allo-HSCT has been shown to be beneficial for patients with intermediate-risk cytogenetics in general,[126,127] and in particular should be considered in patients with higher risk of relapse, including those with *FLT3*-ITD mutations.[120]

Adverse-risk AML

For most patients with adverse-risk cytogenetics, the outcome is extremely poor with conventional consolidation therapies.[129–131] An allo-HSCT from a matched related donor in CR1 is currently considered the treatment of choice for patients with adverse-risk AML[119].

Older AML patients (aged >60 years)

In older patients with AML, the OS remains very poor, hence it is important to select the appropriate up-front therapy for these patients, be it intensive induction chemotherapy, low-dose cytarabine or hypomethylating therapy (decitabine or azacitidine). The best therapeutic option for older patients with AML after achievement of CR remains uncertain. Treatment approaches proposed include the consideration of fit individuals under the age of 75 years for allo-HSCT with reduced-intensity conditioning if deemed appropriate.[132]

Novel agents

The addition of the immunoconjugate gemtuzumab ozogamicin (GO) to standard induction has been shown to be the only novel agent so far to improve outcome in younger[133] and older[134,135] patients with favourable- and intermediate-risk cytogenetics. Currently, GO is only available in the context of clinical trials but it is hoped that it will be approved soon.

A plethora of novel compounds are in development or have just begun clinical trials in AML. Some of these compounds target specific driver mutations, hence eligibility for these compounds should be restricted to patients with the specific genetic lesion. Other compounds in development include monoclonal antibodies, epigenetic modifiers, tyrosine kinase inhibitors and cell-cycle inhibitors.

Monitoring response to therapy (MRD)

The disappearance of a characteristic genetic lesion present at diagnosis indicates a response to therapy. Reappearance of that lesion suggests disease relapse. In some cases, we can exploit this to monitor response to therapy and to detect relapse of leukaemia. The identification of the molecular basis of chromosomal translocations, the development of patient-specific primers and quantitative real-time PCR methodologies have allowed the detection of low levels of a leukaemia using a disease-specific RT-qPCR within a morphologically and cytologically normal marrow. Such technology can be used to follow leukaemia-specific targets such as fusion genes (*PML-RARA, RUNX1-RUNX1T1, CBFB-MYH11*), mutations (mutant *NPM1*) or genes that are commonly over-expressed in AML (*WT1*).[136] The sensitivity limit is dependent on both technique and target.

RT-qPCR assays for *WT1* transcripts typically afford the lowest level of sensitivity because of background amplification due to *WT1* expression in normal progenitors. Flow cytometry might detect one AML blast cell in 10^4 marrow progenitors depending upon the nature of the leukaemia-specific phenotype[137]. For leukaemia-specific targets, such as *RUNX1-RUNX1T1* or canonical *NPM1* mutations, sensitivities of in excess of 1 in 10^6 can be achieved using RT-qPCR[136].

Several studies have shown that *NPM1* is a stable MRD marker which reliably indicates relapse in patients with CN-AML,[138,139] even though recently it has been shown that a small minority of patients lose *NPM1* mutations at relapse.

There is good evidence for the importance of MRD analysis in the setting of APL[140] and in other types of AML that have a molecular marker, e.g. those with a *CBFB-MYH11* or *RUNX1-RUNX1T1* rearrangement.[136,141,142] Quantification of fusion transcripts at diagnosis and early in therapy is able to predict prognosis and the early detection of relapse by the presence of a persistently high level of fusion transcripts following first-line therapy or by a rising transcript number following an initial molecular response.[136] Flow cytometry can also be used for MRD detection in AML if a leukaemia-specific aberrant phenotype was detected at diagnosis.[143]

The genomics of AML

Clonal evolution of AML

Cancer develops through the serial acquisition of somatic mutations over time in cells that undergo selection at a clonal and sub-clonal level.[144] Mutant clones evolve in a process closely resembling Darwinian natural selection whereby mutant stem cells are the unit of selection and somatic mutations accrued over time can either impart a selective advantage to the cell ('driver' lesions) or selective neutrality ('passenger lesions').[145]

In recent years, most, if not all, driver mutations in AML have been identified through the use of next-generation sequencing platforms. AML was the first cancer genome to be subjected to whole genome sequencing.[5] Sequencing of AML genomes and exomes since that time has led to the discovery of many previously unknown driver mutations in AML, which has illuminated our understanding of the molecular pathogenesis of AML, especially in normal karyotype disease.[146–150] Many of these newly discovered driver mutations affect epigenetic regulation of HSCs and since that time the genomic and epigenomic landscape of *de novo* AML has been further defined in a landmark study[150] performed by the AML Cancer Genome Atlas Project, whereby 200 patients with *de novo* AML were either whole genome sequenced (50

Table 3.5 Functional categories of leukaemia mutations as delineated by the Cancer Genome Atlas Research Network.[150]

Functional category of leukaemia mutation	Examples
Transcription factor fusions	*PML-RARA*, *MYH11-CBFB*
NPM1	*NPM1*
Tumour suppressor genes	*TP53*, *WT1*, *PHF6*
DNA methylation	*DNMT3A*, *TET2*, *IDH1*, *IDH1*
Activated signalling	*FLT3*, *KIT*, *NRAS/KRAS*
Myeloid transcription factors	*RUNX1*, *CEBPA*
Chromatin modifiers	*MLL-X* fusions, *MLL-PTD*, *ASXL1*
Cohesin	*STAG2*, *RAD21*, *SMC1A*, *SMC3*
Spliceosome	*SF3B1*, *SRSF2*

patients) or whole exome sequenced (150 patients). This study classified mutations into nine distinct functional groups (Table 3.5), which served to highlight functional collaborations between mutations and molecular heterogeneity, especially within cytogenetically normal AML. Many of these novel mutations, which are described below, have been shown to confer prognostic relevance and therefore many are calling for a shift in classification for CN-AML based mainly on mutational profiling.[151]

The Cancer Genome Atlas Project also found that AML genomes have far fewer mutations than other cancer genomes. Only an average of 13 mutations per individual were identified, of which five were found to be recurrent in AML. The concept that driver mutations belie leukaemic transformation of HSCs is supported by this study as all AML cases were found to have at least one potential driver mutation. Now that the major driver mutations have been discovered, the major challenge is to understand the order of acquisition of mutations AML and the process that belies the clonal evolution of normal HSCs into frank leukaemia.

The essential first step in myeloid leukaemogenesis is the foundation of a pre-leukaemic clone of cells. Leukaemic progression is thought to rely on cells from this pre-leukaemic clone acquiring further cooperating mutations, thus giving rise to sub-clones that in turn acquire further mutations en route to the development of frank leukaemia.[152] Recently, pre-leukaemic HSCs have been shown to survive induction chemotherapy and thus be a reservoir for evolution of relapsed disease.[153,154] These findings are supported by whole genome sequencing studies in paired diagnostic and relapsed samples that have shown that disease relapse originates from the primary clone or a sub-clone thereof.[155] It is therefore important to understand which mutations are 'early' events in clonal evolution and which mutations occur at a 'late' time point. Models of such a paradigm have attributed mutations in 'landscaping genes', involved in global chromatin changes such as DNA methylation, histone modification and chromatin looping, occurring early in the evolution of AML, whereas mutations in 'proliferative' genes such as FLT3 or RAS are seen to occur late.[153]

Established recurrent mutations in AML

Nucleophosmin (NPM1)

NPM1 is a 12-exon, 25 kbp gene situated at chromosome 5q35. Mutations in *NPM1* are the most common mutation in AML, with an overall incidence of 25–35% (CN-AML 45–60%).[120,156] *NPM1* mutations are enriched in normal karyotype AML, such that 85% of AML with *NPM1* mutation display a cytogenetically normal karyotype. NPM1 has roles in many cellular processes, including ribosome biogenesis, centrosome duplication and regulation of apoptosis. Subcellular localization signals on the NPM1 protein allow it to shuttle between the nucleus, nucleolus and the cytoplasm to carry out its roles. Mutations of NPM1 in AML (NPM1c) result in disruption of the localization signals replacing it with a nuclear export signal,[156,157] resulting in an aberrant localization of the NPM1 mutant protein in the cytoplasm.

Some 70–80% of *NPM1* mutations in adults are termed 'type A' where a 4 bp TCTG insertion occurs at position 956–959 leading to a frameshift and the replacement of the last 7αα by 11 different ones at the C-terminus.[158] Additional mutations have similar effects on this region due to a CATG insertion (type B) or CCTG insertion (type D) at the same site, resulting in the loss of one or both of the W288 and W290 tryptophan residues in the mutant NPM C-terminal NLS with abrogation of function and gain of an additional NES[159] (Fig. 3.12b).

A conditional knock-in mouse model of the most common form of Npm1c mutation, type A, has shown that Npm1c mutations are AML-initiating lesions that cause *Hox* gene over-expression, impart increased self-renewal to and prime haematopoietic stem and progenitor cells to leukaemic transformation by activation of pro-proliferative pathways.[160] This mouse model also explained other important features of human NPM1c^{+} AML, including a failure to observe NPM1c mutations in the human germline (embryonic lethality) and a consistent negativity for CD34, a primitive marker, as the

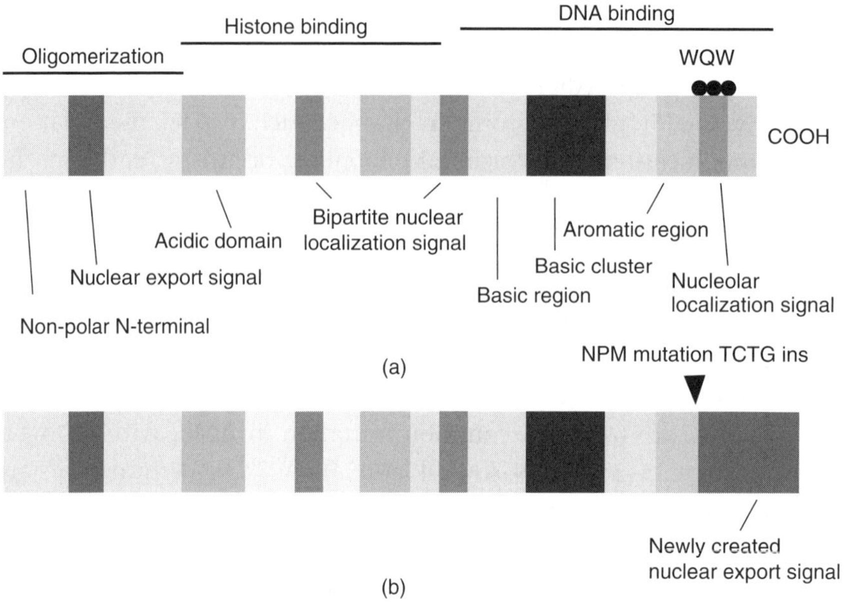

Figure 3.12 Structure of NPM1 and its mutated forms. (a) Schematic representation of the protein structure of NPM illustrating its component parts and functions including its bipartite nuclear localization signals (NLS) and C-terminus nucleolar localization signal that consists of two critical tryptophan residues in a WQW sequence. (b) Type A *NPM1* mutation results in a TCTG insertion where indicated, with the loss of normal C-terminal protein structure and therefore loss of the nucleolar localization signal and gain of a new nuclear export signal (NES).

effects of Npm1c are more noticeable on later progenitors. The long latency observed for these knock-in mice to develop AML reflects the need for additional mutations to be present and cooperate to drive leukaemogenesis.[160]

NPM1 mutations have been found to co-occur with other known AML-associated mutations such as *FLT3*, *DNMT3A*, *IDH1*, *IDH2* and *NRAS*.[120,146,156,161,162]

A high proportion of *NPM1*-mutant AML patients achieve CR with intensive chemotherapy protocols; however, approximately 50% of these patients will relapse within 3 years, especially those with concurrent *FLT3*-ITD mutations.[156,163,164] Relapsed leukaemia is generally much less chemo-sensitive and relapse is therefore the main vehicle for much of the mortality associated with AML.

In relapse of *NPM1*-mutant AML, approximately 10% of patients had been noted to lose the *NPM1* mutation.[138,165,166] A recent study investigating the clonal evolution of relapsed NPM1-mutated AML found a persistence of *DNMT3A* mutations in patients who had lost *NPM1* mutations at relapse.[167] This suggested that *DNMT3A* mutations may precede *NPM1* mutations in AML pathogenesis and that disease relapse in these patients actually originates from an earlier, ancestral *DNMT3A* clone.[167] Indeed, when highly purified HSCs, progenitors and mature cell fractions from the blood of AML patients were examined in a study by Shlush et al.,[154] *DNMT3A* mutations were found to arise earlier than *NPM1*, in the clonal evolution of AML. Furthermore, HSCs with *DNMT3A* mutations were found to survive chemotherapy, providing a potential reservoir for disease relapse. Therefore, the status of *NPM1* as a founder mutation in AML has recently been cast into doubt, although their role in leukaemogenesis remains a critical one.

FLT3 mutations

FLT3 (Fms-like tyrosine kinase 3), fetal liver kinase 2 (flk-2) or stem cell tyrosine kinase 1 (STK-1) is a 24-exon gene situated on chromosome 13q12.2.[168] *FLT3* is a member of the class III receptor tyrosine kinase (RTK) family that includes *KIT, C-FMS* and *PDGFRβ*.[169] All members of this family share amino acid sequence homology consisting of an extracellular component comprised of five immunoglobulin-like domains and a cytoplasmic component with a split tyrosine kinase domain (Fig. 3.13a). Ligand binding by Flt3 ligand (FL) results in receptor dimerization, autophosphorylation and kinase activation with resultant activation of downstream pathways such as STAT5a, RAS and PI3K[170] (Fig. 3.13b).

Mutations in *FLT3* are among the commonest in AML and are classified into the following two types: internal tandem duplications (ITDs) and base substitution mutations in the activation loop of the tyrosine kinase domain (TKDs).

Internal tandem duplications of *FLT3* (*FLT3*-ITDs)

ITDs of *FLT3* (Fig. 3.13c) are common in AML (~20% of all cases), especially in normal karyotype disease (28–34%), where they are found commonly to co-occur with NPM1 mutations and are associated with an inferior outcome.[120,158] ITD mutations are located in exons 14 and 15 of the *FLT3* gene and show a large variation in the number and sizes of duplicated fragments. These mutations result in amino acid sequence changes with intact coding frames, resulting in constitutive activation of the receptor tyrosine kinase and downstream activation of *RAS*, *MAPK* and *STAT5* signalling pathways leading to dysregulated cellular

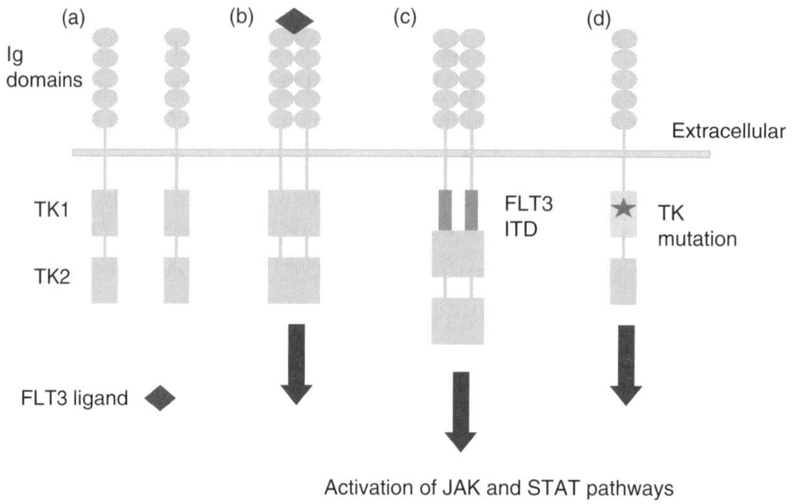

Figure 3.13 Structure of FLT3 and its mutated forms. (a) Diagrammatic representation of the protein structure of FLT3 with five extracellular immunoglobulin-like domains and two intracellular tyrosine kinase domains. (b) Binding of FLT3 ligand leads to FLT3 dimerization and activation of downstream pathways. (c) Acquisition of a FLT3 internal tandem duplication in the juxtamembrane region leads to ligand independent dimerization and activation. (d) Point mutations within the tyrosine kinase domain lead to FLT3 activation and ligand independent activation.

proliferation.[170] In cytogenetically normal AML, *FLT3*-ITDs confer an unfavourable prognosis and inferior overall survival due to a high relapse rate.[120,171,172] Recently published data suggest that in intermediate-risk AML, the negative prognostic association with *FLT3*-ITD is maintained even in the context of other genetic abnormalities such as *DNMT3A* and *TET2*.[151,173]

The mutant to wild-type allelic ratio (AR) and the ITD insertion site (ISS) are the two major prognostic factors associated with *FLT3*-ITDs in CN-AML.[174,175] There is a consistent association between high allelic burden (AR >0.50) and unfavourable outcome.[176,177] An ITD insertion site in the B1 sheet of the tyrosine kinase domain 1 (TKD-1) in approximately one-quarter of cases confers an especially poor prognosis.[175] Evidence is emerging that AML patients with *FLT3*-ITDs benefit from allo-HSCT in first remission, which is recommended for this group.[119] However, recent research has shown differential benefits to allo-HSCT in this group based on AR and ISS. A major benefit of allo-HSCT performed in the first CR was present in patients with an AR of >0.51 with respect to relapse-free survival (RFS) and OS; however, there was no improvement seen in RFS or OS in patients with a low AR.[178] Allo-HSCT also does not seem to alter the dismal prognosis associated with *FLT3*-ITD located in TKD1.[178]

FLT3-ITDs can also occur in AML with abnormal karyotype. In APL, the presence of an *FLT3*-ITD has been shown to have no effect on prognosis, except those with an ITD allele load of <0.5 trended to a better EFS and OS.[179] *FLT3* ITD has been shown to be an adverse factor for OS in core binding factor AML but only in high allelic ratios.[180]

FLT3-TKD mutations

Approximately 11–14% of patients with CN-normal AML show mutations in the activation loop of the tyrosine kinase domain of FLT3, with a mutation hotspot in the aspartic acid codon, D835.[181] Mutations have additionally been observed at codon 836 with I836del being the most common, and also I836M, I836T, I836MA and I836ins LK have been reported.[172] Both classes of mutation result in constitutive activation of the receptor. Such mutation within the activation loop results in conformational change, exposing the active site and resulting in up-regulation of kinase function and ligand-independent activation (Fig. 3.13d). However, the proliferative effects of TKD mutations in FLT3 are not as potent as ITDs and their influence on prognosis remains controversial.[182,183]

Examples of genes that have been identified as targets for dysregulation either by mutation or up-regulation are listed in Table 3.6.

CCAAT/enhancer-binding protein α (*CEBPA)* mutations

CEBPA is a single-exon gene 1077 bp in length sited on 19q13.1 and is a member of the basic leucine zipper transcription factor family, which regulates cell-cycle exit and differentiation in various tissues.[184] In the haematopoietic system, this transcription factor is involved in lineage specification and is crucial for the development of myeloid progenitors to differentiated neutrophils.[185] This protein comprises N-terminus trans-activating domains, a basic region necessary for specific DNA sequence binding, and a C-terminus leucine zipper region necessary for dimerization. They are able to form hetero- and homodimers and can function both as activators and repressors.

Two different isoforms of CEBPA, 42 kDa (p45) and 30 kDa (p30), are produced which differ in transcriptional activity (Fig. 3.14a). The 30 kDa protein is generated from an alternative start codon situated 351 nucleotides downstream of the main ATG start codon and lacks the first trans-activating domain and the antimitotic activity of the

Table 3.6 Examples of genes that have been identified as targets for dysregulation by either mutation or up-regulation.

	Gene symbol (aliases and previous gene symbols)	Gene name	Chromosomal location
Mutation	*FLT3* (*STK1*, *FLK2*, *CD135*)	Fms-related tyrosine kinase 3	13q12
	NPM1 (*B23*, *NPM*, *numatrin*)	Nucleophosmin	5q35.1
	CEBPA (*CEBP*)	CCAAT/enhancer binding protein (CEBP), alpha	19q13.1
	KIT (*c-kit*, *CD117*, *SCFR*)	v-kit Hardy Zuckerman 4 feline sarcoma viral oncogenes homologue	4q11-q12
	MLL (*ALL-1*, *HRX*, *HTRX1*, *CXXC7*, *TRX1*, *MLL1A*, *KMT2A*)	Myeloid/lymphoid or mixed-lineage leukaemia (trithorax homologue, *Drosophila*)	11q23
	TP53	Tumour protein p53	17p13.1
	RUNX1 (*AML1*, *CBF2A*, *AMLCR1*, *PEBP2A2*)	Runt-related transcription factor 1	21q22.3
	WT1 (*GUD*, *AWT1*, *WAGR*, *WIT-2*)	Wilms' tumour 1	11p13
	NRAS	Neuroblastoma RAS viral (v-ras) oncogene homologue	1p13.2
	KRAS	v-Ki-ras2 Kirsten rat sarcoma viral oncogene homologue	12p12.1
	GATA1	GATA binding protein 1	Xp11.23
	GATA2	GATA binding protein 2	3q21
	IDH1	Isocitrate dehydrogenase ($NADP^+$) soluble	2q32-qter
	IDH2	Isocitrate dehydrogenase ($NADP^+$) mitochondrial	15q21-qter
	TET2	Tet methylcytosine dioxygenase 2	4q24
	ASXL1	Additional sex comb like 1 (*Drosophila*)	20q11
	DNMT3A	DNA (cytosine 5) methyltransferase 3 alpha	2p23
	CBL	Cbl proto-oncogene E3 ubiquitin ligase	11q23.3-qter

Table 3.6 (*continued*)

	Gene symbol (aliases and previous gene symbols)	Gene name	Chromosomal location
	EZH2 (*ENX-1*, *EZH1*, *KMT6*, *KMT6A*)	Enhancer of zeste homologue 2 (*Drosophila*)	7q35-q36
	PHF6	PHD finger protein 6	Xq26
Upregulation	*MECOM* (*EVI1*, *MDS1*, *MDS1-EVI1*, *PRDM3*)	MDS1 and EVI1 complex locus	3q26
	BAALC	Brain and acute leukaemia gene, cytoplasmic	8q22.3
	MN1 (*MGCR*)	Meningioma (disrupted in balanced translocation) 1	22q12.1
	ERG (*erg-3*, *p55*)	v-ets erythroblastosis virus E26 oncogene-like (avian)	21q22.3
	SET (*ZPP2A*, *PHAPII*)	SET nuclear oncogene	9q34
	BRE (*BRCC4*, *BRCC5*)	Brain and reproductive organ expressed	2p23

wild-type protein. CEBPA interacts with a variety of transcription factors during haematopoiesis, including *PU.1*[186] and *RUNX1*,[187] and also cellular oncogenes such as *C-MYC* [188]and *RAS*.[189] The role of CEBPA in adult haematopoiesis and in leukaemia has been further elucidated by a conditional knock-out mouse model,[190] which demonstrated that the loss of CEBPA in adult HSCs confers fetal HSC characteristics, including enhanced proliferation, an increased number of functional long-term HSCs (LT-HSCS) and advanced repopulating ability. In this study, transcriptional repression of N-myc by CEBPA was identified as an important effector pathway that is at least partially required for the maintenance of adult HSC quiescence.

Mutations in *CEBPA* occur in AML in 5–14% of normal karyotype cases and lead either to nonsense mutations in the N-terminus or in-frame mutations in the C-terminal basic leucine zipper (bZip region).[120,191] N-terminal nonsense mutations prevent the expression of full-length CEBPA protein, thereby up-regulating the formation of the 30 kDa truncated isoform that has been shown to possess a dominant negative effect on DNA binding to and transactivation of CEBPA target genes such as *CSF3R* (GCSF-R)[191] (Fig. 3.14b). BZip domain mutants are not seen to have a dominant negative function and instead in-frame mutations result in the loss of function of one allele[192] (Fig. 3.14c).

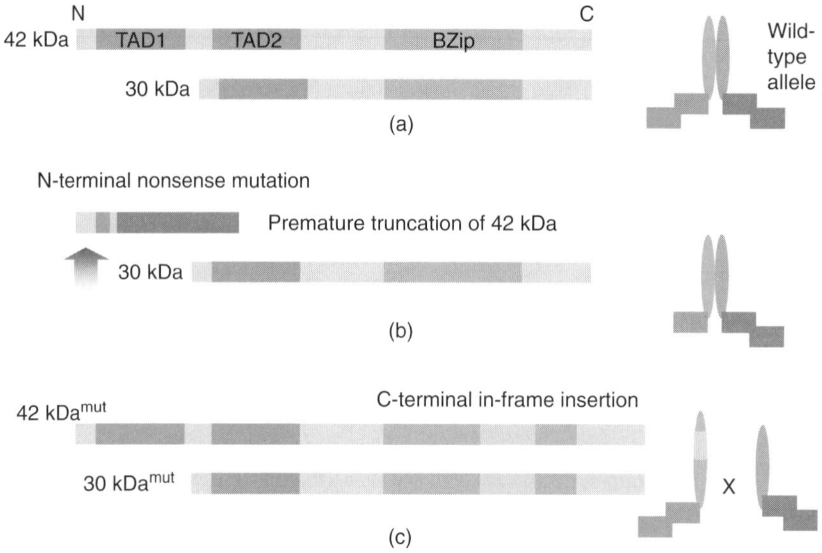

Figure 3.14 Structure of CEBPA and its mutated forms. (a) Normal CEBPA structure illustrated for both the 42 and 30 kDa forms with the transactivating domains (TAD) and basic region and leucine zipper region (bZip) responsible for dimerization highlighted. The shorter form is seen to lack the first TAD. (b) N-terminal CEBPA mutation leads to transcriptional frameshift and premature truncation of the 42 kDa form and therefore predominance of the shorter 30 kDa form and altered activation of downstream pathways. (c) C-terminal in-frame insertions within the BZip domain lead to loss of dimerization ability.

Biallelic mutations of *CEBPA* can occur in AML and typically affect a combination of N- and C-terminal mutations on different alleles – this allows division into single mutated cases ($CEBPA_{SM}$) and double mutated cases ($CEBPA_{DM}$).[123,193] Studies suggest that in CN-AML a single mutation in *CEBPA* is prognostically neutral and therefore other molecular markers such as *NPM1* and *FLT3*-ITD genotype must be taken into account in these patients, whereas biallelic mutations are associated with a favourable prognosis.[123] Accordingly, in current guidelines an allo-HSCT is not recommended in CN-AML patients with biallelic *CEBPA* mutations.[119]

Germline N-terminal frameshift mutations of *CEBPA* have additionally been identified as a predisposing factor in cases of familial AML with acquisition of a second C-terminal mutant allele involved in progression to leukaemogenesis.[194–196] Such cases appear to have a favourable prognosis, although a more aggressive phenotype may be associated with chromosomal deletions of *FLT3* on 13q and *ATM* on 11q.[197]

KIT

KIT (CD117) is a type III tyrosine kinase receptor that binds to stem cell factor (SCF) (KIT ligand). Following ligand binding to the Ig domains, dimerization occurs, thus activating intrinsic kinase activity by the transphosphorylation of specific residues in the JM and TK domains. Mutations in KIT are predominantly associated with the core binding factor leukaemias (CBF-AMLs), t(8;21) or inv(16). In adults, KIT mutations in t(8;21) and inv (16) have generally been associated with a higher incidence of relapse, although whether overall survival is affected is study dependent.[198–203] The most frequent mutations are missense mutations affecting exon 17 encoding the tyrosine kinase 2 domain *(KIT^{TKD})*, the most common observed mutation being a D816V substitution. This residue stabilizes the kinase loop in its active configuration. Such a mutation induces activation of STAT3/STAT5 and PI3K/Akt signalling. Mutations targeting exon 8 are usually in-frame insertions or deletions that affect an extracellular domain of the receptor implicated in receptor dimerization (KIT^{ECD}) and affect MAPK and PI3K/AKT signalling pathways.[204] *KIT* mutations have been found frequently to be lost at relapse, relevant for clinical uses of *KIT* inhibitors and measuring minimal residual disease (MRD).[180]

MLL partial tandem duplication (*MLL* PTD)

The mixed lineage leukaemia *(MLL)* gene is located on chromosome (11q23), which encodes a transcription factor with histone H3 lysine 4 (H3K4) methyltransferase activity. To date, *MLL* has been found to participate in >60 different translocations with distinct fusion partners as described earlier. In addition to these translocations, *MLL* can be mutated via a gain of function, intragenic, partial tandem duplication (PTD) of exons 3–9 or exons 3–12.[205] MLL-PTD cannot be detected by classic karyotypic analysis and is found in ~5% of patients with CN-AML, associated with a poor prognosis and high relapse rates.[206–208] Early allo-HSCT seems prognostically beneficial in patients with a PTD in *MLL*.[119]

TP53

TP53, situated at 17p13.1, has a crucial role in the cell cycle, acting as a tumour suppressor gene. In AML, *TP53* abnormalities are most commonly missense or frameshift mutations in the DNA-binding domain in exons 4–8,[209] or via loss of one allele via 17p monosomy and/or loss of heterozygosity.[210] *TP53* mutations in AML are relatively rare, more common in older than younger patients and closely associated with complex karyotype involving deletions of chromosomes 3, 5, 12 or 17. They confer an extremely poor prognosis even within complex karyotype cohorts.[209]

RUNX1

RUNX1 is a 12-exon gene located at 21q22.12· In addition to the CBF translocations previously outlined as a modality for *RUNX1* deregulation, intragenic mutations of RUNX1 have been found in 5–15% of AML cases.[211–214] There is a predilection for those cases with acquired trisomy 21, trisomy 13 or M0 morphology, where the frequency may be as high as 38, 90 and 65%, respectively.[212] *RUNX1* mutations are associated with secondary AML evolving from myelodysplastic syndrome (MDS), and this is paralleled by an increase in frequency of mutations in AML in older age groups.[128] *RUNX1* mutation predicted a lower CR rate, a shorter DFS and OS rate and in a multivariable analysis was seen to be of independent prognostic significance for OS.[212,213] A frequent co-occurrence in AML with mutations in genes encoding epigenetic modifiers such as *ASXL1*, *MLL-PTD*, *IDH2* and *BCOR* has recently been discovered through genomic sequencing studies.[150]

RUNX1 mutation has additionally been implicated in the syndrome 'familial platelet disorder with predisposition to AML', which is an autosomal dominant condition with a variable severity and age of onset. *RUNX1* has been identified as the candidate gene for this syndrome in at least 11 families.[215–218] All of these mutations occur in the *runt* homology domain of *RUNX1*. A second *RUNX1* mutation has been described in several cases that progressed to develop AML, the mechanism being either an acquired somatic point mutation or duplication of the abnormal allele via acquired trisomy 21.[219]

WT1

The *WT1* gene encodes a zinc-finger transcription factor involved in cell differentiation and tumour suppression in haematopoiesis and nephrogenesis.[220,221]

WT1 mutation is observed in 5–10% of normal karyotype AML associated with a younger age, M6 phenotype, *FLT3*-ITD mutation and a reduced OS, in addition to primary chemotherapy resistance, although prognosis appears to study dependent. [222–225]

The mechanism by which *WT1* exerts a leukaemogenic effect had been elusive until recent research revealed that *WT1* mutations in AML result in a loss of *TET2* function.[226] Further research into understanding this association should reveal important insights into AML pathogenesis.

RAS mutations

Somatic gain of function mutations in *NRAS* and *KRAS* is a common finding in cancer and in AML, occurring in 25 and 15% of AML patients, respectively.[204] RAS proteins are small GTPases that act downstream of tyrosine kinase receptors involved in haematopoiesis. In AML, constitutive activation of RAS occurs when missense mutations (in codons 12, 13 and 61 in *NRAS* and in *KRAS)* impair intrinsic GTPase activity. RAS activation leads to oncogenic deregulation and hyperactive signalling of the downstream effector pathways of RAS that include the PI3K/Akt/mTOR pathway implicated in cellular survival and the pro-proliferative BRAF/MEK/ERK pathway.[227] These pathways downstream of RAS are under investigation as therapeutic targets as, so far, direct targeting of RAS using farnesyltransferase inhibitors have not shown additional clinical benefit in clinical trials when used in older patients with AML.[228]

Novel recurrent mutations in AML

DNA methyltransfersase 3A (DNMT3A)

DNMT3A is a 130 kDa protein encoded by 23 exons on chromosome 2p23. It mediates *de novo* DNA methylation of cytosine residues and thus is seen as an epigenetic regulator. Methylation of DNA refers to an addition of a methyl group (CH_3) to the C5 position of cytosine to form 5-methylcytosine (5mC).[229] The predominant target of DNA methylation is ‘CpG’ sites – dinucleotide pairs, on the same strand of DNA, consisting of a cytosine (C) and a guanine (G) joined by a phosphodiester bond (p).[230] DNA methylation is associated with the silencing of gene expression[231] and hence is an important epigenetic modification that mediates a range of processes, including X-chromosome inactivation, stem cell regulation and genomic imprinting.[232]

Recurring mutations in *DNMT3A* were first found in AML samples using next-generation sequencing (NGS) technology in 2010, when they were found to exist in up to 22% of AML patients,[147–149] making this the most common mutation in AML after mutations in *NPM1* and *FLT3*. There is a mutational hotspot in *DNMT3A* at arginine 882 (R882) where heterozygous missense mutations account for half of all *DNMT3A* mutations in AML. Additional mutations (nonsense, frameshift and splice site) have been found to occur throughout the *DNMT3A* coding sequence.[150]

Research investigating the leukaemogenic effect of mutations in *DNMT3A* suggests that there are two classes of mutations: (i) *R882* mutations and (ii) non-*R882* mutations.

DNMT3A R882 mutations have been shown to lead to a striking reduction of *de novo* DNA methyltransferase activity via a dominant-negative effect of co-expression of the mutant protein with its wild-type counterpart.[233] This dominant negative effect leads to focal hypomethylation throughout the genome that is not seen in AML patients with non-R882 mutations.[233] Murine models have elucidated that when Dnmt3a is conditionally deleted from HSCs, self-renewal is markedly favoured over differentiation, giving these cells a clonal advantage over their wild-type counterparts; however, no leukaemias were observed in these mice.[234]

What has become apparent in human AML through further sequencing studies, including the 200 AML genomes analysed by the Cancer Genome Atlas,[150] is that there is a strong co-occurrence of *DNMT3A* mutations with mutations in *NPM1* and *FLT3*-ITDs. Indeed leukaemias with this trio of mutations have been found to cluster together with distinct mRNA, miRNA and DNA methylation signatures, suggesting that *DNMT3Amut-NPM1mut-FLT3*-ITD leukaemia is a distinct entity.[150] *DNMT3A* mutations have also been found to be associated with older patients, a higher WBC and platelet count, M4/M5 morphology, intermediate-risk cytogenetics and *IDH2* mutations.[147,151,235,236] *DNMT3A* mutations are also found in patients with other haematological malignancies, including MDS, where they are positively correlated with mutations in *SF3B1* and *U2AF1*, but negatively correlated with mutations in another spliceosome gene, *SRSF2*.[230]

A recent important study investigating the clonal evolution of AML demonstrated that *DNMT3A* mutations arise very early in leukaemia evolution, leading to a clonally expanded pool of pre-leukaemic HSCs from which AML evolves.[154] This study went on to demonstrate that in AML patients, HSCs with *DNMT3A* mutations are able to survive induction chemotherapy, persist in the bone marrow at remission and provide a reservoir for disease relapse.[154]

The poorer outcome of AML patients with *DNMT3A* mtuations[147,235] may be a reflection of these findings. However, one study has suggested that the unfavourable effect of *DNMT3A* mutations could be overcome by increasing the dose of daunorubicin during induction therapy.[151]

Isocitrate dehydrogenase (*IDH1* and *IDH2*)

Following whole genome sequencing of a patient with AML M1, a mutation affecting codon 132 of the cytosolic isocitrate dehydrogenase gene, *IDH1*, was noted and subsequently found to occur preferentially in normal karyotype AML at a frequency of 16%.[146] A similar frequency has been reported by other groups, with the *IDH1* R132 mutation occurring typically in those patients with normal cytogenetics and *NPM1* mutation.[151,162,237–239] Mutations have since been identified in the gene *IDH2* at two hotspots, R172 and R140, the latter being more common. *IDH2* mutations similarly affect normal karyotype AML at a frequency of 10–15%.[151,239] Therefore, approximately 15–30% of AML patients have mutations in either *IDH1* or *IDH2*; however, their prognostic relevance is unclear. Some studies have suggested an association with a poorer outcome,[146,240,241] whereas others have found no evidence of an adverse impact.[237,242] The *IDH2* mutation has been found in some studies to be favourable prognostically, but only for the R140Q form.[151,243]

The protein products of isocitrate dehydrogenase genes convert isocitrate to α-ketoglutarate either in the cytosol (IDH1) or in mitochondria (IDH2) (Fig. 3.15).

Leukaemia-associated mutations in IDH1 and IDH2 result in a neomorphic enzyme activity whereby α-ketoglutarate is converted into an oncometabolite 2-hydroxyglutarate (2HG),[244] a competitive inhibitor of histone demethylases and the *TET* family of 5mC hydroxylases.[245] This results in a hypermethylation phenotype and impaired Tet2 function leading to a block in haematopoietic differentiation.[161] This shared pathway of leukaemogenesis is felt to underlie the mutual exclusivity seen between *IDH1/2* mutations and *TET2* mutations in AML patients.[161]

2HG may become an easily accessible biomarker for this form of AML as it can be measured in the serum of peripheral blood. Clinical trials of selective inhibitors of the mutant metabolic enzyme 2HG are in progress.

TET2

The 'ten–eleven-translocation' gene encodes for a protein (Tet2) that exerts epigenetic effects by converting 5mC to 5-hydroxymethylcytosine (5hmC) in DNA.[246] RNAi inhibition of *TET2* in cell lines and $CD34^+$

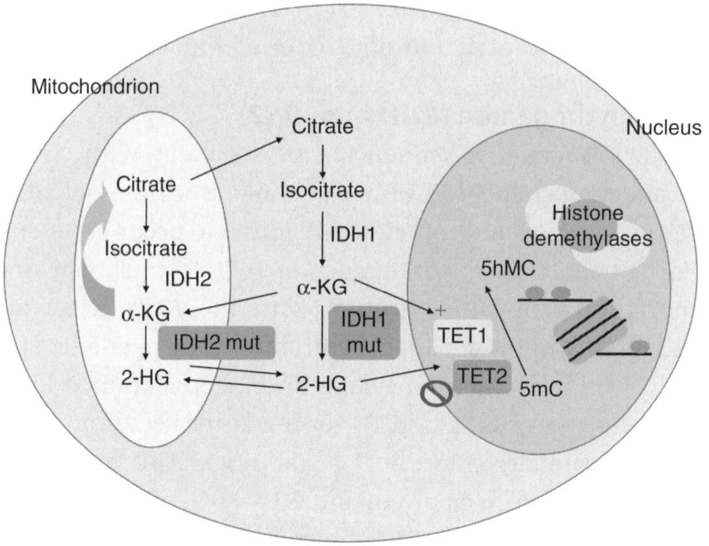

Figure 3.15 Schematic representation of the IDH1, IDH2 and TET2 pathways. Mitochondrial IDH2 and cytosolic IDH1 convert isocitrate to α-ketoglutarate (α-KG), which is a substrate for TET2 hydroxylation of 5-methylcytosine (5mC) DNA residues to produce 5hMC. Mutated IDH1/2 converts α-KG further to 2-hydroxyglutarate (2HG), which acts as an inhibitor of TET function as well as histone demethylases leading to a hypermethylation phenotype and a block in TET2 function. Adapted from Prensner and Chinnaiyan.[279]

stem cells decreases the level of 5hmC and skews differentiation to the granulomonocytic lineages.[247]

A conditional knock-out mouse model revealed that conditional loss of *TET2* in the haematopoietic compartment leads to increased stem cell self-renewal and progressive enlargement of the HSC compartment with eventual resultant myeloproliferation.[248]

Loss of function *TET2* mutations were first reported to occur in a large proportion of myeloid malignancies[249] – MDS (19%), myeloproliferative disorders (MPN) (12%), secondary AML (24%) and chronic myelomonocytic leukaemia (CMML) (22%). Since that time, other studies have shown that *TET2* mutations occur in 10–20% of AML, with some groups reporting them as unfavourable in terms of survival in CN-AML [151,250,251] whereas other studies found no such prognostic effect in either CN-AML patients or within the ELN subsets.[252] Mutations are often homozygous and associated with intermediate-risk cytogenetics, +8 or a normal karyotype and a range of other mutations, especially *ASXL1* and *NPM1* but mutually exclusive of *IDH1* or *IDH2*.[151,161,250]

ASXL1

The *additional sex comb-like 1* (*ASXL1*) gene at 20q11 is mutated in a range of myeloid malignancies including MDS and MPN in addition to AML. It is a member of the enhancer trithorax and polycomb gene family (ETP) transcriptional regulators that can either repress or activate transcription.

ASXL1 mutations promote myeloid transformation by interference with PRC2-mediated histone H3 lysine 27 (K27) trimethylation at specific sites of transcriptional repression, such as the HOXA cluster.[253] The majority of the mutations in AML are an exon 12 truncating mutation, c.1934dupG, but other point mutations also are found within this exon.[254] *ASXL1* mutations occur more commonly in secondary AML and in older patients, with an incidence of 10–16% reported in those over 60 years of age.[255] *ASXL1* mutations are adverse prognostically in terms of both CR and OS in all studies to date and are rarely seen to coexist with *NPM1* or *FLT3*-ITD.[151,254,255]

CBL

Investigation of a site of recurrent UPD at 11q led to the discovery of homozygous mutation in the *Casitas B-cell lymphoma* gene (*CBL*). This is an E3 ubiquitin ligase responsible for terminating tyrosine kinase signalling. Gain of function *CBL* mutations constitutively activate *FLT3* and *KIT* proliferative signalling pathways.[256] Reindl et al. reported exon 8 or 9 *CBL* mutations in three of 279 cases of AML, restricted to CBF leukaemias or those with del(11q).[257] Most of these mutations were homozygous. Such cases had a monoblastic phenotype and a poor outcome. Makishima et al. found a frequency of *CBL* mutation of 9% in secondary AML, suggesting perhaps a role in the malignant evolution of MPD and MDS.[258]

Spliceosome mutations

Introns in protein-coding genes are removed by spliceosomes in the nucleus; these are complexes composed of small nuclear RNA (snRNA) and many protein subunits. Somatic mutations in RNA splicing machinery have been found in the myeloid malignancies and are most prevalent within the myelodysplastic syndromes,[259,260] where they are found to be key early drivers of disease.[261] Mutations in *SF3B1* are strongly associated with MDS with ringed sideroblasts (MDS-RS), whereas mutations in *SRSF2* are strongly associated with CMML.[259] Recent research has identified certain patterns of early mutations which include those in the spliceosome factors *SF3B1, SRSF2, U2AF1* and *ZRSR2* as being highly specific for the diagnosis of secondary AML.[262] This same study also identified these spliceosome mutations in distinct genetic subsets within therapy-related AML and elderly *de novo* AML which had worse clinical outcome, lower complete remission rates and decreased EFS.[262] Further studies will be needed to understand the functional effects of spliceosome mutations in leukaemogenesis. A recent study has suggested that spliceosome mutations only confer a survival advantage to wild-type HSCs in the context of an elderly haematopoietic compartment.[263] It will be important to address whether age-related change in the HSC microenvironment or within the HSC itself contributes towards growth advantage in spliceosome mutant cells in aged individuals.[263]

Cohesin complex mutations

Cohesin is a multiprotein complex involved in the regulation of sister chromatid separation during mitosis. Loss of function somatic mutations have been reported in several of the cohesin factors in ~5% of AML, including *STAG2, RAD21, SMC1A* and *SMC3*.[264] The effect on leukaemogenesis is unclear, but these mutations are felt to represent secondary events in clonal hierarchy and contribute to clonal transformation.[265]

Other mutations

BCOR

Whole exome sequencing of one normal karyotype AML patient identified a mutation in *BCOR* (BCL6 co-repressor) on Xp11.4. Such a mutation had a frequency of 3.8% in unselected normal karyotype AML patients, was associated with *DNMT3A* mutation and showed a trend to inferior outcome.[266]

PHF6

The gene for plant homeodomain finger 6 (*PHF6*) on Xq26 is seen to be mutated in ~3% of cases of adult AML, are more common in males and are of adverse prognostic significance.[151]

PTPN11

PTPN11 mutations have been closely associated with normal karyotype AML occuring with a case prevalence of 5% in adult AML however these mutations have not been shown to effect overall survival or relapse-free survival of patients.[267]

Emerging concepts and future directions

Age-related clonal haematopoiesis (ARCH)

Significant insights into the clonal evolution of myeloid malignancies have recently been gained by four studies showing that clones founded by myeloid leukaemia-associated mutations are relatively common in the blood of healthy individuals and become much commoner with age, a phenomenon referred to as age-related clonal haematopoiesis (ARCH).[263,268–270] Indeed, one study estimated that ARCH is an almost inevitable consequence of ageing.[263] It is likely that ARCH is the precursor of the great majority of myeloid malignancies, although only a small minority of cases will ever culminate in malignancy. Mutations in *DNMT3A*, *TET2*, *JAK2* and *ASXL1* mutations were seen to be the most common drivers of ARCH in each study and were observed in individuals as young as 25 years old.

Distinct patterns of ARCH have been observed to occur during ageing whereby mutations in the spliceosome genes *SRSF2* and *SF3B1* emerged only after the age of 70 years, but thereafter their prevalence increased exponentially in subsequent decades.[263] These findings indicate that spliceosome gene mutations only drive clonal expansion under selection pressures particular to the ageing haematopoietic system and explains the high incidence of MDS and AML associated with these mutations in advanced old age.[263] Convergent evolution, whereby evolution occurs to overcome a shared selective pressure, has been proposed in models of ageing and cancer[145,271] and has been proposed as an explanation of the distinct age-related occurrence of spliceosome mutations found in healthy ageing individuals in this study.

Application of genomic technologies to the diagnosis of AML

NGS in the diagnostic laboratory

Next-generation sequencing can rapidly identify point mutations, SNPs, translocations and chromosomal insertions and deletions at a clonal and subclonal level down to allele frequencies of 1–2%. It is clear that NGS technology will play an important role in the future diagnosis of malignancy, not just in haematology, but throughout pathology services. It is envisaged that NGS will work in parallel with established techniques with the aim that they will work in concert with each other to drive forward a personalized approach to the diagnosis of cancer. To streamline the implementation of NGS in the diagnostic laboratory it will be important to design bespoke assays targeted to the malignancy in question and have validated bioinformatic pipelines tested prior to implementation. It is anticipated that the WHO classification of myeloid malignancies will soon be updated to incorporate more molecular markers. It is likely that risk stratification protocols will follow suit such that molecular profiling of many genes at diagnosis will become the norm in the molecular era of AML. Targeted sequencing assays are also able to determine copy number[272,273] in an accurate quantitative way, which may have added benefits above current cytogenetic techniques.

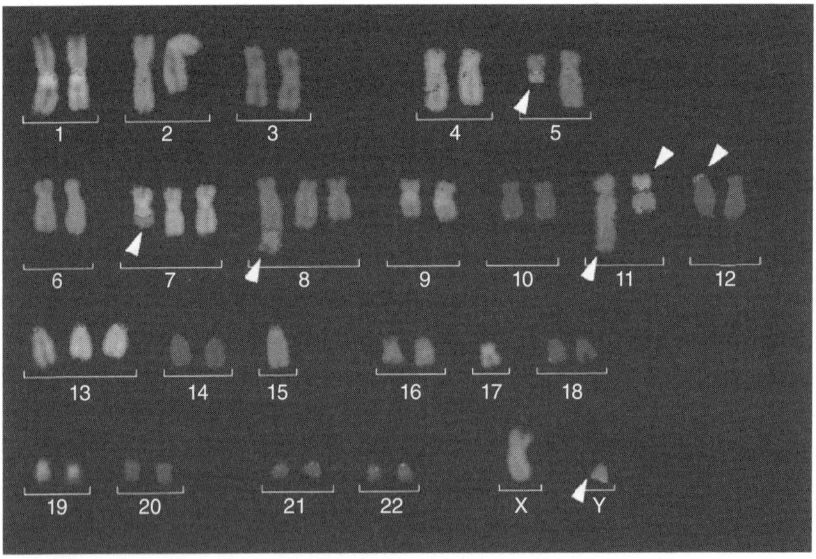

Figure 3.16 Representative colour karyotype after M-FISH. Chromosomes were obtained from an AML-derived cell line, GF-D8. The structural abnormalities identified (arrowheads) are t(5;15), der(7)t(7;15), der(8)ins(8;11);der(11)t(8;11) and t(11;17). In this cell line, the der(Y)t(Y;12) was visible, but difficult to identify accurately. From Tosi et al.[280] Reproduced with permission from John Wiley & Sons.

Advanced karyotyping

Early 'molecular cytogenetic' techniques such as M-FISH and spectral karyotyping (SKY)[274] offered the potential of unravelling complex karyotypes and identifying chromosomal translocations using a 24-colour karyotyping approach (Fig. 3.16). Comparative genomic hybridization (CGH) compares the genomic composition of tumour tissue with that of normal tissue.[275] The application of CGH made it possible to identify regions of chromosomal loss and gain via global genomic analysis at an improved resolution compared with conventional G-banding. Subsequently, array-CGH and single-nucleotide polymorphism (SNP) arrays provided a means of identifying subcytogenetic areas of genomic amplification and deletion, collectively termed copy number aberrations (CNAs) (Fig. 3.17).

The application of this approach to AML was reviewed in Struski et al. in 2002.[276] One study using CGH in 127 cases of AML found copy number alterations and unbalanced translocations with a concordance compared with standard cytogenetics of 72.5%.[277] Additional information was obtained in 20%, butt CGH failed to detect low-level unbalanced translocations in 7.5% of patients. Similarly, 201 acquired copy number abnormalities (CNAs) were found in 86 adult AML genomes when genome-wide copy number analysis was performed with paired normal and tumour DNA using SNP arrays.[94] Copy number abnormalities ranged in size from 35 kbp to 250 Mbp and affected all chromosomes. Approximately 50% had no CNA and the average number of CNAs per genome was 2.34. Normal karyotype cases had a CNA frequency of 24%, which increased to 40% in cases with an abnormal karyotype. Thirty-four of 86 genomes were shown to contain alterations not found by conventional cytogenetic analysis. SNP array analysis additionally demonstrates acquired somatic uniparental disomy (UPD). UPD is now recognized as a mechanism for loss of heterozygosity (LOH) and one study found a UPD frequency of 12% in normal karyotype AML patients, typically at 6p, 11p or 13q.[94]

Conclusion

The rapid development of genomic technology has enabled us to enter a new era of research and therapeutics in AML. Comprehensive molecular

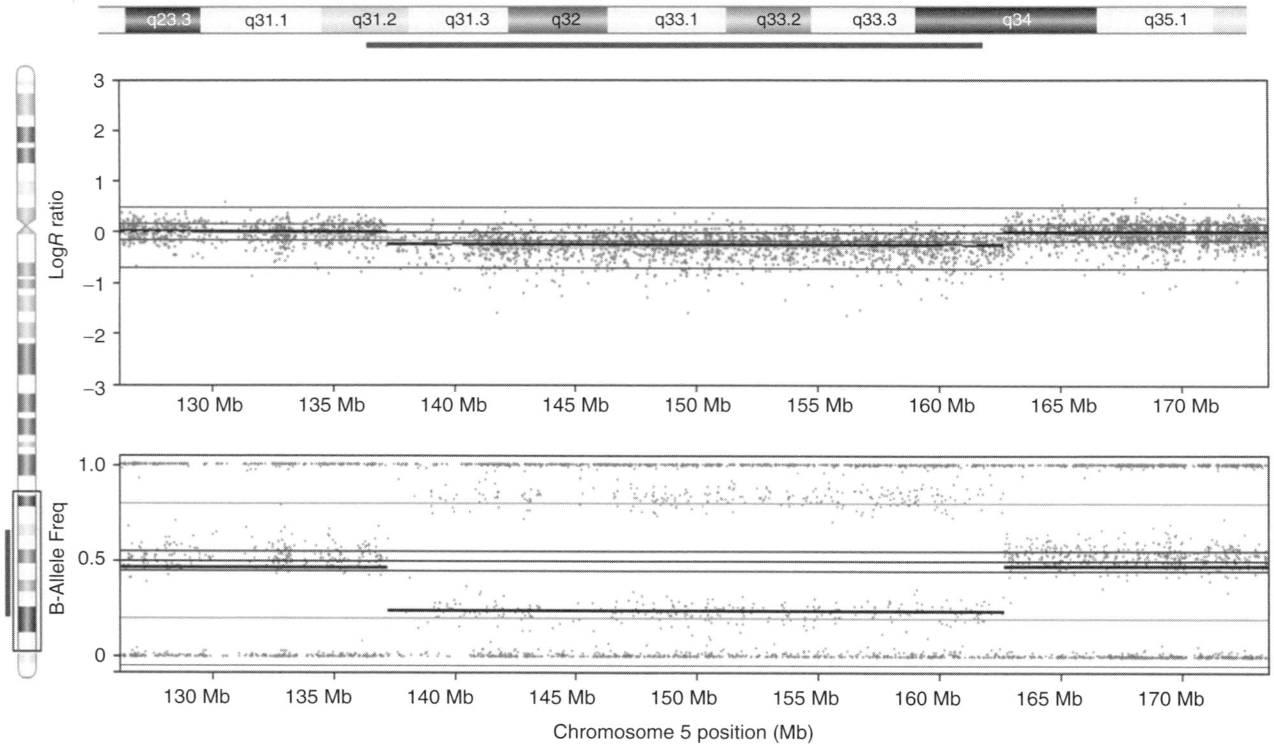

Figure 3.17 Example of high-resolution, genome-wide SNP array data obtained using test DNA from a patient with AML. SNP array analyses differ from conventional aCGH analyses in that only the test DNA is used. The resulting fluorescent signals are then compared with those provided from a standard reference group. In this example, AML patient DNA was tested using a high-resolution SNP array comprising ~300,000 probes. The log*R* ratio plot (above) shows a region of reduced signal in the test DNA corresponding to a genomic loss (indicated by the solid bar). The B-allele Freq (BAF) plot (below) was generated using SNP genotyping information. Across the region of loss, the BAF plot reveals a deviation from the 0, 50 and 100% frequencies that would correspond to AA, AB and BB alleles, respectively, in a diploid region. The plot indicates also that the loss affects only a proportion of cells; if 100% of cells carried the loss then the BAF would be either 0% (corresponding to only A alleles on the non-deleted chromosome) or 100% (only B-alleles on the non-deleted chromosome). In this example, the BAF gives a mixed pattern across the region, reflecting a level of mosaicism in which the majority, but not all, cells carry the loss. No genomic imbalances were reported in the karyotype. Image courtesy of Dr Samantha J.L. Knight, NIHR Biomedical Research Centre and Wellcome Trust Centre For Human Genetics, University of Oxford, Oxford, UK.

profiling involving the use of gene panels and targeted NGS will allow a more precise diagnosis and identification of patient subsets, especially amongst the patients with intermediate-risk cytogenetics. It is hoped that a better understanding of the molecular pathogenesis of AML will translate into improvement of patient outcomes through the development of novel therapeutic approaches.

Mini-glossary

Allele burden: The fraction of alleles with a specific sequence in relation to the total number of alleles for the same region of the genome. For example, a heterozygous mutation in a pure population of leukaemia cells has an allele burden of 0.5. If 80% of cells are leukaemic and 20% of cells are normal, the mutant allele burden of the heterozygous mutation would be $0.8 \times 0.5 = 0.4$.

Branching evolution (of cancer): A form of clonal evolution of cancer that leads to the generation of more than one clone of cells characterized by distinct somatic mutations, but which share at least one mutation traceable back to a single ancestral cell.

Clonal evolution (of cancer): The stepwise acquisition of mutations in a founder cell and its progeny leading towards the development of a cancer.

Co-occurrence (of cancer mutations): The occurrence of two or more mutations in the same cancer more often than would be expected by chance.

Convergent evolution (of cancer): A pattern of cancer evolution during which independent clones expand after acquiring the same or a very similar mutation. This is unlikely to reflect an increased likelihood of acquiring the mutation *per se* and more likely to reflect the fact that such a mutation is particularly advantageous to the specific cancer and gives a marked growth advantage when acquired by chance.

Dominant negative mutation: A heterozygous mutation that leads to marked or complete loss of function of the coded protein and of the normal protein coded by the other (wild-type) copy of the gene.

Driver gene or driver mutation: A mutated gene that confers a selective growth advantage to a cancer cell.

Exome sequencing: Sequencing of all exons in protein-coding genes within the genome.

Gain-of-function mutation: A mutation that gives the coded protein a novel or markedly enhanced function.

Loss-of-function mutation: A mutation that leads to marked or complete loss of function of the coded protein.

Loss-of-heterozygosity (LOH): A genetic modification leading to the loss of the maternally or paternally derived copy of a genomic region. This can happen as a result of deletion or uniparental disomy (uPD).

Missense mutation: A nucleotide substitution (e.g. G to T) that results in an amino acid change (e.g. valine to phenylalanine).

Next-generation sequencing (NGS): DNA sequencing using one of the methodologies developed since 2005 and which allow massively parallel sequencing of thousands or millions of fragments of DNA simultaneously.

Nonsense mutation: A nucleotide substitution that results in the production of a stop codon (i.e. TAA, TGA or TAG).

Non-synonymous mutation: A mutation that alters the encoded amino acid sequence of a protein. These include missense, nonsense, splice site, gain of translation start, loss of translation stop and indel mutations.

References

1 Juliusson, G., et al. Age and acute myeloid leukemia: real world data on decision to treat and outcomes from the Swedish Acute Leukemia Registry. *Blood* 2009; 113: 4179–4187.

2 Dohner, K., Dohner, H. Molecular characterization of acute myeloid leukemia. *Haematologica* 2008; 93: 976–982.

3 Craddock, C., Tauro, S., Moss, P., Grimwade, D. Biology and management of relapsed acute myeloid leukaemia. *Br J Haematol* 2005; 129: 18–34.

4 Schlenk, R.F., Dohner, H. Genomic applications in the clinic: use in treatment paradigm of acute myeloid leukemia. *Hematology Am Soc Hematol Educ Program* 2013; 201(3): 324–330.

5 Ley, T.J., et al. DNA sequencing of a cytogenetically normal acute myeloid leukaemia genome. *Nature* 2008; 456: 66–72.

6 Bennett, J.M., et al. Proposals for the classification of the acute leukaemias. French–American–British (FAB) co-operative group. *Br J Haematol* 1976; 33: 451–458.

7 Bennett, J.M., et al. Proposed revised criteria for the classification of acute myeloid leukemia. A report of the French–American–British Cooperative Group. *Ann Intern Med* 1985; 103: 620–625.

8 Vardiman, J.W., et al. The 2008 revision of the World Health Organization (WHO) classification of myeloid neoplasms and acute leukemia: rationale and important changes. *Blood* 2009; 114: 937–951.

9 Speck, N.A., Gilliland, D.G. Core-binding factors in haematopoiesis and leukaemia. *Nat Rev Cancer* 2002; 2: 502–513.

10 Lutterbach, B., Hiebert, S.W. Role of the transcription factor AML-1 in acute leukemia and hematopoietic differentiation. *Gene* 2000; 245: 223–235.

11 Otto, F., Lubbert, M., Stock, M. Upstream and downstream targets of RUNX proteins. *J Cell Biochem* 2003; 89: 9–18.

12 Wang, Q., et al. Disruption of the Cbfa2 gene causes necrosis and hemorrhaging in the central nervous system and blocks definitive hematopoiesis. *Proc Natl Acad Sci U S A* 1996; 93: 3444–3449.

13 Okuda, T., van Deursen, J., Hiebert, S.W., Grosveld, G., Downing, J.R. AML1, the target of multiple chromosomal translocations in human leukemia, is essential for normal fetal liver hematopoiesis. *Cell* 1996; 84: 321–330.

14 Downing, J.R. The AML1-ETO chimaeric transcription factor in acute myeloid leukaemia: biology and clinical significance. *Br J Haematol* 1999; 106: 296–308.

15 Yuan, Y., et al. AML1-ETO expression is directly involved in the development of acute myeloid leukemia in the presence of additional mutations. *Proc Natl Acad Sci U S A* 2001; 98: 10398–10403.

16 Okuda, T., et al. Expression of a knocked-in AML1-ETO leukemia gene inhibits the establishment of normal definitive hematopoiesis and directly generates dysplastic hematopoietic progenitors. *Blood* 1998; 91: 3134–3143.

17 Licht, J.D. AML1 and the AML1-ETO fusion protein in the pathogenesis of t(8;21) AML. *Oncogene* 2001; 20: 5660–5679.

18 Shigesada, K., van de Sluis, B., Liu, P.P. Mechanism of leukemogenesis by the inv(16) chimeric gene CBFB/PEBP2B-MHY11. *Oncogene* 2004; 23: 4297–4307.

19 Hyde, R.K., Liu, P.P. RUNX1 repression-independent mechanisms of leukemogenesis by fusion genes CBFB-MYH11 and AML1-ETO (RUNX1-RUNX1T1). *J Cell Biochem* 2010; 110: 1039–1045.

20 Lutterbach, B., Hou, Y., Durst, K.L., Hiebert, S.W. The inv(16) encodes an acute myeloid leukemia 1 transcriptional corepressor. *Proc Natl Acad Sci U S A* 1999; 96: 12822–12827.

21 Marcucci, G., et al. Overexpression of the ETS-related gene, ERG, predicts a worse outcome in acute myeloid leukemia with normal karyotype: a Cancer and Leukemia Group B study. *J Clin Oncol* 2005; 23: 9234–9242.

22 Paschka, P., et al. Secondary genetic lesions in acute myeloid leukemia with inv(16) or t(16;16): a study of the German–Austrian AML Study Group (AMLSG). *Blood* 2013; 121: 170–177.

23 Asou, N. The role of a Runt domain transcription factor AML1/RUNX1 in leukemogenesis and its clinical implications. *Crit Rev Oncol Hematol* 2003; 45: 129–150.

24 De Braekeleer, E., et al. RUNX1 translocations and fusion genes in malignant hemopathies. *Future Oncol* 2011; 7: 77–91.

25 de The, H., Chomienne, C., Lanotte, M., Degos, L., Dejean, A. The t(15;17) translocation of acute promyelocytic leukaemia fuses the retinoic acid receptor alpha gene to a novel transcribed locus. *Nature* 1990; 347: 558–561.

26 Borrow, J., Goddard, A.D., Sheer, D., Solomon, E. Molecular analysis of acute promyelocytic leukemia breakpoint cluster region on chromosome 17. *Science* 1990; 249: 1577–1580.

27 Guglielmi, C., et al. Immunophenotype of adult and childhood acute promyelocytic leukaemia: correlation with morphology, type of PML gene breakpoint and clinical outcome. A cooperative Italian study on 196 cases. *Br J Haematol* 1998; 102: 1035–1041.

28 Wang, Z.G., et al. Role of PML in cell growth and the retinoic acid pathway. *Science* 1998; 279: 1547–1551.

29 Brown, N.J., et al. PML nuclear bodies in the pathogenesis of acute promyelocytic leukemia: active players or innocent bystanders? *Front Biosci* 2009; 14: 1684–1707.

30 Chambon, P. A decade of molecular biology of retinoic acid receptors. *FASEB J* 1996; 10: 940–954.

31 Grimwade, D. The pathogenesis of acute promyelocytic leukaemia: evaluation of the role of molecular diagnosis and monitoring in the management of the disease. *Br J Haematol* 1999; 106: 591–613.

32 Welch, J.S., Yuan, W., Ley, T.J. PML-RARA can increase hematopoietic self-renewal without causing a myeloproliferative disease in mice. *J Clin Invest* 2011;121: 1636–1645.

33 Wartman, L.D., et al. Sequencing a mouse acute promyelocytic leukemia genome reveals genetic events relevant for disease progression. *J Clin Invest* 2011; 121: 1445–1455.

34 Gale, R.E., et al. Relationship between FLT3 mutation status, biologic characteristics, and response to targeted therapy in acute promyelocytic leukemia. *Blood* 2005; 106: 3768–3776.

35 Golomb, H.M., Rowley, J.D., Vardiman, J.W., Testa, J.R., Butler, A. "Microgranular" acute promyelocytic leukemia: a distinct clinical, ultrastructural, and cytogenetic entity. *Blood* 1980; 55: 253–259.

36 Sanz, M.A., et al. Management of acute promyelocytic leukemia: recommendations from an expert panel on behalf of the European LeukemiaNet. *Blood* 2009; 113: 1875–1891.

37 Swerdlow, S.H., Campo, E., Harris, N.L., et al. *WHO Classification of Tumours of Haematopoietic and Lymphoid Tissues*, 4th edn. Lyon: IARC Press, 2008.

38 Dimov, N.D., et al. Rapid and reliable confirmation of acute promyelocytic leukemia by immunofluorescence staining with an antipromyelocytic leukemia antibody: the M.D. Anderson Cancer Center experience of 349 patients. *Cancer* 2010; 116: 369–376.

39 Grimwade, D., et al. Characterization of acute promyelocytic leukemia cases lacking the classic t(15;17): results of the European Working Party. Groupe Français de Cytogénétique Hématologique, Groupe de Français d'Hématologie Cellulaire, UK Cancer Cytogenetics Group and BIOMED 1 European Community-Concerted Action "Molecular Cytogenetic Diagnosis in Haematological Malignancies". *Blood* 2000; 96: 1297–1308.

40 Sainty, D., et al. A new morphologic classification system for acute promyelocytic leukemia distinguishes cases with underlying PLZF/RARA gene rearrangements. *Blood* 2000; 96: 1287–1296.

41 Meyer, C., et al. New insights to the MLL recombinome of acute leukemias. *Leukemia* 2009; 23: 1490–1499.

42 Archimbaud, E., et al. Clinical and biological characteristics of adult de novo and secondary acute myeloid leukemia with balanced 11q23 chromosomal anomaly or MLL gene rearrangement compared to cases with unbalanced 11q23 anomaly: confirmation of the existence of different entities with 11q23 breakpoint. *Leukemia* 1998; 12: 25–33.

43 Chandra, P., et al. Acute myeloid leukemia with t(9;11)(p21–22;q23): common properties of dysregulated ras pathway signaling and genomic progression characterize de novo and therapy-related cases. *Am J Clin Pathol* 2010; 133: 686–693.

44 Shih, L.Y., et al. Characterization of fusion partner genes in 114 patients with de novo acute myeloid leukemia and MLL rearrangement. *Leukemia* 2006; 20: 218–223.

45 Somervaille, T.C. and Cleary, M.L. Grist for the MLL: how do MLL oncogenic fusion proteins generate leukemia stem cells? *Int J Hematol* 2010; 91: 735–741.

46 Goyama, S. and Kurokawa, M. Evi-1 as a critical regulator of leukemic cells. *Int J Hematol* 2010; 91: 753–757.

47 Lugthart, S., et al. Aberrant DNA hypermethylation signature in acute myeloid leukemia directed by EVI1. *Blood* 2011; 117: 234–241.

48 Kataoka, K., et al. Evi1 is essential for hematopoietic stem cell self-renewal, and its expression marks hematopoietic cells with long-term multilineage repopulating activity. *J Exp Med* 2011; 208: 2403–2416.

49 Yuasa, S., et al. Transient inhibition of BMP signaling by Noggin induces cardiomyocyte differentiation of mouse embryonic stem cells. *Nat Biotechnol* 2005; 23: 607–611.

50 Kataoka, K. and Kurokawa, M. Ecotropic viral integration site 1, stem cell self-renewal and leukemogenesis. *Cancer Sci* 2012; 103: 1371–1377.

51 Senyuk, V., Premanand, K., Xu, P., Qian, Z., Nucifora, G. The oncoprotein EVI1 and the DNA methyltransferase Dnmt3 co-operate in binding and de novo methylation of target DNA. *PloS One* 2011; 6: e20793.

52 Gomez-Benito, M., et al. EVI1 controls proliferation in acute myeloid leukaemia through modulation of miR-1-2. *Br J Cancer* 2010; 103: 1292–1296.

53 Eppert, K., et al. Stem cell gene expression programs influence clinical outcome in human leukemia. *Nat Med* 2011; 17: 1086–1093.

54 Valk, P.J., et al. Prognostically useful gene-expression profiles in acute myeloid leukemia. *N Engl J Med* 2004; 350: 1617–1628.

55 Grimwade, D., et al. Refinement of cytogenetic classification in acute myeloid leukemia: determination of prognostic significance of rare recurring chromosomal abnormalities among 5876 younger adult patients treated in the United Kingdom Medical Research Council trials. *Blood* 2010; 116: 354–365.

56 Lugthart, S., et al. Clinical, molecular, and prognostic significance of WHO type inv(3)(q21q26.2)/t(3;3)(q21;q26.2) and various other 3q abnormalities in acute myeloid leukemia. *J Clin Oncol* 2010; 28: 3890–3898.

57 Secker-Walker, L.M., Mehta, A. and Bain, B. Abnormalities of 3q21 and 3q26 in myeloid malignancy: a United Kingdom Cancer Cytogenetic Group study. *Br J Haematol* 1995; 91: 490–501.

58 Slovak, M.L., et al. A retrospective study of 69 patients with t(6;9)(p23;q34) AML emphasizes the need for a prospective, multicenter initiative for rare 'poor prognosis' myeloid malignancies. *Leukemia* 2006; 20: 1295–1297.

59 Chi, Y., Lindgren, V., Quigley, S., Gaitonde, S. Acute myelogenous leukemia with t(6;9)(p23;q34) and marrow basophilia: an overview. *Arch Pathol Lab Med* 2008; 132: 1835–1837.

60 Oyarzo, M.P., et al. Acute myeloid leukemia with t(6;9)(p23;q34) is associated with dysplasia and a high frequency of flt3 gene mutations. *Am J Clin Pathol* 2004; 122: 348–358.

61 Gupta, M., et al. The t(6;9)(p22;q34) in myeloid neoplasms: a retrospective study of 16 cases. *Cancer Genet Cytogenet* 2010; 203: 297–302.

62 Keung, Y.K., et al. Philadelphia chromosome positive myelodysplastic syndrome and acute myeloid leukemia – retrospective study and review of literature. *Leukemia Res* 2004; 28: 579–586.

63 Cuneo, A., et al. Philadelphia chromosome-positive acute myeloid leukemia: cytoimmunologic and cytogenetic features. *Haematologica* 1996; 81: 423–427.

64 Paietta, E., et al. Biologic heterogeneity in Philadelphia chromosome-positive acute leukemia with myeloid morphology: the Eastern Cooperative Oncology Group experience. *Leukemia* 1998; 12: 1881–1885.

65 Bacher, U., et al. Subclones with the t(9;22)/BCR-ABL1 rearrangement occur in AML and seem to cooperate with distinct genetic alterations. *Br J Haematol* 2011; 152: 713–720.

66 Cirmena, G., et al. A BCR-JAK2 fusion gene as the result of a t(9;22)(p24;q11) in a patient with acute myeloid leukemia. *Cancer Genet Cytogenet* 2008; 183: 105–108.

67 Torres, L., et al. Acute megakaryoblastic leukemia with a four-way variant translocation originating the RBM15-MKL1 fusion gene. *Pediatr Blood Cancer* 2011; 56: 846–849.

68 Champagne, N., Pelletier, N., Yang, X.J. The monocytic leukemia zinc finger protein MOZ is a histone acetyltransferase. *Oncogene* 2001; 20: 404–409.

69 Coulthard, S., et al. Two cases of inv(8)(p11q13) in AML with erythrophagocytosis: a new cytogenetic variant. *Br J Haematol* 1998; 100: 561–563.

70 Jekarl, D.W., et al. CD56 antigen expression and hemophagocytosis of leukemic cells in acute myeloid leukemia with t(16;21)(p11;q22). *Int J Hematol* 2010; 92: 306–313.

71 Block, A.W., et al. Rare recurring balanced chromosome abnormalities in therapy-related myelodysplastic syndromes and acute leukemia: report from an international workshop. *Genes Chromosomes Cancer* 2002; 33: 401–412.

72 De Braekeleer, E., et al. ETV6 fusion genes in hematological malignancies: a review. *Leukemia Res* 2012; 36: 945–961.

73 Ballabio, E., et al. Ectopic expression of the HLXB9 gene is associated with an altered nuclear position in t(7;12) leukaemias. *Leukemia* 2009; 23: 1179–1182.

74 von Bergh, A.R., et al. High incidence of t(7;12)(q36;p13) in infant AML but not in infant ALL, with a dismal outcome and ectopic expression of HLXB9. *Genes Chromosomes Cancer* 2006; 45: 731–739.
75 Duhoux, F.P., et al. PRDM16 (1p36) translocations define a distinct entity of myeloid malignancies with poor prognosis but may also occur in lymphoid malignancies. *Br J Haematol* 2012; 156: 76–88.
76 Raza, S., et al. Autosomal monosomies among 24,262 consecutive cytogenetic studies: prevalence, chromosomal distribution and clinicopathologic correlates of sole abnormalities. *Am J Hematol* 2011; 86: 353–356.
77 Mrozek, K., Heerema, N.A., Bloomfield, C.D. Cytogenetics in acute leukemia. *Blood Rev* 2004; 18: 115–136.
78 Rucker, F.G., et al. TP53 alterations in acute myeloid leukemia with complex karyotype correlate with specific copy number alterations, monosomal karyotype, and dismal outcome. *Blood* 2012; 119: 2114–2121.
79 Breems, D.A., et al. Monosomal karyotype in acute myeloid leukemia: a better indicator of poor prognosis than a complex karyotype. *J Clin Oncol* 2008; 26: 4791–4797.
80 Cuneo, A., et al. Detection and monitoring of trisomy 8 by fluorescence in situ hybridization in acute myeloid leukemia: a multicentric study. *Haematologica* 1998; 83: 21–26.
81 Schoch, C., et al. Impact of trisomy 8 on expression of genes located on chromosome 8 in different AML subgroups. *Genes Chromosomes Cancer* 2006; 45: 1164–1168.
82 Ganmore, I., Smooha, G., Izraeli, S. Constitutional aneuploidy and cancer predisposition. *Hum Mol Genet* 2009; 18: R84–R93.
83 Leon, E., Jamal, S.M., Zou, Y.S., Milunsky, J.M. Partial trisomy 8 mosaicism due to a pseudoisodicentric chromosome 8. *Am J Med Genet A* 2011; 155A: 1740–1744.
84 Mitelman, F., Heim, S., Mandahl, N. Trisomy 21 in neoplastic cells. *Am J Med Genet Suppl* 1990; 7: 262–266.
85 Larsson, N., Lilljebjorn, H., Lassen, C., Johansson, B., Fioretos, T. Myeloid malignancies with acquired trisomy 21 as the sole cytogenetic change are clinically highly variable and display a heterogeneous pattern of copy number alterations and mutations. *Eur J Haematol* 2012; 88: 136–143.
86 Mehta, A.B., Bain, B.J., Fitchett, M., Shah, S., Secker-Walker, L.M. Trisomy 13 and myeloid malignancy – characteristic blast cell morphology: a United Kingdom Cancer Cytogenetics Group survey. *Br J Haematol* 1998; 101: 749–752.
87 Dicker, F., Haferlach, C., Kern, W., Haferlach, T., Schnittger, S. Trisomy 13 is strongly associated with AML1/RUNX1 mutations and increased FLT3 expression in acute myeloid leukemia. *Blood* 2007; 110: 1308–1316.
88 Thomas, L., Stamberg, J., Gojo, I., Ning, Y., Rapoport, A.P. Double minute chromosomes in monoblastic (M5) and myeloblastic (M2) acute myeloid leukemia: two case reports and a review of literature. *Am J Hematol* 2004; 77: 55–61.
89 Bruyere, H., Sutherland, H., Chipperfield, K., Hudoba, M. Concomitant and successive amplifications of MYC in APL-like leukemia. *Cancer Genet Cytogenet* 2010; 197: 75–80.

90 Receveur, A., et al. Trisomy 4 associated with double minute chromosomes and MYC amplification in acute myeloblastic leukemia. *Ann Genet* 2004; 47: 423–427.
91 Rowe, D., et al. Cytogenetically cryptic AML1-ETO and CBF beta-MYH11 gene rearrangements: incidence in 412 cases of acute myeloid leukaemia. *Br J Haematol* 2000; 111: 1051–1056.
92 Grimwade, D. Impact of cytogenetics on clinical outcome in AML. In *Acute Myelogenous Leukemia*, ed. Karp, J.E. Totowa, NJ: Humana Press, 2007, pp. 177–192.
93 Cuneo, A., et al. Incidence and significance of cryptic chromosome aberrations detected by fluorescence in situ hybridization in acute myeloid leukemia with normal karyotype. *Leukemia* 2002; 16: 1745–1751.
94 Walter, M.J., et al. Acquired copy number alterations in adult acute myeloid leukemia genomes. *Proc Natl Acad Sci U S A* 2009; 106: 12950–12955.
95 Ballabio, E., et al. Genomic imbalances are confined to non-proliferating cells in paediatric patients with acute myeloid leukaemia and a normal or incomplete karyotype. *PloS One* 2011; 6: e20607.
96 Mrozek, K., Marcucci, G., Paschka, P., Whitman, S.P., Bloomfield, C.D. Clinical relevance of mutations and gene-expression changes in adult acute myeloid leukemia with normal cytogenetics: are we ready for a prognostically prioritized molecular classification? *Blood* 2007; 109: 431–448.
97 Lugthart, S., et al. High EVI1 levels predict adverse outcome in acute myeloid leukemia: prevalence of EVI1 overexpression and chromosome 3q26 abnormalities underestimated. *Blood* 2008; 111: 4329–4337.
98 Groschel, S., et al. High EVI1 expression predicts outcome in younger adult patients with acute myeloid leukemia and is associated with distinct cytogenetic abnormalities. *J Clin Oncol* 2010; 28: 2101–2107.
99 Vazquez, I., et al. Down-regulation of EVI1 is associated with epigenetic alterations and good prognosis in patients with acute myeloid leukemia. *Haematologica* 2011; 96: 1448–1456.
100 Haferlach, C., et al. Gene expression of BAALC, CDKN1B, ERG, and MN1 adds independent prognostic information to cytogenetics and molecular mutations in adult acute myeloid leukemia. *Genes Chromosomes Cancer* 2012; 51: 257–265.
101 Langer, C., et al. High BAALC expression associates with other molecular prognostic markers, poor outcome, and a distinct gene-expression signature in cytogenetically normal patients younger than 60 years with acute myeloid leukemia: a Cancer and Leukemia Group B (CALGB) study. *Blood* 2008; 111: 5371–5379.
102 Schwind, S., et al. BAALC and ERG expression levels are associated with outcome and distinct gene and microRNA expression profiles in older patients with de novo cytogenetically normal acute myeloid leukemia: a Cancer and Leukemia Group B study. *Blood* 2010; 116: 5660–5669.
103 Langer, C., et al. Prognostic importance of MN1 transcript levels, and biologic insights from MN1-associated gene and microRNA expression signatures in cytogenetically normal acute myeloid leukemia: a Cancer and Leukemia Group B study. *J Clin Oncol* 2009; 27: 3198–3204.

104 Schwind, S., et al. Low expression of MN1 associates with better treatment response in older patients with de novo cytogenetically normal acute myeloid leukemia. *Blood* 2011; 118: 4188–4198.
105 Metzeler, K.H., et al. ERG expression is an independent prognostic factor and allows refined risk stratification in cytogenetically normal acute myeloid leukemia: a comprehensive analysis of ERG, MN1, and BAALC transcript levels using oligonucleotide microarrays. *J Clin Oncol* 2009; 27: 5031–5038 .
106 Cristobal, I., et al. Overexpression of SET is a recurrent event associated with poor outcome and contributes to protein phosphatase 2A inhibition in acute myeloid leukemia. *Haematologica* 2012; 97: 543–550.
107 Noordermeer, S.M., et al. High BRE expression predicts favorable outcome in adult acute myeloid leukemia, in particular among MLL-AF9-positive patients. *Blood* 2011; 118: 5613–5621.
108 Inoue, K., et al. Aberrant overexpression of the Wilms tumor gene (WT1) in human leukemia. *Blood* 1997; 89: 1405–1412.
109 Cilloni, D., et al. Real-time quantitative polymerase chain reaction detection of minimal residual disease by standardized WT1 assay to enhance risk stratification in acute myeloid leukemia: a European LeukemiaNet study. *J Clin Oncol* 2009; 27: 5195–5201.
110 Marcucci, G., Mrozek, K., Radmacher, M.D., Garzon, R., Bloomfield, C.D. The prognostic and functional role of microRNAs in acute myeloid leukemia. *Blood* 2011; 117: 1121–1129.
111 Garzon, R., Croce, C.M. MicroRNAs in normal and malignant hematopoiesis. *Curr Opin Hematol* 2008; 15: 352–358.
112 Marcucci, G., et al. MicroRNA expression in cytogenetically normal acute myeloid leukemia. *N Engl J Med* 2008; 358: 1919–1928.
113 Garzon, R., et al. MicroRNA signatures associated with cytogenetics and prognosis in acute myeloid leukemia. *Blood* 2008; 111: 3183–3189.
114 Whitman, S.P., et al. FLT3 internal tandem duplication associates with adverse outcome and gene- and microRNA-expression signatures in patients 60 years of age or older with primary cytogenetically normal acute myeloid leukemia: a Cancer and Leukemia Group B study. *Blood* 2010; 116: 3622–3626.
115 Schwind, S., et al. Prognostic significance of expression of a single microRNA, miR-181a, in cytogenetically normal acute myeloid leukemia: a Cancer and Leukemia Group B study. *J Clin Oncol* 2010; 28: 5257–5264.
116 Vardiman, J.W., Harris, N.L., Brunning, R.D. The World Health Organization (WHO) classification of the myeloid neoplasms. *Blood* 2002; 100: 2292–2302.
117 Pedersen-Bjergaard, J., Andersen, M.K., Andersen, M.T., Christiansen, D.H. Genetics of therapy-related myelodysplasia and acute myeloid leukemia. *Leukemia* 2008; 22: 240–248.
118 Beaumont, M., et al. Therapy-related acute promyelocytic leukemia. *J Clin Oncol* 2003; 21: 2123–2137.
119 Dohner, H., et al. Diagnosis and management of acute myeloid leukemia in adults: recommendations from an international expert panel, on behalf of the European LeukemiaNet. *Blood* 2010; 115: 453–474.

120 Schlenk, R.F., et al. Mutations and treatment outcome in cytogenetically normal acute myeloid leukemia. *N Engl J Med* 2008; 358: 1909–1918.
121 Dufour, A., et al. Acute myeloid leukemia with biallelic CEBPA gene mutations and normal karyotype represents a distinct genetic entity associated with a favorable clinical outcome. *J Clin Oncol* 2010; 28: 570–577.
122 Schnittger, S., et al. Nucleophosmin gene mutations are predictors of favorable prognosis in acute myelogenous leukemia with a normal karyotype. *Blood* 2005; 106: 3733–3739.
123 Wouters, B.J., et al. Double CEBPA mutations, but not single CEBPA mutations, define a subgroup of acute myeloid leukemia with a distinctive gene expression profile that is uniquely associated with a favorable outcome. *Blood* 2009; 113: 3088–3091.
124 Green, C.L., et al. Prognostic significance of CEBPA mutations in a large cohort of younger adult patients with acute myeloid leukemia: impact of double CEBPA mutations and the interaction with FLT3 and NPM1 mutations. *J Clin Oncol* 2010; 28: 2739–2747.
125 Taskesen, E., et al. Prognostic impact, concurrent genetic mutations, and gene expression features of AML with CEBPA mutations in a cohort of 1182 cytogenetically normal AML patients: further evidence for CEBPA double mutant AML as a distinctive disease entity. *Blood* 2011; 117: 2469–2475.
126 Cornelissen, J.J., et al. Results of a HOVON/SAKK donor versus no-donor analysis of myeloablative HLA-identical sibling stem cell transplantation in first remission acute myeloid leukemia in young and middle-aged adults: benefits for whom? *Blood* 2007; 109: 3658–3666.
127 Koreth, J., et al. Allogeneic stem cell transplantation for acute myeloid leukemia in first complete remission: systematic review and meta-analysis of prospective clinical trials. *JAMA* 2009; 301: 2349–2361.
128 Dohner, K., Paschka, P. Intermediate-risk acute myeloid leukemia therapy: current and future. *Hematology Am Soc Hematol Educ Program* 2014; 201(4): 34–43.
129 Grimwade, D., et al. The importance of diagnostic cytogenetics on outcome in AML: analysis of 1,612 patients entered into the MRC AML 10 trial. The Medical Research Council Adult and Children's Leukaemia Working Parties. *Blood* 1998; 92: 2322–2333.
130 Slovak, M.L., et al. Karyotypic analysis predicts outcome of preremission and postremission therapy in adult acute myeloid leukemia: a Southwest Oncology Group/Eastern Cooperative Oncology Group Study. *Blood* 2000; 96: 4075–4083.
131 Byrd, J.C., et al. Pretreatment cytogenetic abnormalities are predictive of induction success, cumulative incidence of relapse, and overall survival in adult patients with de novo acute myeloid leukemia: results from Cancer and Leukemia Group B (CALGB 8461). *Blood* 2002; 100: 4325–4336.
132 Wang, E.S. Treating acute myeloid leukemia in older adults. *Hematology Am Soc Hematol Educ Program* 2014; 201(4): 14–20.
133 Burnett, A., Wetzler, M., Lowenberg, B. Therapeutic advances in acute myeloid leukemia. *J Clin Oncol* 2011; 29: 487–494.

134 Burnett, A.K., et al. Addition of gemtuzumab ozogamicin to induction chemotherapy improves survival in older patients with acute myeloid leukemia. *J Clin Oncol* 2012; 30: 3924–3931.
135 Castaigne, S., et al. Effect of gemtuzumab ozogamicin on survival of adult patients with de-novo acute myeloid leukaemia (ALFA-0701): a randomised, open-label, phase 3 study. *Lancet* 2012; 379: 1508–1516.
136 Freeman, S.D., Jovanovic, J.V., Grimwade, D. Development of minimal residual disease-directed therapy in acute myeloid leukemia. *Semin Oncol* 2008; 35: 388–400.
137 Cilloni, D. Is WT1 helping the molecular monitoring of minimal residual disease to get easier in acute myeloid leukaemia? *Leuk Res* 2009; 33: 603–604.
138 Kronke, J., et al. Monitoring of minimal residual disease in NPM1-mutated acute myeloid leukemia: a study from the German–Austrian Acute Myeloid Leukemia Study Group. *J Clin Oncol* 2011; 29: 2709–2716.
139 Schnittger, S., et al. Minimal residual disease levels assessed by NPM1 mutation-specific RQ-PCR provide important prognostic information in AML. *Blood* 2009; 114: 2220–2231.
140 Grimwade, D., Lo Coco, F. Acute promyelocytic leukemia: a model for the role of molecular diagnosis and residual disease monitoring in directing treatment approach in acute myeloid leukemia. *Leukemia* 2002; 16: 1959–1973.
141 Venditti, A., et al. Level of minimal residual disease after consolidation therapy predicts outcome in acute myeloid leukemia. *Blood* 2000; 96: 3948–3952.
142 Maurillo, L., et al. Toward optimization of postremission therapy for residual disease-positive patients with acute myeloid leukemia. *J Clin Oncol* 2008; 26: 4944–4951.
143 Freeman, S.D., et al. Prognostic relevance of treatment response measured by flow cytometric residual disease detection in older patients with acute myeloid leukemia. *J Clin Oncol* 2013; 31: 4123–4131.
144 Nowell, P.C. The clonal evolution of tumor cell populations. *Science* 1976; 194: 23–28.
145 Greaves, M. and Maley, C.C. Clonal evolution in cancer. *Nature* 2012; 481: 306–313.
146 Mardis, E.R., et al. Recurring mutations found by sequencing an acute myeloid leukemia genome. *N Engl J Med* 2009; 361: 1058–1066.
147 Ley, T.J., et al. DNMT3A mutations in acute myeloid leukemia. *N Engl J Med* 2010; 363: 2424–2433.
148 Yamashita, Y., et al. Array-based genomic resequencing of human leukemia. *Oncogene* 2010; 29: 3723–3731.
149 Yan, X.J., et al. Exome sequencing identifies somatic mutations of DNA methyltransferase gene DNMT3A in acute monocytic leukemia. *Nat Genet* 2011; 43: 309–315.
150 Cancer Genome Atlas Research Network. Genomic and epigenomic landscapes of adult de novo acute myeloid leukemia. *N Engl J Med* 2013; 368: 2059–2074.
151 Patel, J.P., et al. Prognostic relevance of integrated genetic profiling in acute myeloid leukemia. *N Engl J Med* 2012; 366: 1079–1089.

152 Welch, J.S., et al. The origin and evolution of mutations in acute myeloid leukemia. *Cell* 2012; 150: 264–278.

153 Corces-Zimmerman, M.R., Hong, W.J., Weissman, I.L., Medeiros, B.C., Majeti, R. Preleukemic mutations in human acute myeloid leukemia affect epigenetic regulators and persist in remission. *Proc Natl Acad Sci U S A* 2014; 111: 2548–2553.

154 Shlush, L.I., et al. Identification of pre-leukaemic haematopoietic stem cells in acute leukaemia. *Nature* 2014; 506: 328–333.

155 Ding, L., et al. Clonal evolution in relapsed acute myeloid leukaemia revealed by whole-genome sequencing. *Nature* 2012; 481: 506–510.

156 Falini, B., et al. Cytoplasmic nucleophosmin in acute myelogenous leukemia with a normal karyotype. *N Engl J Med* 2005; 352: 254–266.

157 Liso, A., et al. In human genome, generation of a nuclear export signal through duplication appears unique to nucleophosmin (NPM1) mutations and is restricted to AML. *Leukemia* 2008; 22: 1285–1289.

158 Thiede, C., Creutzig, E., Reinhardt, D., Ehninger, G., Creutzig, U. Different types of *NPM1* mutations in children and adults: evidence for an effect of patient age on the prevalence of the TCTG-tandem duplication in *NPM1*-exon 12. *Leukemia* 2007; 21: 366–367.

159 Rau, R., Brown, P. Nucleophosmin (NPM1) mutations in adult and childhood acute myeloid leukaemia: towards definition of a new leukaemia entity. *Hematol Oncol* 2009; 27: 171–181.

160 Vassiliou, G.S., et al. Mutant nucleophosmin and cooperating pathways drive leukemia initiation and progression in mice. *Nat Genet* 2011; 43: 470–475.

161 Figueroa, M.E., et al. Leukemic IDH1 and IDH2 mutations result in a hypermethylation phenotype, disrupt TET2 function, and impair hematopoietic differentiation. *Cancer Cell* 2010; 18: 553–567.

162 Paschka, P., et al. IDH1 and IDH2 mutations are frequent genetic alterations in acute myeloid leukemia and confer adverse prognosis in cytogenetically normal acute myeloid leukemia with NPM1 mutation without FLT3 internal tandem duplication. *J Clin Oncol* 2010; 28: 3636–3643.

163 Dohner, K., et al. Mutant nucleophosmin (NPM1) predicts favorable prognosis in younger adults with acute myeloid leukemia and normal cytogenetics: interaction with other gene mutations. *Blood* 2005; 106: 3740–3746.

164 Verhaak, R.G., et al. Mutations in nucleophosmin (NPM1) in acute myeloid leukemia (AML): association with other gene abnormalities and previously established gene expression signatures and their favorable prognostic significance. *Blood* 2005; 106: 3747–3754.

165 Papadaki, C., et al. Monitoring minimal residual disease in acute myeloid leukaemia with NPM1 mutations by quantitative PCR: clonal evolution is a limiting factor. *Br J Haematol* 2009; 144: 517–523.

166 Suzuki, T., et al. Clinical characteristics and prognostic implications of NPM1 mutations in acute myeloid leukemia. *Blood* 2005; 106: 2854–2861.

167 Kronke, J., et al. Clonal evolution in relapsed NPM1-mutated acute myeloid leukemia. *Blood* 2013; 122: 100–108.

168 Carow, C.E., et al. Localization of the human stem cell tyrosine kinase-1 gene (FLT3) to 13q12→q13. *Cytogenet Cell Genet* 1995; 70: 255–257.
169 Gupta, R., Knight, C.L., Bain, B.J. Receptor tyrosine kinase mutations in myeloid neoplasms. *Br J Haematol* 2002; 117: 489–508.
170 Zhang, S., et al. Essential role of signal transducer and activator of transcription (Stat)5a but not Stat5b for Flt3-dependent signaling. *J Exp Med* 2000; 192: 719–728.
171 Frohling, S., et al. Prognostic significance of activating FLT3 mutations in younger adults (16 to 60 years) with acute myeloid leukemia and normal cytogenetics: a study of the AML Study Group Ulm. *Blood* 2002; 100: 4372–4380.
172 Thiede, C., et al. Analysis of FLT3-activating mutations in 979 patients with acute myelogenous leukemia: association with FAB subtypes and identification of subgroups with poor prognosis. *Blood* 2002; 99: 4326–4335.
173 Gaidzik, V.I., et al. Clinical impact of DNMT3A mutations in younger adult patients with acute myeloid leukemia: results of the AML Study Group (AMLSG). *Blood* 2013; 121: 4769–4777.
174 Whitman, S.P., et al. Absence of the wild-type allele predicts poor prognosis in adult de novo acute myeloid leukemia with normal cytogenetics and the internal tandem duplication of FLT3: a cancer and leukemia group B study. *Cancer Res* 2001; 61: 7233–7239.
175 Kayser, S., et al. Insertion of FLT3 internal tandem duplication in the tyrosine kinase domain-1 is associated with resistance to chemotherapy and inferior outcome. *Blood* 2009; 114: 2386–2392.
176 Gale, R.E., et al. The impact of FLT3 internal tandem duplication mutant level, number, size, and interaction with NPM1 mutations in a large cohort of young adult patients with acute myeloid leukemia. *Blood* 2008; 111: 2776–2784.
177 Pratcorona, M., et al. Favorable outcome of patients with acute myeloid leukemia harboring a low-allelic burden FLT3-ITD mutation and concomitant NPM1 mutation: relevance to post-remission therapy. *Blood* 2013; 121: 2734–2738.
178 Schlenk, R.F., et al. Differential impact of allelic ratio and insertion site in FLT3-ITD-positive AML with respect to allogeneic transplantation. *Blood* 2014; 124: 3441–3449.
179 Schnittger, S., et al. Clinical impact of FLT3 mutation load in acute promyelocytic leukemia with t(15;17)/PML-RARA. *Haematologica* 2011; 96: 1799–1807.
180 Allen, C., et al. The importance of relative mutant level for evaluating impact on outcome of KIT, FLT3 and CBL mutations in core-binding factor acute myeloid leukemia. *Leukemia* 2013; 27: 1891–1901.
181 Yamamoto, Y., et al. Activating mutation of D835 within the activation loop of FLT3 in human hematologic malignancies. *Blood* 2001; 97: 2434–2439.
182 Mead, A.J., et al. FLT3 tyrosine kinase domain mutations are biologically distinct from and have a significantly more favorable prognosis than FLT3 internal tandem duplications in patients with acute myeloid leukemia. *Blood* 2007; 110: 1262–1270.

183 Whitman, S.P., et al. FLT3 D835/I836 mutations are associated with poor disease-free survival and a distinct gene-expression signature among younger adults with de novo cytogenetically normal acute myeloid leukemia lacking FLT3 internal tandem duplications. *Blood* 2008; 111: 1552–1559.
184 Nerlov, C. The C/EBP family of transcription factors: a paradigm for interaction between gene expression and proliferation control. *Trends Cell Biol* 2007; 17: 318–324.
185 Zhang, D.E., et al. Absence of granulocyte colony-stimulating factor signaling and neutrophil development in CCAAT enhancer binding protein alpha-deficient mice. *Proc Natl Acad Sci U S A* 1997; 94: 569–574.
186 Reddy, V.A., et al. Granulocyte inducer C/EBPalpha inactivates the myeloid master regulator PU.1: possible role in lineage commitment decisions. *Blood* 2002; 100: 483–490.
187 Puig-Kroger, A., et al. RUNX/AML and C/EBP factors regulate CD11a integrin expression in myeloid cells through overlapping regulatory elements. *Blood* 2003; 102: 3252–3261.
188 Johansen, L.M., et al. c-Myc is a critical target for c/EBPalpha in granulopoiesis. *Mol Cell Biol* 2001; 21: 3789–3806.
189 Behre, G., et al. Ras signaling enhances the activity of C/EBP alpha to induce granulocytic differentiation by phosphorylation of serine 248. *J Biol Chem* 2002; 277: 26293–26299.
190 Ye, M., et al. C/EBPa controls acquisition and maintenance of adult haematopoietic stem cell quiescence. *Nat Cell Biol* 2013; 15: 385–394.
191 Pabst, T., et al. Dominant-negative mutations of CEBPA, encoding CCAAT/enhancer binding protein-alpha (C/EBPalpha), in acute myeloid leukemia. *Nat Genet* 2001; 27: 263–270.
192 Gombart, A.F., et al. Mutations in the gene encoding the transcription factor CCAAT/enhancer binding protein alpha in myelodysplastic syndromes and acute myeloid leukemias. *Blood* 2002; 99: 1332–1340.
193 Pabst, T., Eyholzer, M., Fos, J., Mueller, B.U. Heterogeneity within AML with CEBPA mutations; only CEBPA double mutations, but not single CEBPA mutations are associated with favourable prognosis. *Br J Cancer* 2009; 100: 1343–1346.
194 Smith, M.L., Cavenagh, J.D., Lister, T.A., Fitzgibbon, J. Mutation of CEBPA in familial acute myeloid leukemia. *N Engl J Med* 2004; 351: 2403–2407.
195 Sellick, G.S., Spendlove, H.E., Catovsky, D., Pritchard-Jones, K., Houlston, R.S. Further evidence that germline CEBPA mutations cause dominant inheritance of acute myeloid leukaemia. *Leukemia* 2005; 19: 1276–1278.
196 Pabst, T., Eyholzer, M., Haefliger, S., Schardt, J., Mueller, B.U. Somatic CEBPA mutations are a frequent second event in families with germline CEBPA mutations and familial acute myeloid leukemia. *J Clin Oncol* 2008; 26: 5088–5093.
197 Carmichael, C.L., et al. Poor prognosis in familial acute myeloid leukaemia with combined biallelic CEBPA mutations and downstream events affecting the ATM, FLT3 and CDX2 genes. *Br J Haematol* 2010; 150: 382–385.

198 Care, R.S., et al. Incidence and prognosis of c-KIT and FLT3 mutations in core binding factor (CBF) acute myeloid leukaemias. *Br J Haematol* 2003; 121: 775–777.
199 Nanri, T., et al. Mutations in the receptor tyrosine kinase pathway are associated with clinical outcome in patients with acute myeloblastic leukemia harboring t(8;21)(q22;q22). *Leukemia* 2005; 19: 1361–1366.
200 Cairoli, R., et al. Prognostic impact of c-KIT mutations in core binding factor leukemias: an Italian retrospective study. *Blood* 2006; 107: 3463–3468.
201 Boissel, N., et al. Incidence and prognostic impact of c-Kit, FLT3, and Ras gene mutations in core binding factor acute myeloid leukemia (CBF-AML). *Leukemia* 2006; 20: 965–970.
202 Schnittger, S., et al. KIT-D816 mutations in AML1-ETO-positive AML are associated with impaired event-free and overall survival. *Blood* 2006; 107: 1791–1799.
203 Paschka, P., et al. Adverse prognostic significance of KIT mutations in adult acute myeloid leukemia with inv(16) and t(8;21): a Cancer and Leukemia Group B Study. *J Clin Oncol* 2006; 24: 3904–3911.
204 Meyer, S.C., Levine, R.L. Translational implications of somatic genomics in acute myeloid leukaemia. *Lancet Oncol* 2014; 15: e382–e394.
205 Basecke, J., Whelan, J.T., Griesinger, F., Bertrand, F.E. The MLL partial tandem duplication in acute myeloid leukaemia. *Br J Haematol* 2006; 135: 438–449.
206 Strout, M.P., Marcucci, G., Bloomfield, C.D., Caligiuri, M.A. The partial tandem duplication of ALL1 (MLL) is consistently generated by Alu-mediated homologous recombination in acute myeloid leukemia. *Proc Natl Acad Sci U S A* 1998; 95: 2390–2395.
207 Dohner, K., et al. Prognostic significance of partial tandem duplications of the MLL gene in adult patients 16 to 60 years old with acute myeloid leukemia and normal cytogenetics: a study of the Acute Myeloid Leukemia Study Group Ulm. *J Clin Oncol* 2002; 20: 3254–3261.
208 Schnittger, S., et al. Screening for MLL tandem duplication in 387 unselected patients with AML identify a prognostically unfavorable subset of AML. *Leukemia* 2000; 14: 796–804.
209 Grossmann, V., et al. A novel hierarchical prognostic model of AML solely based on molecular mutations. *Blood* 2012; 120: 2963–2972.
210 Christiansen, D.H., Andersen, M.K., Pedersen-Bjergaard, J. Mutations with loss of heterozygosity of p53 are common in therapy-related myelodysplasia and acute myeloid leukemia after exposure to alkylating agents and significantly associated with deletion or loss of 5q, a complex karyotype, and a poor prognosis. *J Clin Oncol* 2001; 19: 1405–1413.
211 Tang, J.L., et al. AML1/RUNX1 mutations in 470 adult patients with de novo acute myeloid leukemia: prognostic implication and interaction with other gene alterations. *Blood* 2009; 114: 5352–5361.
212 Schnittger, S., et al. RUNX1 mutations are frequent in de novo AML with noncomplex karyotype and confer an unfavorable prognosis. *Blood* 2011; 117: 2348–2357.

213 Gaidzik, V.I., et al. RUNX1 mutations in acute myeloid leukemia: results from a comprehensive genetic and clinical analysis from the AML study group. *J Clin Oncol* 2011; 29: 1364–1372.
214 Mendler, J.H., et al. RUNX1 mutations are associated with poor outcome in younger and older patients with cytogenetically normal acute myeloid leukemia and with distinct gene and microRNA expression signatures. *J Clin Oncol* 2012; 30: 3109–3118.
215 Buijs, A., et al. A novel CBFA2 single-nucleotide mutation in familial platelet disorder with propensity to develop myeloid malignancies. *Blood* 2001; 98: 2856–2858.
216 Michaud, J., et al. In vitro analyses of known and novel RUNX1/AML1 mutations in dominant familial platelet disorder with predisposition to acute myelogenous leukemia: implications for mechanisms of pathogenesis. *Blood* 2002; 99: 1364–1372.
217 Song, W.J., et al. Haploinsufficiency of CBFA2 causes familial thrombocytopenia with propensity to develop acute myelogenous leukaemia. *Nat Genet* 1999; 23: 166–175.
218 Walker, L.C., et al. A novel inherited mutation of the transcription factor RUNX1 causes thrombocytopenia and may predispose to acute myeloid leukaemia. *Br J Haematol* 2002; 117: 878–881.
219 Preudhomme, C., et al. High frequency of RUNX1 biallelic alteration in acute myeloid leukemia secondary to familial platelet disorder. *Blood* 2009; 113: 5583–5587.
220 Huff, V. Wilms' tumours: about tumour suppressor genes, an oncogene and a chameleon gene. *Nat Rev Cancer* 2011; 11: 111–121.
221 Rivera, M.N. and Haber, D.A. Wilms' tumour: connecting tumorigenesis and organ development in the kidney. *Nat Rev Cancer* 2005; 5: 699–712.
222 Paschka, P., et al. Wilms' tumor 1 gene mutations independently predict poor outcome in adults with cytogenetically normal acute myeloid leukemia: a Cancer and Leukemia Group B study. *J Clin Oncol* 2008; 26: 4595–4602.
223 Hou, H.A., et al. WT1 mutation in 470 adult patients with acute myeloid leukemia: stability during disease evolution and implication of its incorporation into a survival scoring system. *Blood* 2010; 115: 5222–5231.
224 Becker, H., et al. Mutations of the Wilms tumor 1 gene (WT1) in older patients with primary cytogenetically normal acute myeloid leukemia: a Cancer and Leukemia Group B study. *Blood* 2010; 116: 788–792.
225 Virappane, P., et al. Mutation of the Wilms' tumor 1 gene is a poor prognostic factor associated with chemotherapy resistance in normal karyotype acute myeloid leukemia: the United Kingdom Medical Research Council Adult Leukaemia Working Party. *J Clin Oncol* 2008; 26: 5429–5435.
226 Rampal, R., et al. DNA hydroxymethylation profiling reveals that WT1 mutations result in loss of TET2 function in acute myeloid leukemia. *Cell Rep* 2014; 9: 1841–1855.
227 Schubbert, S., Shannon, K., Bollag, G. Hyperactive Ras in developmental disorders and cancer. *Nat Rev Cancer* 2007; 7: 295–308.

228 Burnett, A.K., et al. The addition of the farnesyl transferase inhibitor, tipifarnib, to low dose cytarabine does not improve outcome for older patients with AML. *Br J Haematol* 2012; 158: 519–522.

229 Holliday, R. and Grigg, G.W. DNA methylation and mutation. *Mutat Res* 1993; 285: 61–67.

230 Yang, L., Rau, R. and Goodell, M.A. DNMT3A in haematological malignancies. *Nat Rev Cancer* 2015; 15: 152–165.

231 You, J.S. and Jones, P.A. Cancer genetics and epigenetics: two sides of the same coin? *Cancer Cell* 2012; 22: 9–20.

232 Bird, A. DNA methylation patterns and epigenetic memory. *Genes Dev* 2002; 16: 6–21.

233 Russler-Germain, D.A., et al. The R882H DNMT3A mutation associated with AML dominantly inhibits wild-type DNMT3A by blocking its ability to form active tetramers. *Cancer Cell* 2014; 25: 442–454.

234 Challen, G.A., et al. Dnmt3a is essential for hematopoietic stem cell differentiation. *Nat Genet* 2012; 44: 23–31.

235 Thol, F., et al. Incidence and prognostic influence of DNMT3A mutations in acute myeloid leukemia. *J Clin Oncol* 2011; 29: 2889–2896.

236 Renneville, A., et al. Prognostic significance of DNA methyltransferase 3A mutations in cytogenetically normal acute myeloid leukemia: a study by the Acute Leukemia French Association. *Leukemia* 2012; 26: 1247–1254.

237 Ho, P.A., et al. Molecular alterations of the IDH1 gene in AML: a Children's Oncology Group and Southwest Oncology Group study. *Leukemia* 2010; 24: 909–913.

238 Green, C.L., et al. The prognostic significance of IDH1 mutations in younger adult patients with acute myeloid leukemia is dependent on FLT3/ITD status. *Blood* 2010; 116: 2779–2782.

239 Abbas, S., et al. Acquired mutations in the genes encoding IDH1 and IDH2 both are recurrent aberrations in acute myeloid leukemia: prevalence and prognostic value. *Blood* 2010; 116: 2122–2126.

240 Marcucci, G., et al. IDH1 and IDH2 gene mutations identify novel molecular subsets within de novo cytogenetically normal acute myeloid leukemia: a Cancer and Leukemia Group B study. *J Clin Oncol* 2010; 28: 2348–2355.

241 Boissel, N., et al. Prognostic impact of isocitrate dehydrogenase enzyme isoforms 1 and 2 mutations in acute myeloid leukemia: a study by the Acute Leukemia French Association group. *J Clin Oncol* 2010; 28: 3717–3723.

242 Thol, F., et al. Prognostic impact of IDH2 mutations in cytogenetically normal acute myeloid leukemia. *Blood* 2010; 116: 614–616.

243 Green, C.L., et al. The prognostic significance of IDH2 mutations in AML depends on the location of the mutation. *Blood* 2011; 118: 409–412.

244 Ward, P.S., et al. The common feature of leukemia-associated IDH1 and IDH2 mutations is a neomorphic enzyme activity converting alpha-ketoglutarate to 2-hydroxyglutarate. *Cancer Cell* 2010; 17: 225–234.

245 Xu, W., et al. Oncometabolite 2-hydroxyglutarate is a competitive inhibitor of alpha-ketoglutarate-dependent dioxygenases. *Cancer Cell* 2011; 19: 17–30.

246 Tahiliani, M., et al. Conversion of 5-methylcytosine to 5-hydroxymethylcytosine in mammalian DNA by MLL partner TET1. *Science* 2009; 324: 930–935.
247 Pronier, E., et al. Inhibition of TET2-mediated conversion of 5-methylcytosine to 5-hydroxymethylcytosine disturbs erythroid and granulomonocytic differentiation of human hematopoietic progenitors. *Blood* 2011; 118: 2551–2555.
248 Moran-Crusio, K., et al. Tet2 loss leads to increased hematopoietic stem cell self-renewal and myeloid transformation. *Cancer Cell* 2011; 20: 11–24.
249 Delhommeau, F., et al. Mutation in TET2 in myeloid cancers. *N Engl J Med* 2009; 360: 2289–2301.
250 Chou, W.C., et al. TET2 mutation is an unfavorable prognostic factor in acute myeloid leukemia patients with intermediate-risk cytogenetics. *Blood* 2011; 118: 3803–3810.
251 Metzeler, K.H., et al. TET2 mutations improve the new European LeukemiaNet risk classification of acute myeloid leukemia: a Cancer and Leukemia Group B study. *J Clin Oncol* 2011; 29: 1373–1381.
252 Gaidzik, V.I., et al. TET2 mutations in acute myeloid leukemia (AML): results from a comprehensive genetic and clinical analysis of the AML study group. *J Clin Oncol* 2012; 30: 1350–1357.
253 Abdel-Wahab, O., et al. ASXL1 mutations promote myeloid transformation through loss of PRC2-mediated gene repression. *Cancer Cell* 2012; 22: 180–193.
254 Pratcorona, M., et al. Acquired mutations in ASXL1 in acute myeloid leukemia: prevalence and prognostic value. *Haematologica* 2012; 97: 388–392.
255 Metzeler, K.H., et al. ASXL1 mutations identify a high-risk subgroup of older patients with primary cytogenetically normal AML within the ELN Favorable genetic category. *Blood* 2011; 118: 6920–6929.
256 Fernandes, M.S., et al. Novel oncogenic mutations of CBL in human acute myeloid leukemia that activate growth and survival pathways depend on increased metabolism. *J Biol Chem* 2010; 285: 32596–32605.
257 Reindl, C., et al. CBL exon 8/9 mutants activate the FLT3 pathway and cluster in core binding factor/11q deletion acute myeloid leukemia/myelodysplastic syndrome subtypes. *Clin Cancer Res* 2009; 15: 2238–2247.
258 Makishima, H., et al. Mutations of e3 ubiquitin ligase cbl family members constitute a novel common pathogenic lesion in myeloid malignancies. *J Clin Oncol* 2009; 27: 6109–6116.
259 Yoshida, K., et al. Frequent pathway mutations of splicing machinery in myelodysplasia. *Nature* 2011; 478: 64–69.
260 Papaemmanuil, E., et al. Somatic SF3B1 mutation in myelodysplasia with ring sideroblasts. *N Engl J Med* 2011; 365: 1384–1395.
261 Papaemmanuil, E., et al. Clinical and biological implications of driver mutations in myelodysplastic syndromes. *Blood* 2013; 122: 3616–3627; quiz 3699.
262 Lindsley, R.C., et al. Acute myeloid leukemia ontogeny is defined by distinct somatic mutations. *Blood* 2015; 125: 1367–1376.

263 McKerrell, T., et al. Leukemia-associated somatic mutations drive distinct patterns of age-related clonal hemopoiesis. *Cell Rep* 2015; 10: 1239–1245.
264 Solomon, D.A., et al. Mutational inactivation of STAG2 causes aneuploidy in human cancer. *Science* 2011; 333: 1039–1043.
265 Thota, S., et al. Genetic alterations of the cohesin complex genes in myeloid malignancies. *Blood* 2014; 124: 1790–1798.
266 Grossmann, V., et al. Whole-exome sequencing identifies somatic mutations of BCOR in acute myeloid leukemia with normal karyotype. *Blood* 2011; 118: 6153–6163.
267 Hou, H.A., et al. Characterization of acute myeloid leukemia with PTPN11 mutation: the mutation is closely associated with NPM1 mutation but inversely related to FLT3/ITD. *Leukemia* 2008; 22: 1075–1078.
268 Xie, M., et al. Age-related mutations associated with clonal hematopoietic expansion and malignancies. *Nat Med* 2014; 20: 1472–1478.
269 Genovese, G., et al. Clonal hematopoiesis and blood-cancer risk inferred from blood DNA sequence. *N Engl J Med* 2014; 371: 2477–2487.
270 Jaiswal, S., et al. Age-related clonal hematopoiesis associated with adverse outcomes. *N Engl J Med* 2014; 371: 2488–2498.
271 Rossi, D.J., Jamieson, C.H., Weissman, I.L. Stems cells and the pathways to aging and cancer. *Cell* 2008; 132: 681–696.
272 Conte, N., et al. Detailed molecular characterisation of acute myeloid leukaemia with a normal karyotype using targeted DNA capture. *Leukemia* 2013; 27: 1820–1825.
273 Fischer, A., Vazquez-Garcia, I., Illingworth, C.J., Mustonen, V. High-definition reconstruction of clonal composition in cancer. *Cell Rep* 2014; 7: 1740–1752.
274 Kearney, L., Horsley, S.W. Molecular cytogenetics in haematological malignancy: current technology and future prospects. *Chromosoma* 2005; 114: 286–294.
275 Tachdjian, G., Aboura, A., Lapierre, J.M., Viguie, F. Cytogenetic analysis from DNA by comparative genomic hybridization. *Ann Genet* 2000; 43: 147–154.
276 Struski, S., Doco-Fenzy, M., Cornillet-Lefebvre, P. Compilation of published comparative genomic hybridization studies. *Cancer Genet Cytogenet* 2002; 135: 63–90.
277 Casas, S., et al. Genetic diagnosis by comparative genomic hybridization in adult de novo acute myelocytic leukemia. *Cancer Genet Cytogenet* 2004; 153: 16–25.
278 Smith, M.L., Hills, R.K., Grimwade, D.. Independent prognostic variables in acute myeloid leukaemia. *Blood Rev* 2011; 25(1): 39–51.
279 Prensner, J.R., Chinnaiyan, A.M. Metabolism unhinged: IDH mutations in cancer. *Nat Med* 2011; 17: 291–293.
280 Tosi, S, et al. Characterization of the human myeloid leukemia-derived cell line GF-D8 by multiplex fluorescence in situ hybridization, subtelomeric probes, and comparative genomic hybridization. *Genes Chromosomes Cancer* 1999; 24(3): 213–221.

281 Bain, B.J. *Leukaemia Diagnosis,* 3rd edn. Oxford: Blackwell Publishing, 2003.

282 Harris, N.L., Jaffe, E.S., Diebold, J., et al. World Health Organization classification of neoplastic diseases of the hematopoietic and lymphoid tissues: report of the Clinical Advisory Committee meeting – Airlie House, Virginia, November 1997. *J Clin Oncol* 1999; 17: 3835–3849.

283 Naiel, A., Vetter, M., Plekhanova, O., et al. A novel three-colour fluorescence in situ hybridization approach for the detection of t(7;12)(q36;p13) in acute myeloid leukaemia reveals new cryptic three way translocation t(7;12;16). *Cancers (Basel)* 2013; 5(1): 281–295.

CHAPTER 4

Molecular genetics of paediatric acute myeloid leukaemia*

Marry van den Heuvel-Eibrink, Jasmijn D.E. de Rooij and Christian Michel Zwaan

Clinical introduction

Epidemiology of AML

In children, the most frequently occurring haematological malignancies are acute leukaemias, of which 80% are classified as acute lymphoblastic leukaemia (ALL) and 15–20% as acute myeloid leukaemia (AML). The incidence of AML in infants is 1.5 per 100,000 individuals per year; thereafter the incidence decreases to 0.4 per 100,000 individuals aged 5–9 years, after which it gradually increases into adulthood, up to an incidence of 16.2 per 100,000 individuals aged over 65 years (SEER Cancer Statistics Registry USA, 2005–2009).[1]

The underlying cause of AML is unknown, and childhood AML generally occurs *de novo*. Rare cases are preceded by clonal evolution of preleukaemic myeloproliferative diseases such as myelodysplastic syndrome (MDS) and juvenile myelomonocytic leukaemia (JMML). In addition, AML may occur after previous radiotherapy or chemotherapy (alkylating agents or epipodophyllotoxins) as secondary neoplasm.[2,3] In elderly AML patients, environmental factors have been suggested in the pathogenesis of AML, which probably is less relevant in children as exposure time is usually short, unless the carcinogenic impact of an environmental exposure is truly significant.[4,5] In rare cases, AML occurs in children with underlying genetic diseases, such as chromosomal-breakage syndromes (e.g. Fanconi anaemia, Bloom

*This chapter is adapted from de Rooij, J.D.E., Zwaan, C.M., van den Heuvel-Eibrink, M. Pediatric AML: from biology to clinical management. *Journal of Clinical Medicine* 2015; 4: 127–149.

The Genetic Basis of Haematological Cancers, First Edition. Edited by Sabrina Tosi and Alistair G. Reid.

syndrome) or diseases with disturbed myelopoiesis (e.g. severe congenital neutropenia, Diamond–Blackfan anaemia, dyskeratosis congenita).[6,7] Excessive telomere erosion might also induce chromosomal instability, which could predispose to malignant transformation.[8] Loss-of function mutations in the telomerase complex genes *TERT* and *TERC* were found in aplastic anaemia and AML (9%) in adults.[9,10] However, in paediatric AML, the frequency of *TERT* and *TERC* gene variants was not increased compared with a geographically matched control group.[11] Children with Down syndrome classically present with a unique subtype of AML, following a transient myeloproliferative disorder in the neonatal period, which is characterized by mutations in the *GATA1* gene, occurring as a first hit.[12] Recently, germline mutations in several genes, such as *TP53, RUNX1, GATA2* and *CEBPA*, have been found in families with an unexplained high risk of AML, suggesting a familial predisposition to develop AML.[13–17]

Diagnostic approach

AML is a heterogeneous disease with respect to morphology, underlying germline and somatic genetic abnormalities and clinical behaviour. The standard diagnostic process of AML is based on a combination of morphology in combination with cytochemistry, immunophenotyping and cytogenetic and molecular–genetic characterization of the leukaemic blasts derived from the bone marrow. Thereby, each AML patient can be classified according to clinically relevant subgroups. The older morphology-based French–American–British (FAB) classification is nowadays replaced by the World Health Organization (WHO) classification, which also takes genetic abnormalities into account (Table 4.1).[18,19]

Immunophenotyping is generally used to distinguish AML from ALL and subclassifies paediatric AML according to the cell lineage of origin and differentiation stage at which the differentiation arrest occurs. Especially for the diagnosis of FAB types M0 and M7, immunophenotyping is indispensable.[20,21] The majority of chromosomal abnormalities are detected by conventional karyotyping and complemented with fluorescence *in situ* hybridization (FISH) to detect, for instance, relevant (cryptic) translocations, fusion genes or loss of chromosome material.[22] In young children under 2 years of age, it is important to search for paediatric AML-specific translocations that are not yet acknowledged in the WHO classification as separate entities, such as t(7;12)(q36;p13) and t(1;22)(p13;q13).[21]

Table 4.1 The WHO classification of AML and related neoplasms.[19]

Category	Morphology
Acute myeloid leukaemia with recurrent genetic abnormalities	AML with t(8;21)(q22;q22); *RUNX1-RUNX1T1*
	AML with inv(16)(p13.1q22) or t(16;16)(p13.1;q22); *CBFB-MYH11*
	APL with t(15;17)(q22;q12); *PML-RARA*
	AML with t(9;11)(p22;q23); *MLLT3-MLL*
	AML with t(6;9)(p23;q34); *DEK-NUP214*
	AML with inv(3)(q21q26.2) or t(3;3)(q21;q26.2); *RPN1-EVI1*
	AML (megakaryoblastic) with t(1;22)(p13;q13); *RBM15-MKL1*
	Provisional entity: AML with mutated *NPM1*
	Provisional entity: AML with mutated *CEBPA*
Acute myeloid leukaemia with myelodysplasia-related changes	
Therapy-related myeloid neoplasms	
Acute myeloid leukaemia, not otherwise specified	AML with minimal differentiation
	AML without maturation
	AML with maturation
	Acute myelomonocytic leukaemia
	Acute monoblastic/monocytic leukaemia
	Acute erythroid leukaemia
	Pure erythroid leukaemia
	Erythroleukaemia, erythroid/myeloid
	Acute megakaryoblastic leukaemia
	Acyte basophilic leukaemia
	Acute panmyelosis with myelofibrosis
Myeloid sarcoma	
Myeloid proliferations related to Down syndrome	Transient abnormal myelopoiesis
	Myeloid leukaemia associated with Down syndrome
Blastic plasmacytoid dendritic cell neoplasm	

Treatment and outcome

The clinical outcome of paediatric AML has improved significantly over the past decades, with current long-term survival rates of ~70%.[23–30] This improvement is mainly due to intensification of chemotherapeutic regimens, enhanced risk-group stratification and improved supportive care. Risk-group stratification is usually based on cytogenetic abnormalities present in the leukaemic blasts and on early response to treatment, either measured as the complete remission (CR) rate after one or two courses or applying minimal-residual disease measurements. The backbone of most chemotherapeutic regimens consists of four or five cycles of intensive chemotherapy consisting of cytarabine plus an anthracycline, to which other drugs are added. The added value of haematopoietic stem cell transplantation (SCT) in paediatric AML is under discussion, as in general the reduction in relapse risk is counterbalanced by procedure-related deaths and is also dependent on the intensity or prior induction chemotherapy.[31] SCT in first CR is therefore currently only recommended for certain high-risk cases in most European protocols, whereas SCT plays a more prominent role in most North American treatment protocols.[32] Despite intensive treatment, 30–40% of patients relapse and their outcome is poor, with 38% of patients surviving in the largest and most recent series reported to date.[33]

Because of the high frequency of treatment-related deaths (5–10%) both in treatment protocols for newly diagnosed and for relapsed disease and because of long-term side effects such as anthracycline-induced cardiomyopathy, further intensification of chemotherapy seems no longer feasible.[34] Therefore, knowledge of the molecular and genetic background is of utmost relevance in order to detect novel, leukaemia-specific treatment targets.

Relevant molecular and genetic aberrations in paediatric AML

Type I/II aberrations and their non-random associations

AML is thought to arise from at least two classes of cooperating genetic events (Fig. 4.1).[35] Type I abnormalities result in increased, uncontrolled proliferation and/or survival and are often activating mutations of genes involved in signal transduction pathways, such as *FLT3*, *C-KIT*, *N-RAS*, *K-RAS* and *PTPN11*. Type II abnormalities impair differentiation and mainly result from genetic aberrations in haematopoietic transcription factors, such as the AML-characteristic translocations t(15;17)(q22;q21)/*PML-RARA*, t(8;21)(q22;q22)/*AML1-ETO*, inv(16)(p13;q22)/*CBFB-MYH11* and 11q23/*MLL* rearrangements, or from mutations in genes such as

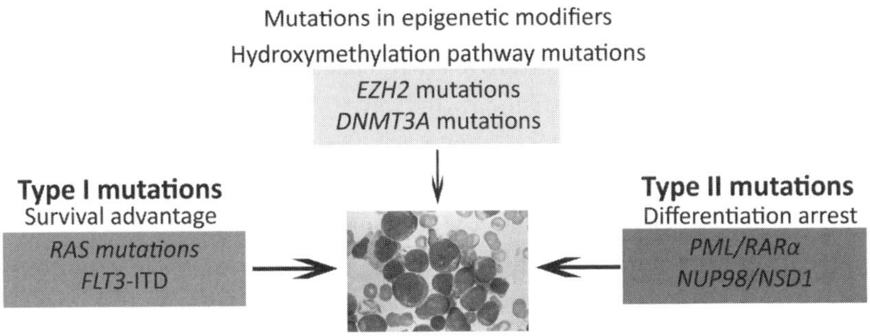

Figure 4.1 Model of cooperating (epi-)genetic events in AML. Different types of genetic and epigenetic events collaborate in leukaemogenesis.

NPM1 and *CEBPA*.[36–38] Translocations involving haematopoietic transcription factors often lead to dysregulated gene expression, as a result of either the fusion partner itself or the recruitment of different co-factors to the transcription complex. For example, the *MLL* gene has histone methyltransferase activity and is part of a chromatin modifying complex. More than 60 fusion partners have been identified in AML, but the breakpoint of the *MLL* gene is highly conserved. Fusion proteins lead to a gain of function of the *MLL* complex, resulting in inappropriate histone modification and increased expression of *MEIS1* and specific *HOX* genes, maintain a stem cell phenotype. In addition, the presence of DOT1L in the *MLL* complex has been shown to be required for the leukaemogenic activity of several *MLL* rearrangements and may be a target for treatment.[39–41] Mutations in epigenetic regulators (e.g. *EZH2, ASXL1, DNMT3A*) have been found that add another level of complexity and contribute to both the maturation arrest and proliferative capacity which are needed to develop AML.

In paediatric AML, several type I abnormalities are mutually exclusive, although mutated *N-RAS* can be found in combination with mutations of *FLT3, C-KIT, K-RAS* and *PTPN11*, and also the combination of *WT1* and *FLT3*. The cytogenetic type II aberrations, including *MLL* rearrangements, CBF-AML and t(15;17)(q22;q21), are mutually exclusive with mutations of *NPM1, CEBPA* (biallelic) and *MLL*-PTD.[42] Most type I abnormalities are non-randomly distributed over the different type II-defined AML subtypes (Figure 4.2).

The most common cytogenetic abnormalities in children are t(8;21)(q22;q22), inv(16)(p13.1q22) [together referred to as core binding factor (CBF)-AML], t(15;17)(q22;q21) and 11q23/*MLL*-rearranged abnormalities (up to 25%) (Table 4.2).[43–46] Together these account for approximately half of all paediatric AML cases, a much higher frequency than in adults. Some translocations, for example t(1;22)(p13;q13), t(7;12)(q36;p13) and t(5;11)(q35;p15.5), are specific to children and are seldom or never found in adults.[47–54]

Only 20–25% of paediatric AML cases are cytogenetically normal, whereas in adults approximately 50% of cases do not have any cytogenetic abnormalities.[55,56] Of interest, recent work identified single gene mutations (such as *NPM1* mutations and biallelic *CEBPA* mutations),

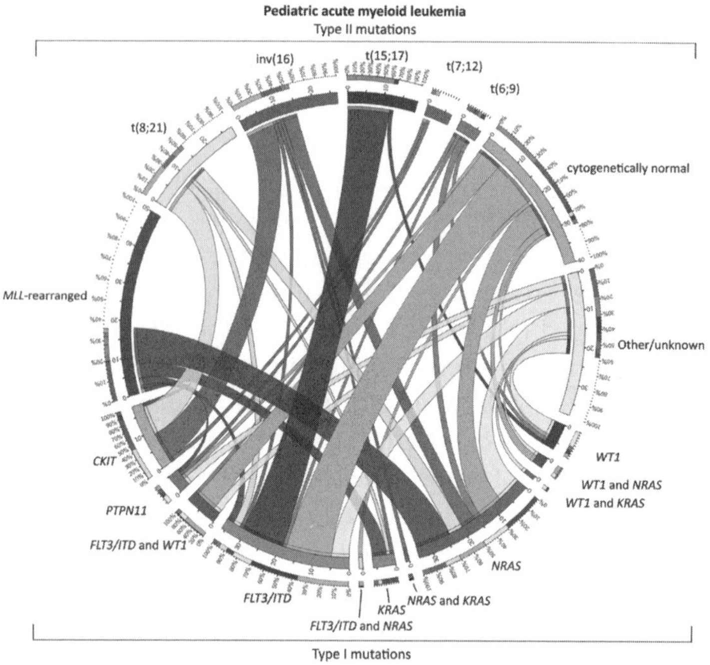

(a)

Figure 4.2 Distribution type I/II abnormalities in paediatric AML. (a) The circos plot depicts the frequency of the type II mutations and co-occurrence of type I mutations in patients with *de novo* paediatric AML. The length of the arch corresponds to the frequency of the type II mutation and the width of the ribbon with the percentage of patients with a specific type I mutation or combination of type I mutations. *FLT3*-ITD denotes *FLT3* internal tandem duplication. (b) The circos plot depicts the frequency of the type II mutations and co-occurrence of type I mutations in patients with *de novo* paediatric cytogenetically normal AML. The length of the arch corresponds to the frequency of the type II mutation and the width of the ribbon with the percentage of patients with a specific type I mutation or combination of type I mutations. *FLT3*-ITD denotes *FLT3* internal tandem duplication. *(See plate section for color representation of this figure.)*

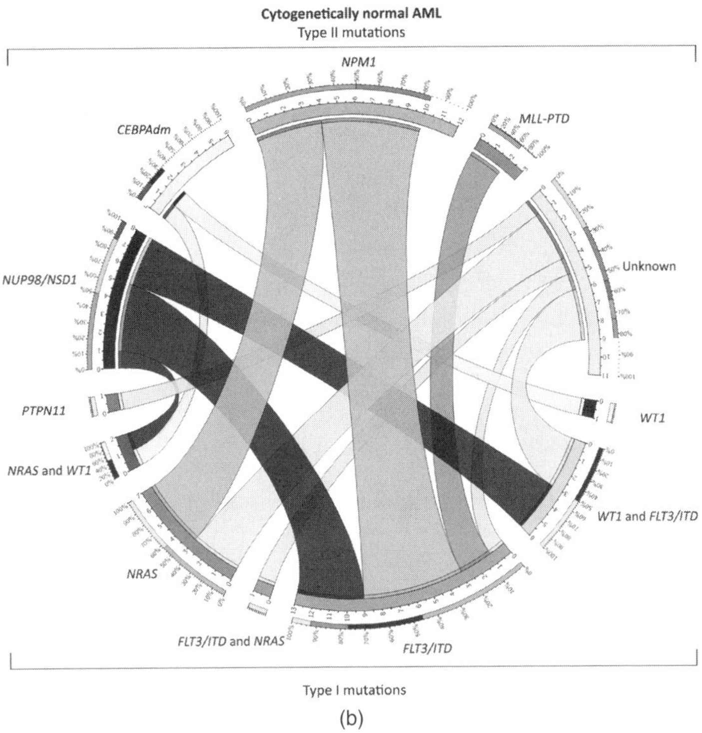

Figure 4.2 (*continued*)

cryptic translocations (*NUP98/NSD1*) and deregulated gene expression (e.g. of the *BAALC* and *ERG* genes) underlying cytogenetically normal AML.[37,53,57]

The distribution of genetic abnormalities differed significantly between children below and above the age of 2 years. Very young children are characterized by a high incidence of *MLL* rearrangements and a low incidence of t(8;21)(q22;q22), t(15;17)(q22;q21) and CN-AML. In addition, complex karyotypes are more frequently observed and t(7;12)(q36;p13) and t(1;22)(p13;q13) are almost exclusively found in this age group.[42]

Relevance of type I/II aberrations for outcome and stratification of paediatric AML treatment

Although the classification and prognosis of paediatric AML are roughly similar to those of adult AML, some clear differences exist, especially in the frequency of the underlying genetic changes.[44–46] In general, good risk abnormalities are more frequent in children. Table 4.3 gives

Table 4.2 Recurrent cytogenetic aberrations in paediatric and adult AML.

Cytogenetic aberration	Involved gene(s)/ fusion gene	Paediatric AML (%)	Adult AML (%)	Prognostic relevance
None	–	20–25	45	Intermediate
t(15;17)(q22;q21)	*PML-RARA*	6–10	8	Favourable
inv(16)(p13q22)/t(16;16)(p13;q22)	*CBFB-MYH11*	6–9	5	Favourable
t(8;21)(q22;q22)	*AML1-ETO*	12–15	6	Favourable
t/inv(11q23)	*MLL*-diverse partner genes	16–23	3	Dependent on partner gene
t(6;9)(p23;q34)	*DEK-NUP214*	1–2	1	Unfavourable
t(9;22)(q34;q11)	*BCR-ABL*	<1	1	Unfavourable
t(1;22)(p13;q13)	*RBM15-MKL1*	<1	0	Intermediate/unfavourable
t(7;12)(q36;p13)	*HLXB9-ETV6*	1	0	Unfavourable
t(8;16)(p11;p13)/inv(8)(p11q13)	*MOZ-CBP/-TIF2*	1	NA	Unknown
inv(3)(q21q26)/t(3;3)(q21;q26)	*EVI1*	<1	1–2	Unfavourable
t(5;11)(q35;p15.5)	*NUP98/NSD1*	4	1	Unfavourable
abn(12p)*	?	2–4	3	Unfavourable
abn(17p)*	?	2–3	2	Intermediate/unfavourable
del(9q)*	?	3–5	2	Intermediate
+21*	?	5–6	2	Intermediate
+8*	?	9–13	9	Intermediate
–Y*	?	4–5	4	Intermediate
–7/7q–*	?	2–7	8	Unfavourable(in paediatric AML restricted to –7)
–5/5q–*	?	1–2	7	Intermediate/unfavourable
Complex (≥3 aberrations)	?	8–15	11	Intermediate/unfavourable

*Percentage reflects the mentioned aberration as sole aberration or in combination with other cytogenetic aberrations.
Based on references 45, 46, 50, 53, 106 and 107.

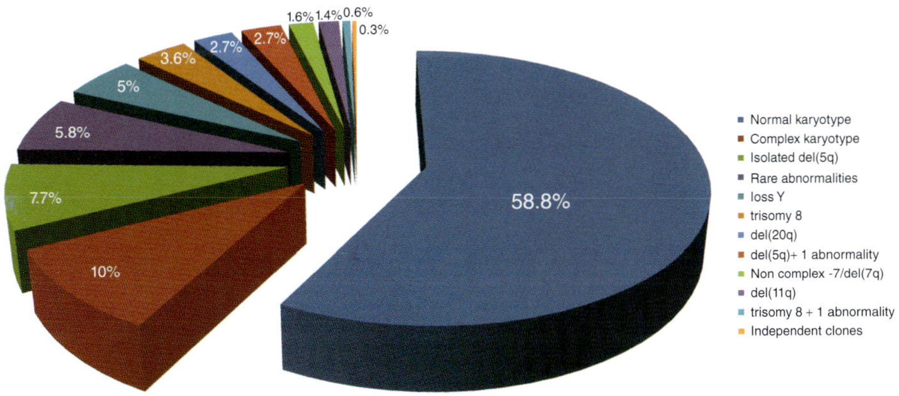

Figure 1.2 Distribution of cytogenetic aberrations in 364 cases of MDS.

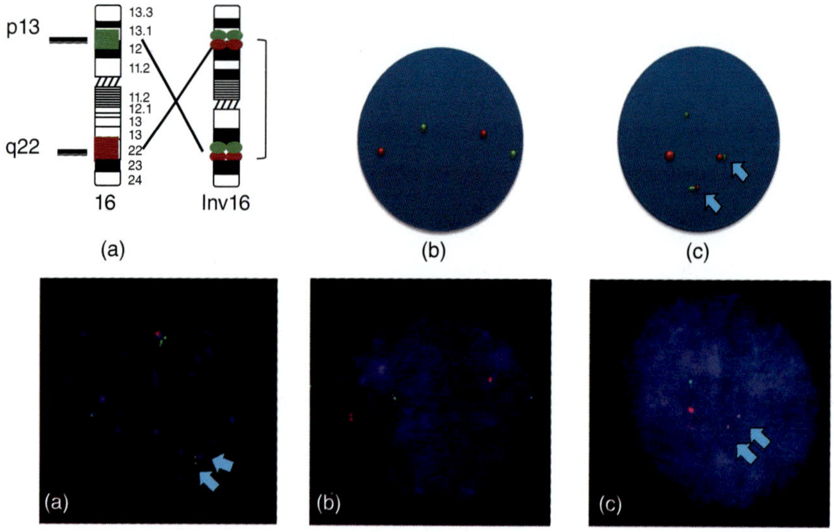

Figure 3.4 Rearrangement inv(16) in AML. Fluorescence *in situ* hybridization performed on the bone marrow metaphase spread of a patient with inv(16) using a dual colour probe set (Metasystems, Altlussheim, Germany). Green and red signals correspond to the *MYH11* and the *CBFB* regions, respectively. A schematic representation of the distribution of FISH signals is shown in the upper row, on both normal and rearranged chromosomes (a), on a normal interphase nucleus (b) and on an interphase nucleus carrying the inv(16) (c). Arrows indicate the fusion signals in (c) and (C). The corresponding photomicrographs are shown in the bottom row (A, B and C). Image courtesy of S. Tosi and A. Naiel, Leukaemia and Chromosome Research Laboratory, Brunel University, London, UK; the patient sample was provided by Professor Jochen Harbott, Department of Paediatric Haematology and Oncology, Justus Liebig University, Giessen, Germany.

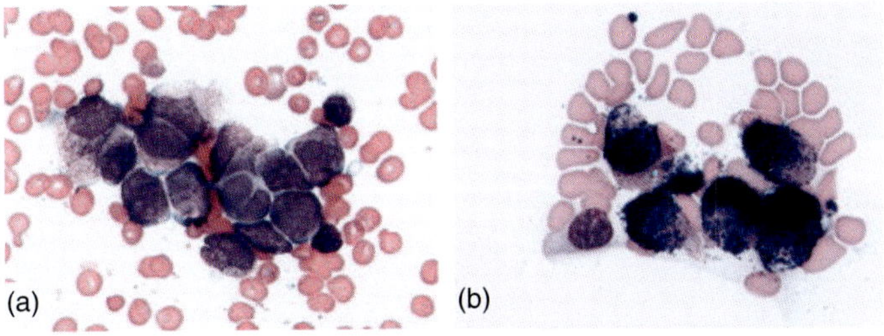

Figure 3.6 A composite of the methodologies for diagnostic evaluation of APL. (a) Bone marrow aspirate smear shows many promyelocytes. Compared with the adjacent normal small lymphocytes, promyelocytes are of medium to large size with irregularly shaped to convoluted to bilobed nuclei, open chromatin, visible to prominent nucleoli and a moderate amount of basophilic cytoplasm with abundant granularity. (b) Cytochemical stain for myeloperoxidase. Compared with the negative small lymphocyte, the promyelocytes are strongly positive, with numerous granules covering the outlines of the nuclei. (c) Negative (macrogranular) immunofluorescent stain for promyelocytic leukaemia (PML). The PML oncogenic domains are observed as several distinct particles in each nucleus. (d) Positive (microgranular) immunofluorescent stain for PML. Numerous (too many to count) fine, dusty granules are present in each nucleus. Panels (a)–(d) reproduced with permission-from Dimov et al. (e, f) Examples of the application of FISH probes specific for and spanning, the PML gene on chromosome 15 (in green) and the RARA gene on chromosome 17 (in red; Kreatech Diagnostics, The Netherlands) to metaphase (e) and interphase (f) cells from a patient with APL. Fusion signals are present on both the der(15) and the der(17), in addition to single green and red signals marking the normal chromosomes 15 and 17, respectively (e). The same hybridization signal pattern is visible in interphase nuclei (f) and is readily distinguishable from normal cells which show two single green and two single red signals (not shown). Panels (e) and (f) courtesy of A Reid and I. Ortiz de Mendibil, Imperial Molecular Pathology, Imperial College Healthcare Trust, London, UK. (g) Example of G-banded karyotype from a bone marrow metaphase of an APL patient. Note that in this case additional abnormalities are present in addition to the typical t(15;17) rearrangement. The full karyotype reads 47,XX,+8,i(8)(q10), t(15;17)(q24;q21). Panel (g) courtesy of John Swansbury, Clinical Cytogenetics; McElwain Laboratories, The Royal Marsden Hospital and the Institute of Cancer Research, Sutton, Surrey, UK.

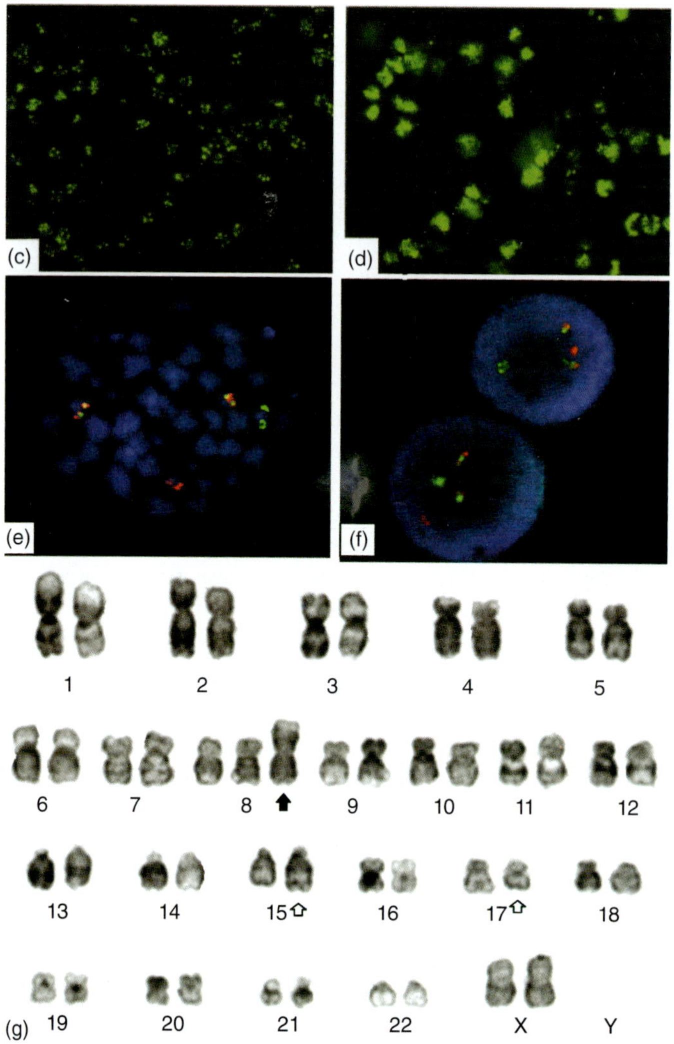

Figure 3.6 (*continued*)

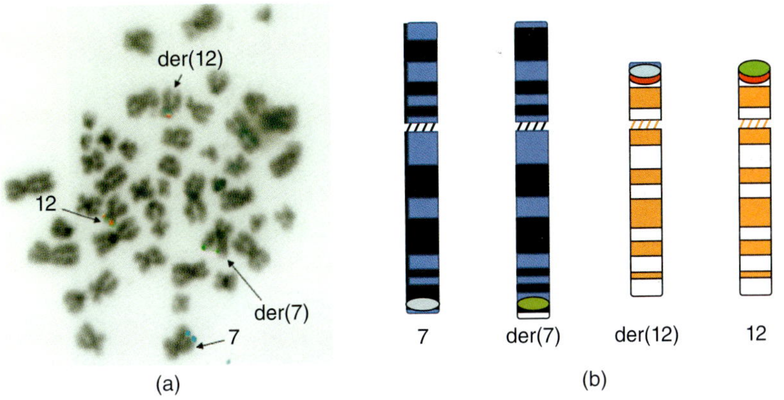

Figure 3.7 Translocation t(7;12) in infant AML. Fluorescence *in situ* hybridization performed on the bone marrow metaphase spread of a patient with t(7;12)(q36;p13) using a three-colour probe set (Metasystems, Altlussheim, Germany). Note localization of FISH signals on chromosome 7 (blue signals), der(7) (green signals), chromosome 12 (green and orange signals) and der(12) (blue and orange signals). The DAPI counterstain used to visualize the chromosomes has been converted into greyscale to simulate a G-like banding pattern (a). The schematic representation of the hybridization pattern is also shown on the ideograms (b). From Naiel et al.[283]

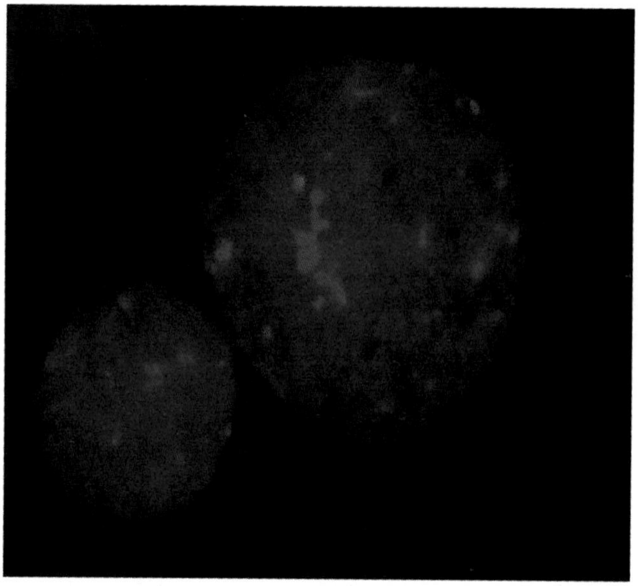

Figure 3.9 Visualization of trisomy 8 in AML. Example of interphase FISH using a probe specific for the centromeric alphoid sequences of chromosome 8. The interphase nuclei were obtained from the bone marrow of a patient with AML. Three hybridization signals (in red) are visible in each of the two nuclei represented here. DAPI was used to counterstain the nuclei in blue. Image capture and analysis were performed using a Zeiss microscope (Axioplan2 Imaging) equipped with a Sensys cooled CCD camera and Smart Capture v2 imaging software (Digital Scientific UK). Image courtesy of S. Tosi, A. Naiel and H. Al-Badri, Leukaemia and Chromosome Research Laboratory, Brunel University, London, UK.

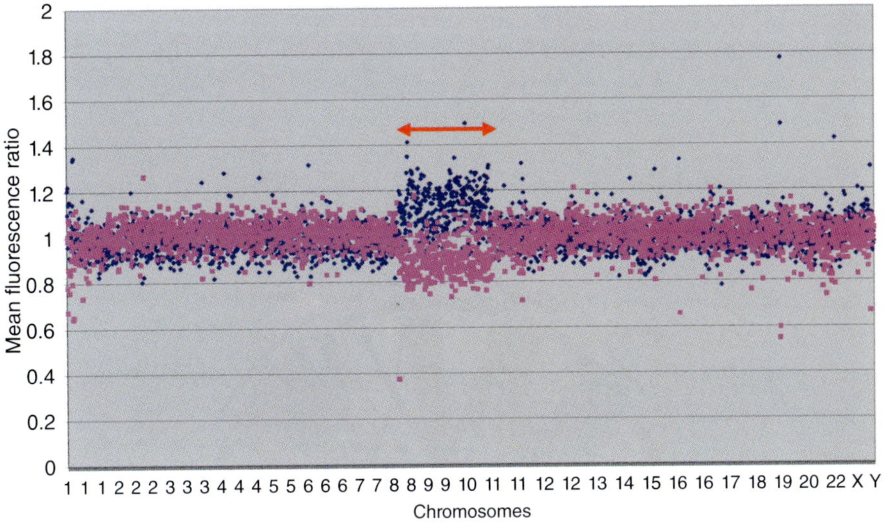

Figure 3.10 Early example of genome-wide array-CGH test data obtained using test DNA from a patient with AML. The earliest genome-wide arrays used for detecting genome imbalance were made up of ~3000 probes spaced at 1 Mb intervals along the genome. Data were obtained by comparing patient versus reference DNAs, each detected with different fluorescent dyes. In this example, dye-swap experiments are shown in blue and pink, respectively, and the mean fluorescent ratios (patient versus reference) reveal genomic gains involving chromosomes 7 and 8 indicated by the red double-ended arrow. Reproduced with permission from Ballabio et al.[95]

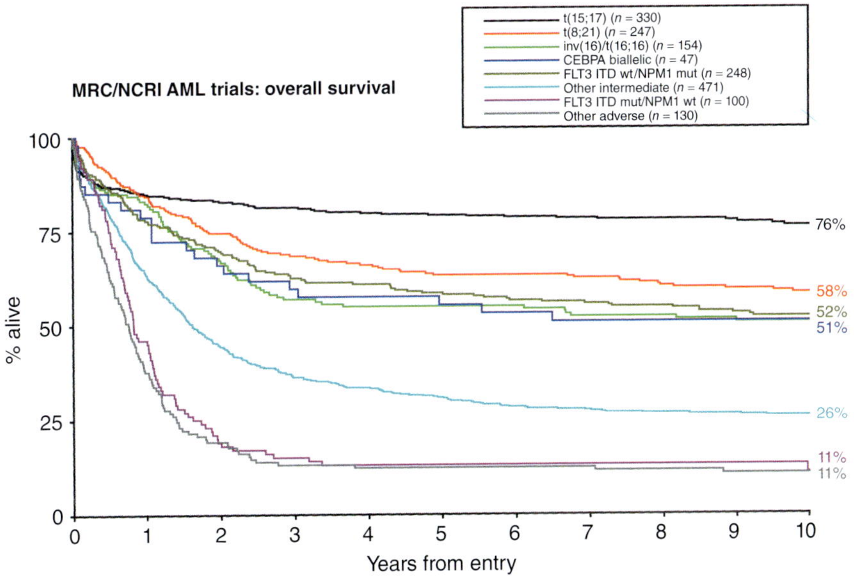

Figure 3.11 Kaplan–Meier plot of overall survival for patients aged under 60 years entered into recent MRC/NCRI clinical studies divided according to karyotype and molecular data. From Smith et al. Reproduced with permission from Elsevier.

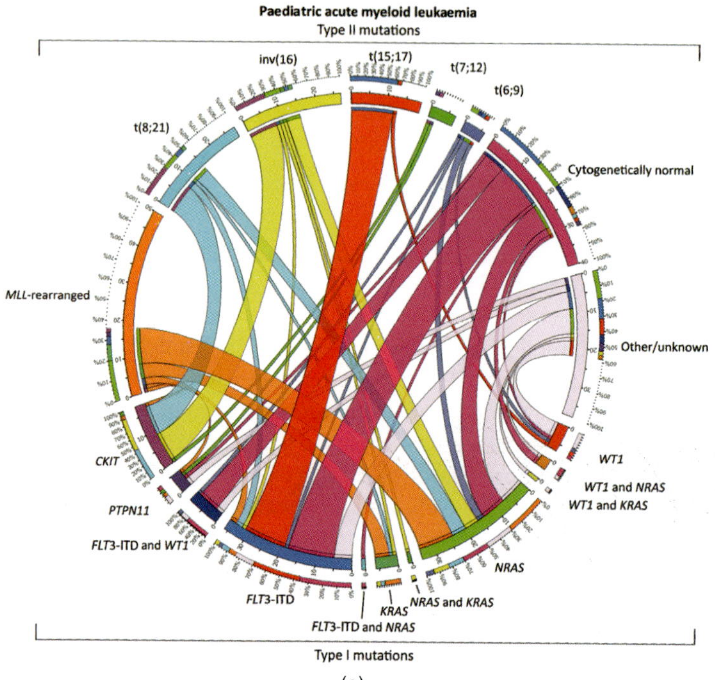

Figure 4.2 Distribution type I/II abnormalities in paediatric AML. (a) The circos plot depicts the frequency of the type II mutations and co-occurrence of type I mutations in patients with *de novo* paediatric AML. The length of the arch corresponds to the frequency of the type II mutation and the width of the ribbon with the percentage of patients with a specific type I mutation or combination of type I mutations. *FLT3*-ITD denotes *FLT3* internal tandem duplication. (b) The circos plot depicts the frequency of the type II mutations and co-occurrence of type I mutations in patients with *de novo* paediatric cytogenetically normal AML. The length of the arch corresponds to the frequency of the type II mutation and the width of the ribbon with the percentage of patients with a specific type I mutation or combination of type I mutations. *FLT3*-ITD denotes *FLT3* internal tandem duplication.

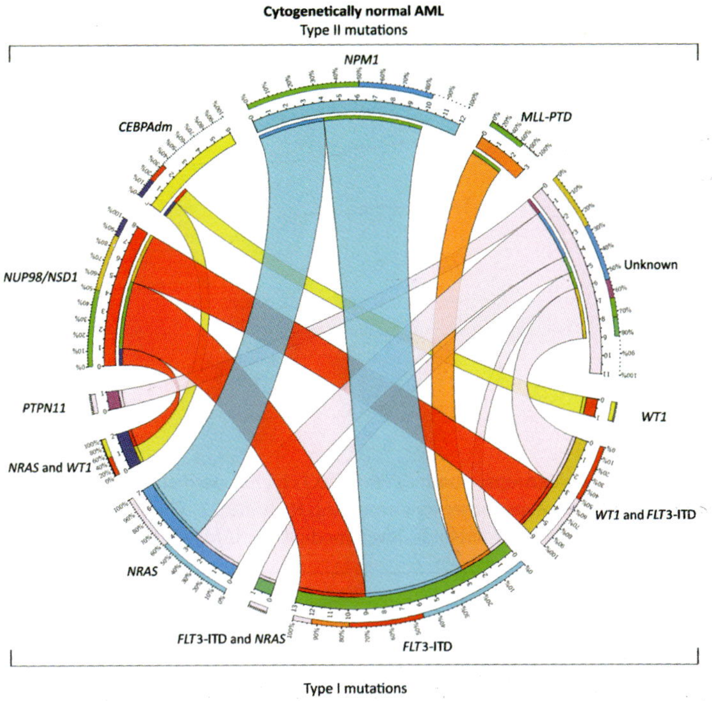

(b)

Figure 4.2 (*continued*)

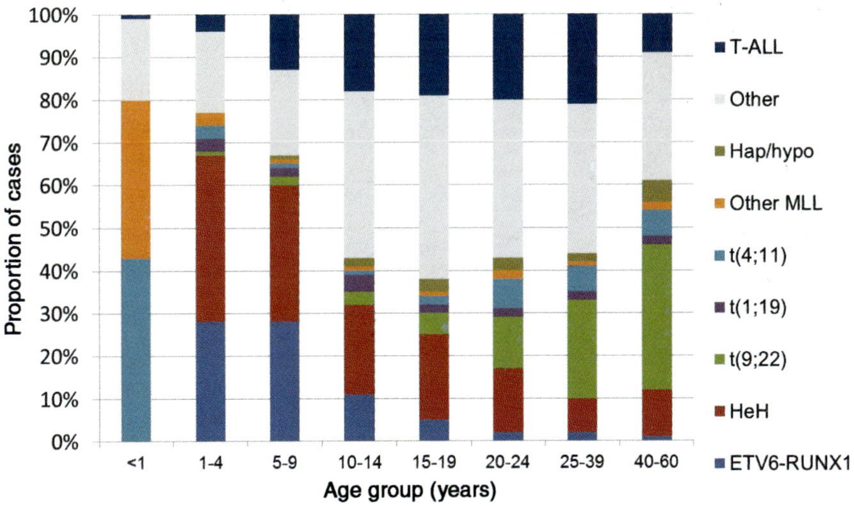

Figure 5.2 Distribution of the most common chromosomal abnormalities according to age. The abnormalities are colour coded according to the key on the right. Other, other chromosomal abnormalities; Hap/hypo, hypodiploidy with less than 40 chromosomes; Other *MLL*, other rearrangements involving the *MLL* gene; t(4;11), t(4;11)(q21;q23); t(1;19), t(1;19)(q23;p13); t(9;22), t(9;22)(q34;q11); HeH, high hypodiploidy; *ETV6-RUNX1*, fusion from t(12;21)(p13;q22).

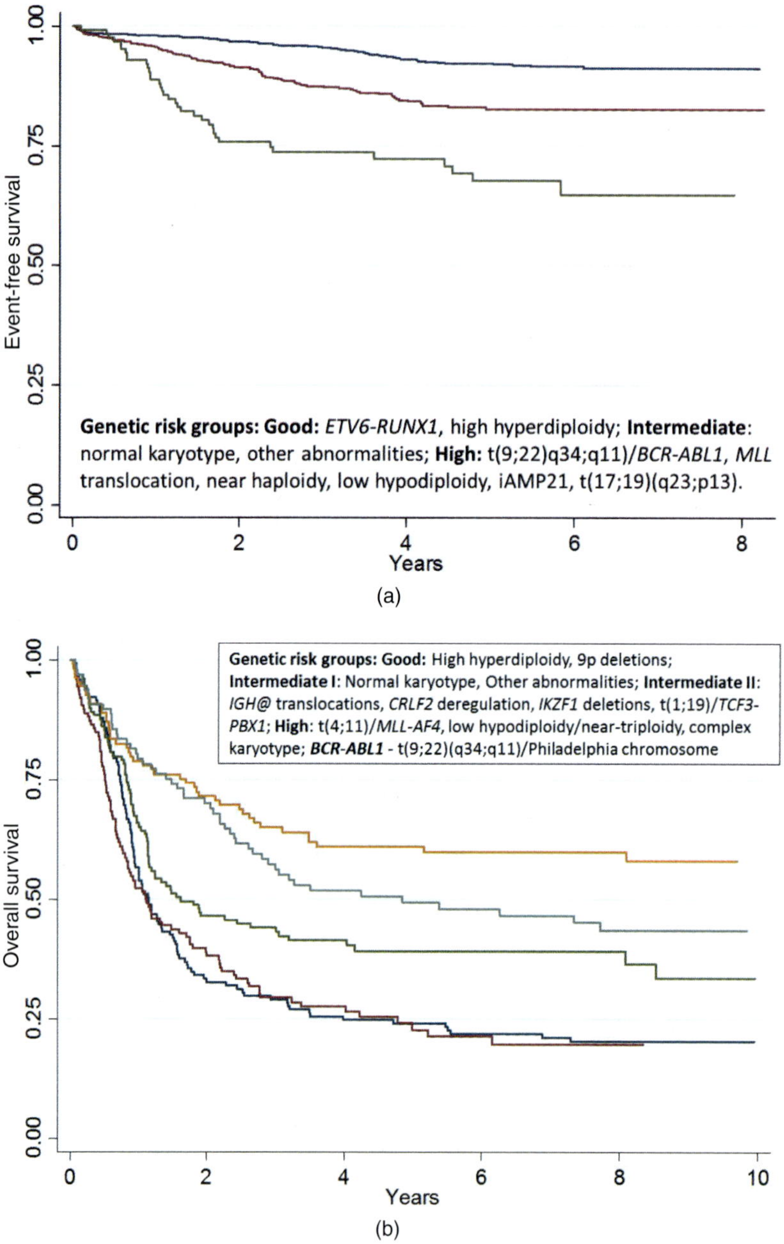

Figure 5.3 Kaplan–Meier survival curves. (a) Event-free survival of childhood BCP-ALL classified according to genetic risk group as indicated. Blue, good risk; red, intermediate risk; green, high risk. (b) Overall survival of adult BCP-ALL classified according to genetic risk group as indicated. Yellow, good risk, green, two intermediate risk groups; red, *BCR-ABL1* positive; blue, high risk.

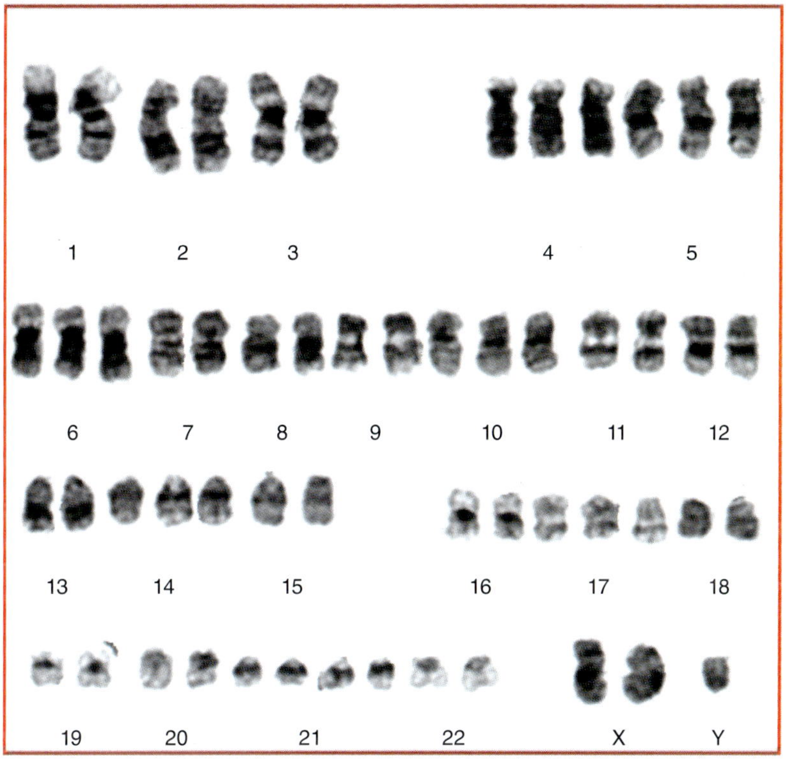

(a)

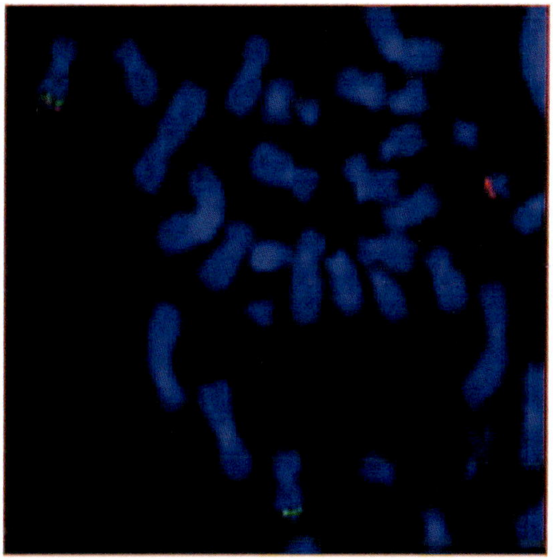

(b)

Figure 5.4 Common cytogenetic abnormalities found in BCP-ALL. (a) A high hyperdiploid karyogram. (b) FISH of *MLL* rearrangement. The normal chromosome 11 shows the closely apposed red and green signals, the abnormal chromosome 11 shows the green signal only with the red signal translocated to the abnormal chromosome 19. (c) Karyogram showing the translocation t(9;22)(q34;q11).

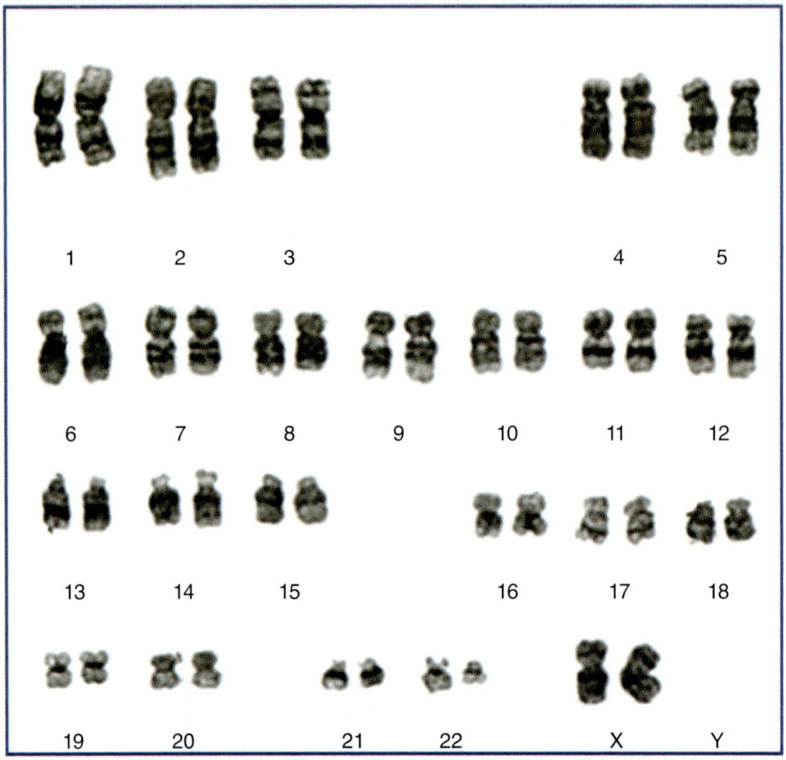

(c)

Figure 5.4 (*continued*)

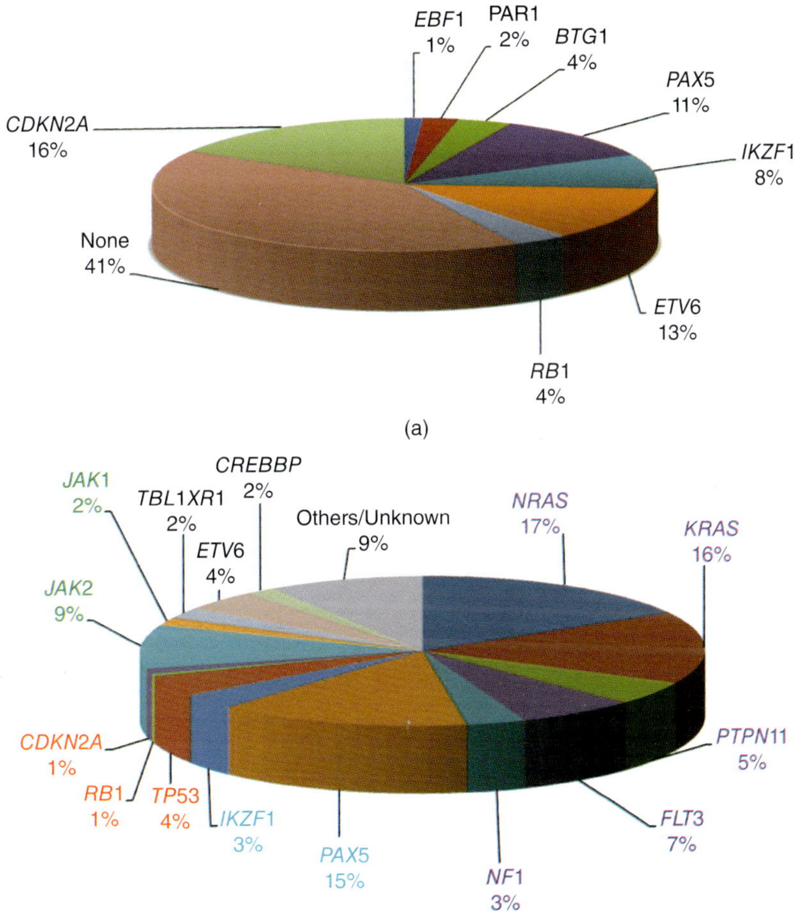

Figure 5.6 Distribution of copy number abnormalities and mutations in BCP-ALL. (a) Deletions of significant genes, PAR1, deletions within the pseudoautosomal region of the sex chromosomes. (b) Mutations in significant genes colour coded according to the signalling pathway to which they belong: purple, RAS signalling; blue, B-cell development genes; red, cell cycle control; green, JAK-STAT pathway, black, others.

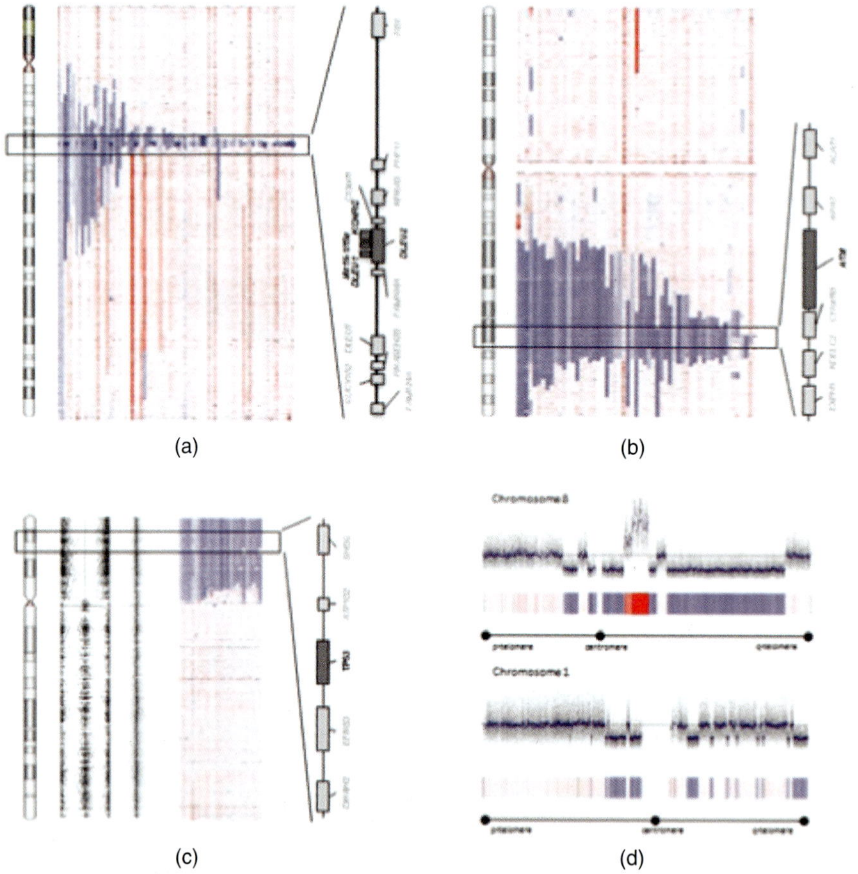

Figure 6.2 Copy-number changes in patients with CLL. (a), (b) and (c) show copy-number deletions of chromosomes 13, 11 and 17, respectively. Each chromosome runs vertically from p telomere (top), through the centromere to the q telomere (bottom) An idiogram is shown on the left, with a copy number heatmap of multiple patients, where blue, white and red show deletions, normal copy number and duplications, respectively. The location of each MDR is highlighted with an expanded view of the genes within these regions. For chromosome 17 an allelic ratio (left) and copy number (right) profile is also shown for a *TP53* mutated patient with copy number neutral LOH. (d) Two examples of chromothrypsis targeting chromosomes 8 (top) and 1 (bottom). The chromosome is positioned horizontally, running from p telomere (left) to q telomere (right). For each example, a copy-number profile and a heat map are shown.

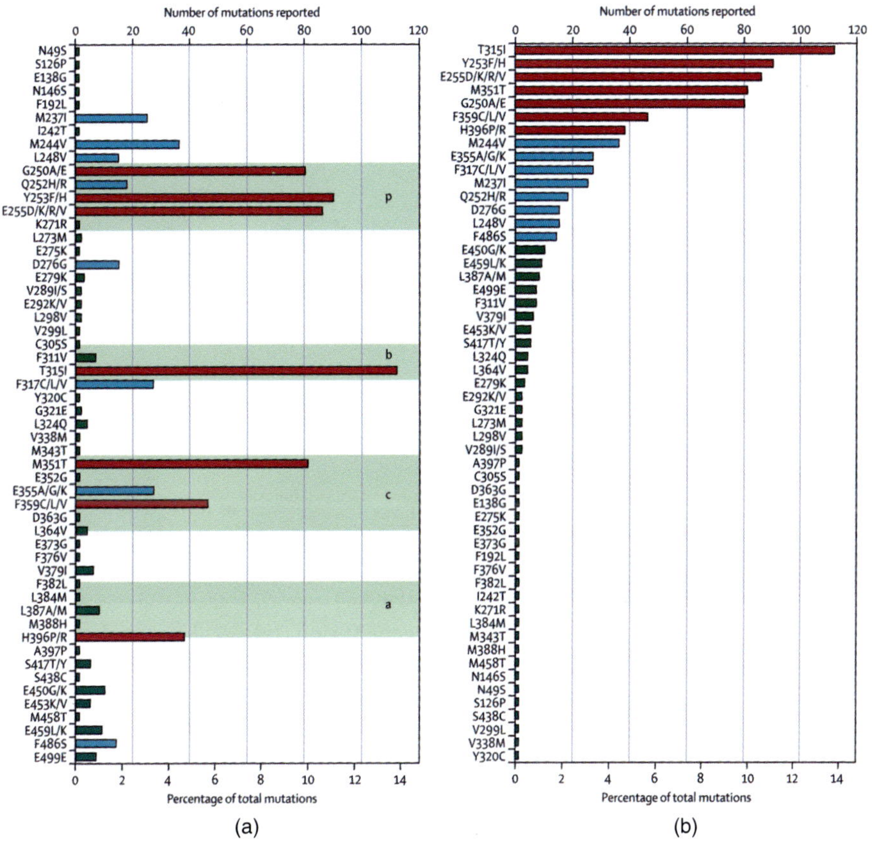

Figure 7.4 Relative frequency of different *BCR-ABL1* point mutations. Note that the most common 15 substitutions account for over 85% of cases. Reproduced from Apperley, with permission.

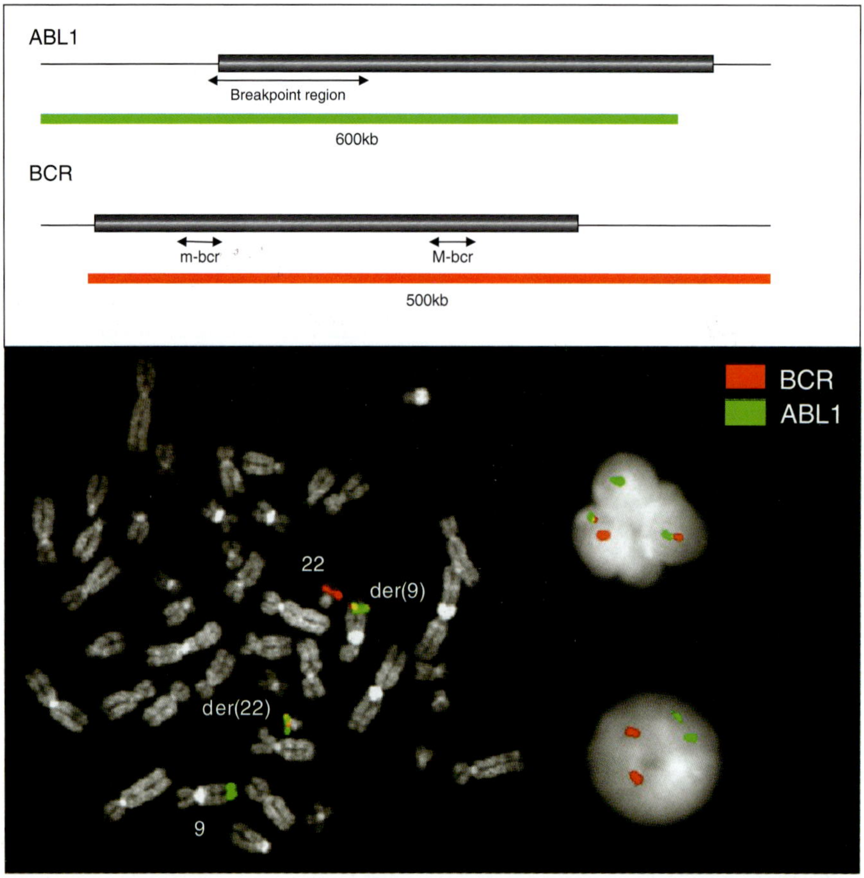

Figure 7.5 FISH detection of the BCR-ABL1 gene fusion. (a) Coverage of the probes used in a standard dual-colour, dual-fusion fluorescence *in situ* hybridization (FISH) probe system with regard to the *ABL1* and *BCR* loci. At the *ABL1* locus the probe typically consists of a single large red-labelled contig of at least 3–400 kb, spanning the *ABL* breakpoint region, covering the majority of the gene and extending beyond ABL in a 5′ (centromeric) direction. At the *BCR* locus, the probe is represented by a second large contig specific for the majority of the *BCR* gene, usually extending beyond the 3′ (telomeric) end of the gene. The BCR component is designed to span all common breakpoints, with both probes producing two hybridization signals of roughly equal size when split by a BCR-ABL1 rearrangement. Note that the colour scheme may differ between manufacturers. (b) Application of a dual-fusion *BCR-ABL1* FISH probe to a bone marrow metaphase and interphase cells from a patient with CML. Single green and red signals signal mark the unrearranged ABL1 and BCR genes, respectively. Fused red–green doublets are present on the der(9) and der(22), marking the presence of ABL1-BCR and BCR-ABL1 genes, respectively. The same hybridization pattern of one red, one green and two fusion signals (1R1G2F) is apparent on interphase cells (LHS, upper cell) and is readily distinguishable from a normal cell showing two red and two green signals (2R2G) only (LHS, lower cell).

Table 4.3 Genetically defined prognostic groups in paediatric AML as proposed by Creutzig et al.[21].

Prognosis	Genetics
Favourable	t(8;21)(q22;q22)/*RUNX1-RUNX1T1*
	inv(16)(p13.1q22) or t(16;16)(p13.1;q22)/*CBFB-MYH11*
	t(15;17)(q22;q21)/*PML-RARA**
	Molecular (in CN-AML)
	NPM1-mutated AML
	CEBPA double mutation
	t(1;11)(q21;q23)/*MLL-MLLT11(AF1Q)*
	GATA1s†
Intermediate‡	Cytogenetic abnormalities not classified as favourable or adverse§
Adverse	−7,¶ −5 or del(5q)
	inv(3)(q21q26.2) or
	t(3;3)(q21;q26.2)/*RPN1-MECOM(EVI1-MDS1-EAP)*
	t(6;9)(p23;q34)/*DEK-NUP214*
	t(7;12)(q36;p13)/*ETV6(TEL)-HLXB9(MNX1)*
	t(4;11)(q21;q23)/*MLL-MLLT2(AF4)*
	t(6;11)(q27;q23)/*MLL-MLLT4(AF6)*
	t(5;11)(q35;p15.5)/*NUP98-NSD1*
	t(10;11)(p12;q23)/*MLL-MLLT10(AF10)*#
	Complex karyotype**
	*WT1*mut/*FLT3*-ITD††
	t(9;22)(q34;q11.2)‡‡

Frequencies, response rates and outcome measures should be reported by genetic group and, if sufficient numbers are available, by specific subsets indicated.

*t(15;17)/PML-RARA is treated separately from other AMLs.

†In particular in DS patients and infants with acute megakaryoblastic leukaemia, analysis of GATA1s mutations should be included. Identification of GATA1s-associated leukaemia in trisomy 21 mosaicism can prevent over-treatment.

‡Includes all AMLs with normal karyotype, except for those included in the favourable subgroup; most of these cases are associated with poor prognosis, but they should be reported separately as they may respond differently to treatment.

§For most abnormalities, adequate numbers have not been studied to draw firm conclusions on their prognostic significance.

¶Excluding recurrent genetic aberrations, as defined in the WHO 2008 classification.

#Results in t(10;11)(p12;q23) are heterogeneous; therefore, intermediate prognosis may also be adequate.

**Three or more chromosome abnormalities in the absence of one of the WHO-designated recurring translocations or inversions.

††There are differences in the risk allocation of *FLT3*-ITD considering the allelic ratio.

‡‡t(9;22) is rare, but it is included because its poor prognostic impact is known.

an overview of genetically defined prognostic groups in paediatric AML as proposed by Creutzig et al.,[21] based on large studies in Europe and North America.

Outcome in *MLL*-rearranged AML is variable and depends on the translocation partner. For example, the *MLL* translocation t(1;11)(q21;q23) is associated with very favourable outcome in paediatric AML. However, the translocations t(6;11)(q27;q23) and t(10;11)(p12;q23) confer a poor prognosis.[58,59] Complex karyotype (more than three chromosomal abnormalities, excluding recurrent changes) is associated with poor outcome in adults, but data in children are inconsistent, due in part to differences in definition and because it is rare in children.[45,46] Monosomy 7 is a well-known poor prognostic factor and confers worse outcome than del(7q) cases.[60] Abnormalities such as the monosomal karyotype and monosomy 5 or del(5q) are rare in paediatric AML.[61]

The clinical outcome of CN-AML is highly dependent on the presence of single-gene mutations or cryptic translocations. Of special interest are *NPM1* and biallelic *CEBPA* mutations, conferring favourable prognosis. A high allelic ratio of *FLT3*-ITD versus wild type is considered an adverse prognostic factor. Other mutations, i.e. of *WT1* and *NPM1* and the cryptic translocation t(5;11)(q35;p15.5)/*NUP98-NSD1*, can modify the prognostic relevance of *FLT3*-ITD.[53,62,63]

Epigenetic modifiers and hydroxymethylation pathway mutations

Several genetic abnormalities found in AML affect histone modification or DNA methylation (e.g. *IDH1/2, TET2, DNMT3A, MLL*), which suggests that epigenetic changes also contribute to leukemogenesis.[64–67] Epigenetic profiling was able to distinguish cytogenetic subtypes of adult AML.[68] So far, the frequency of epigenetic regulator mutations in paediatric AML has been remarkably low,[69–74] but differences in promoter hypermethylation of selected genes between paediatric and adult AML warrant the profiling of DNA methylation in paediatric AML.[75] Of note, these studies may point out subsets of patients eligible for treatment with demethylating agents or histone modification inhibitors, as was shown for paediatric ALL.[76]

Differences in microRNA expression levels can classify several types of cancer.[77] Profiling studies in adult AML have shown that variations in microRNA expression patterns are associated with subtypes of AML and that specific microRNAs target genes of interest for the biology of AML.[78–80] In paediatric AML, microRNA expression patterns also vary among subtypes of AML, although some differences in expression patterns of specific microRNAs were observed between children and adults.[81]

Further strategies

Further genomic approaches to unravelling the biology of paediatric AML

In order to provide more insight into the heterogeneity and biology of AML, genome-wide approaches have recently been employed, although the success rate is variable. Array-based comparative genomic hybridization (array-CGH) and single nucleotide polymorphism (SNP) arrays identified several regions of loss-of-heterozygosity and recurrent copy number aberrations (CNAs), albeit with low frequency in AML.[82] These CNAs included aberrations in *WT1*, *NF1* and *TET2*, the last being more common in adults than in children.[38,67,72,83] Gene expression profiling could predict the cytogenetic subtypes of AML with high accuracy and identified a novel subtype characterized by mutation or epigenetic silencing of *CEBPA*, although its value for diagnostic purposes remains limited.[37,84–87] However, novel genes involved in the pathogenesis of AML subtypes were identified using this method, such as *BRE* and *IGSF4*.[88,89] High-throughput sequencing led to the identification of recurrent mutations in *IDH1*, *IDH2* and *DNMT3A* in adult AML, although these are rare events in children.[64,65,69,71,73,74] Whole exome sequencing revealed a different spectrum of gene mutations in paediatric AML compared with adult AML.[90] Novel techniques such as whole exome (DNA) and RNA sequencing will allow further elucidation of novel relevant molecular markers that may be targeted for therapy in the future.

In addition to discovering novel gene mutations, next-generation sequencing has also proved to be a powerful tool in the study of clonal evolution in both adult and paediatric AML.[91,92] By comparing the mutation spectrum of diagnosis-relapse pairs, it was shown that the founding clone gained novel mutations and evolved into the relapse clone. Moreover, minor subclones present at diagnosis can survive chemotherapy, gain mutations and present as dominant clones at relapse. Such data were already available using a candidate-gene approach in paediatric

AML.[93] Therapeutic targeting of newly identified mutations to prevent relapse may provide improved outcomes for patients.

Next-generation sequencing can also be applied in order to discover mutations involved in the progression from predisposing conditions to AML. For example, comparison of serial samples from a severe congenital neutropenia (SCN) patient who later progressed to AML revealed 12 acquired mutations during disease progression. The sequential gain of two *CSF3R* mutations in this case of SCN implicated the *CSF3R* gene in leukaemic transformation.[94]

Studies to determine the value of next-generation sequencing techniques in AML diagnostics are ongoing.[95,96] As the number of gene mutations is constantly increasing, it will be relevant to translate this knowledge into prognostically relevant gene panels that may be sequenced in a massive parallel way, providing information to guide therapy.

Molecularly targeted therapy

New therapeutic approaches that are more tumour specific and cause less severe side effects are urgently needed. Some new compounds directed at specific molecular targets have already been investigated in early clinical trials in paediatric AML. Tyrosine kinase inhibitors directed at inhibiting the constitutive activation of the *FLT3* gene are among the best studied drugs in this respect in paediatric AML and include trials using PKC412, CEP701, AC220 and sorafenib.[97,98] Of these drugs, AC220 is the most potent and selective inhibitor.[97] Recent data suggest a potentially generic mechanism of drug resistance when combining these inhibitors with chemotherapy due to FLT3-ligand upregulation, which questions their use in this fashion.[99] In addition, it is debated whether FLT3 mutations are tumour driving, but the occurrence of secondary FLT3 mutations in patients treated with inhibitors may imply tumour dependency.[100]

Other potential targets in AML consist of *KIT* and *RAS* gene mutations. *KIT* mutations include the imatinib-resistant *D816V/Y* mutation, which is, however, sensitive to dasatinib.[42,101] There is an ongoing trial in adults using dasatinib together with chemotherapy in CBF-AML. No trials have yet been reported using small-molecule RAS-pathway inhibitors, e.g. MEK inhibitors, after studies using farnesyl transferase inhibitors failed to show benefit in older patients with AML.[102]

Recent data show that inhibition of Aurora kinase B may be beneficial in paediatric AML and initial data on the selective Aurora kinase

B inhibitor AZD1152 warrant further study.[103] Other potential targets comprise Polo-like kinase inhibitors and Pim kinase inhibitors.[104,105] In *MLL*-rearranged AML, efforts are directed at developing targeted therapy, for instance by inhibiting DOT1L, which is part of the MLL complex.[39] A first-in-man clinical trial with a DOT1L inhibitor, EPZ-5676, is currently being conducted (NCT01684150).

Conclusion

The heterogeneity of AML is illustrated by the various prognostically relevant molecular and cytogenetic aberrations that have been discovered in recent years. However, many cooperating events in leukaemogenesis still remain unknown. The application of new techniques, especially next-generation sequencing, will contribute to our understanding of the genetic landscape of AML and allow the development of targeted therapy in the near future. To achieve such goals for a rare disease such as paediatric AML, international collaboration is crucial.

References

1 Howlader, N., Noone, A.M., Krapcho, et al. *SEER Cancer Statistics Review, 1975–2009 (Vintage 2009 Populations)*. Bethesda, MD: National Cancer Institute, 2012, http://seer.cancer.gov/csr/1975_2009_pops09/, based on November 2011 SEER data submission, posted to the SEER web site April 2012 (accessed 4 November 2015).

2 Sandler, E.S., Friedman, D.J., Mustafa, M.M., Winick, N.J., Bowman, W.P., Buchanan, G.R. Treatment of children with epipodophyllotoxin-induced secondary acute myeloid leukemia. *Cancer* 1997; 79: 1049–1054.

3 Weiss, B., Vora, A., Huberty, J., Hawkins, R.A., Matthay, K.K. Secondary myelodysplastic syndrome and leukemia following ^{131}I-metaiodobenzylguanidine therapy for relapsed neuroblastoma. *J Pediatr Hematol Oncol* 2003; 25: 543–547.

4 Smith, M.T., Zhang, L., McHale, C.M., Skibola, C.F., Rappaport, S.M. Benzene, the exposome and future investigations of leukemia etiology. *Chem Biol Interact* 2011; 192: 155–159.

5 Hsu, W.L., Preston, D.L., Soda, M., et al. The incidence of leukemia, lymphoma and multiple myeloma among atomic bomb survivors: 1950–2001. *Radiat Res* 2013; 179: 361–382.

6 Seif, A.E. Pediatric leukemia predisposition syndromes: clues to understanding leukemogenesis. *Cancer Genet* 2011; 204: 227–244.

7 Tonnies, H., Huber, S., Kuhl, J.S., Gerlach, A., Ebell, W., Neitzel, H. Clonal chromosomal aberrations in bone marrow cells of Fanconi anemia patients: gains of

the chromosomal segment 3q26q29 as an adverse risk factor. *Blood* 2003; 101: 3872–3874.

8 Aalbers, A.M., Calado, R.T., Young, N.S., et al. Telomere length and telomerase complex mutations in pediatric acute myeloid leukemia. *Leukemia* 2013; 27: 1786–1789.

9 Calado, R.T., Young, N.S. Telomere diseases. *N Engl J Med* 2009; 361: 2353–2365.

10 Calado, R.T., Regal, J.A., Hills, M., et al. Constitutional hypomorphic telomerase mutations in patients with acute myeloid leukemia. *Proc Natl Acad Sci U S A* 2009; 106: 1187–1192.

11 Aalbers, A.M., Calado, R.T., Young, N.S., et al. Telomere length and telomerase complex mutations in pediatric acute myeloid leukemia. *ASH Annu Meet Abstr* 2012; 120: 1482.

12 Zwaan, M.C., Reinhardt, D., Hitzler, J., Vyas, P. Acute leukemias in children with Down syndrome. *Pediatr Clin North Am* 2008; 55: 53–70, x.

13 Hahn, C.N., Chong, C.E., Carmichael, C.L., et al. Heritable GATA2 mutations associated with familial myelodysplastic syndrome and acute myeloid leukemia. *Nat Genet* 2011; 43: 1012–1017.

14 Link, D.C., Schuettpelz, L.G., Shen, D., et al. Identification of a novel TP53 cancer susceptibility mutation through whole-genome sequencing of a patient with therapy-related AML. *JAMA* 2011; 305: 1568–1576.

15 Owen, C., Barnett, M., Fitzgibbon, J. Familial myelodysplasia and acute myeloid leukaemia – a review. *Br J Haematol* 2008; 140: 123–132.

16 Smith, M.L., Cavenagh, J.D., Lister, T.A., Fitzgibbon, J. Mutation of CEBPA in familial acute myeloid leukemia. *N Engl J Med* 2004; 351: 2403–2407.

17 Song, W.J., Sullivan, M.G., Legare, R.D., et al. Haploinsufficiency of CBFA2 causes familial thrombocytopenia with propensity to develop acute myelogenous leukaemia. *Nat Genet* 1999; 23: 166–175.

18 Bennett, J.M., Catovsky, D., Daniel, M.T., et al. Proposals for the classification of the acute leukaemias. French–American–British (FAB) Co-operative Group. *Br J Haematol* 1976; 33: 451–458.

19 Vardiman, J.W., Thiele, J., Arber, D.A., et al. The 2008 revision of the World Health Organization (WHO) classification of myeloid neoplasms and acute leukemia: rationale and important changes. *Blood Rev* 2009; 114: 937–951.

20 Bennett, J.M., Catovsky, D., Daniel, M.T., et al. Criteria for the diagnosis of acute leukemia of megakaryocyte lineage (M7). A report of the French–American–British Cooperative Group. *Ann Intern Med* 1985; 103: 460–462.

21 Creutzig, U., van den Heuvel-Eibrink, M.M., Gibson, B., et al. Diagnosis and management of acute myeloid leukemia in children and adolescents: recommendations from an international expert panel. *Blood Rev* 2012; 120: 3187–3205.

22 Mrozek, K., Heinonen, K., Bloomfield, C.D. Clinical importance of cytogenetics in acute myeloid leukaemia. *Best Pract Res Clin Haematol* 2001; 14: 19–47.

23 Abrahamsson, J., Forestier, E., Heldrup, J., et al. Response-guided induction therapy in pediatric acute myeloid leukemia with excellent remission rate. *J Clin Oncol* 2011; 29: 310–315.

24 Creutzig, U., Zimmermann, M., Dworzak, M., et al. Study AML-BFM 2004: Improved survival in childhood acute myeloid leukemia without increased toxicity. *ASH Annu Meet Abstr* 2010; 116: 181.
25 Entz-Werle, N., Suciu, S., van der Werff ten Bosch, J., et al. Results of 58872 and 58921 trials in acute myeloblastic leukemia and relative value of chemotherapy vs allogeneic bone marrow transplantation in first complete remission: the EORTC Children Leukemia Group report. *Leukemia* 2005; 19: 2072–2081.
26 Gibson, B.E., Wheatley, K., Hann, I.M., et al. Treatment strategy and long-term results in paediatric patients treated in consecutive UK AML trials. *Leukemia* 2005; 19: 2130–2138.
27 Perel, Y., Auvrignon, A., Leblanc, T., et al. Treatment of childhood acute myeloblastic leukemia: dose intensification improves outcome and maintenance therapy is of no benefit – multicenter studies of the French LAME (Leucémie Aiguë Myéloblastique Enfant) Cooperative Group. *Leukemia* 2005; 19: 2082–2089.
28 Pession, A., Rondelli, R., Basso, G., et al. Treatment and long-term results in children with acute myeloid leukaemia treated according to the AIEOP AML protocols. *Leukemia* 2005; 19: 2043–2053.
29 Rubnitz, J.E., Inaba, H., Dahl, G., et al. Minimal residual disease-directed therapy for childhood acute myeloid leukaemia: results of the AML02 multicentre trial. *Lancet Oncol* 2010; 11: 543–552.
30 Tsukimoto, I., Tawa, A., Horibe, K., et al. Risk-stratified therapy and the intensive use of cytarabine improves the outcome in childhood acute myeloid leukemia: the AML99 trial from the Japanese Childhood AML Cooperative Study Group. *J Clin Oncol* 2009; 27: 4007–4013.
31 Niewerth, D., Creutzig, U., Bierings, M.B., Kaspers, G.J. A review on allogeneic stem cell transplantation for newly diagnosed pediatric acute myeloid leukemia. *Blood Rev* 2010; 116: 2205–2214.
32 Reinhardt, D., Kremens, B., Zimmermann, M., et al. No improvement of overall survival in children with high-risk acute myeloid leukemia by stem cell transplantation in 1st complete remission. *ASH Annu Meet Abstr* 2006; 108: 320.
33 Kaspers, G.J., Zimmermann, M., Reinhardt, D., et al. Improved outcome in pediatric relapsed acute myeloid leukemia: results of a randomized trial on liposomal daunorubicin by the International BFM Study Group. *J Clin Oncol* 2013; 31: 599–607.
34 Slats, A.M., Egeler, R.M., van der Does-van den Berg, A., et al. Causes of death – other than progressive leukemia – in childhood acute lymphoblastic (ALL) and myeloid leukemia (AML): the Dutch Childhood Oncology Group experience. *Leukemia* 2005; 19: 537–544.
35 Kelly, L.M., Gilliland, D.G. Genetics of myeloid leukemias. *Annu Rev Genomics Hum Genet* 2002; 3: 179–198.
36 Ahmed, M., Sternberg, A., Hall, G., et al. Natural history of GATA1 mutations in Down syndrome. *Blood Rev* 2004; 103: 2480–2489.
37 Hollink, I.H., van den Heuvel-Eibrink, M.M., Arentsen-Peters, S.T., et al. Characterization of CEBPA mutations and promoter hypermethylation in pediatric acute myeloid leukemia. *Haematologica* 2011; 96: 384–392.

38 Hollink, I.H., van den Heuvel-Eibrink, M.M., Zimmermann, M., et al. Clinical relevance of Wilms tumor 1 gene mutations in childhood acute myeloid leukemia. *Blood Rev* 2009; 113: 5951–5960.

39 Bernt, K.M., Zhu, N., Sinha, A.U., et al. MLL-rearranged leukemia is dependent on aberrant H3K79 methylation by DOT1L. *Cancer Cell* 2011; 20: 66–78.

40 Marschalek, R. Mechanisms of leukemogenesis by MLL fusion proteins. *Br J Haematol* 2011; 152: 141–154.

41 McLean, C.M., Karemaker, I.D., van Leeuwen, F. The emerging roles of DOT1L in leukemia and normal development. *Leukemia* 2014; 28: 2131–2138.

42 Balgobind, B.V., Hollink, I.H., Arentsen-Peters, S.T., et al. Integrative analysis of type-I and type-II aberrations underscores the genetic heterogeneity of pediatric acute myeloid leukemia. *Haematologica* 2011; 96: 1478–1487.

43 Betts, D.R., Ammann, R.A., Hirt, A., et al. The prognostic significance of cytogenetic aberrations in childhood acute myeloid leukaemia. A study of the Swiss Paediatric Oncology Group (SPOG). *Eur J Haematol* 2007; 78: 468–476.

44 Grimwade, D. The clinical significance of cytogenetic abnormalities in acute myeloid leukaemia. *Best Pract Res Clin Haematol* 2001; 14: 497–529.

45 Harrison, C.J., Hills, R.K., Moorman, A.V., et al. Cytogenetics of childhood acute myeloid leukemia: United Kingdom Medical Research Council Treatment Trials AML 10 and 12. *J Clin Oncol* 2010; 28: 2674–2681.

46 von Neuhoff, C., Reinhardt, D., Sander, A., et al. Prognostic impact of specific chromosomal aberrations in a large group of pediatric patients with acute myeloid leukemia treated uniformly according to trial AML-BFM 98. *J Clin Oncol* 2010; 28: 2682–2689.

47 Park, J., Kim, M., Lim, J., et al. Three-way complex translocations in infant acute myeloid leukemia with t(7;12)(q36;p13): the incidence and correlation of a HLXB9 overexpression. *Cancer Genet Cytogenet* 2009; 191: 102–105.

48 Simmons, H.M., Oseth, L., Nguyen, P., O'Leary, M., Conklin, K.F., Hirsch, B. Cytogenetic and molecular heterogeneity of 7q36/12p13 rearrangements in childhood AML. *Leukemia* 2002; 16: 2408–2416.

49 Slater, R.M., von Drunen, E., Kroes, W.G., et al. t(7;12)(q36;p13) and t(7;12)(q32;p13) – translocations involving ETV6 in children 18 months of age or younger with myeloid disorders. *Leukemia* 2001; 15: 915–920.

50 von Bergh, A.R., van Drunen, E., van Wering, E.R., et al. High incidence of t(7;12)(q36;p13) in infant AML but not in infant ALL, with a dismal outcome and ectopic expression of HLXB9. *Genes Chromosomes Cancer* 2006; 45: 731–739.

51 Mercher, T., Busson-Le Coniat, M., Nguyen Khac, F., et al. Recurrence of OTT-MAL fusion in t(1;22) of infant AML-M7. *Genes Chromosomes Cancer* 2002; 33: 22–28.

52 Torres, L., Lisboa, S., Vieira, J., et al. Acute megakaryoblastic leukemia with a four-way variant translocation originating the RBM15-MKL1 fusion gene. *Pediatr Blood Rev Cancer* 2011; 56: 846–849.

53 Hollink, I.H., van den Heuvel-Eibrink, M.M., Arentsen-Peters, S.T., et al. NUP98/NSD1 characterizes a novel poor prognostic group in acute myeloid leukemia with a distinct HOX gene expression pattern. *Blood Rev* 2011; 118: 3645–3656.

54 Tosi, S., Harbott, J., Teigler-Schlegel, A., et al. t(7;12)(q36;p13), a new recurrent translocation involving *ETV6* in infant leukaemia. *Genes Chromosomes Cancer* 2000; 29: 325–332.
55 Balgobind, B.V., Zwaan, C.M., Pieters, R., van den Heuvel-Eibrink, M.M. The heterogeneity of pediatric MLL-rearranged acute myeloid leukemia. *Leukemia* 2011; 25: 1239–1248.
56 Marcucci, G., Haferlach, T., Dohner, H. Molecular genetics of adult acute myeloid leukemia: prognostic and therapeutic implications. *J Clin Oncol* 2011; 29: 475–486.
57 Hermkens, M.C., van den Heuvel-Eibrink, M.M., Arentsen-Peters, S.T., et al. The clinical relevance of BAALC and ERG expression levels in pediatric AML. *Leukemia* 2013; 27: 735–737.
58 Balgobind, B.V., Raimondi, S.C., Harbott, J., et al. Novel prognostic subgroups in childhood 11q23/MLL-rearranged acute myeloid leukemia: results of an international retrospective study. *Blood Rev* 2009; 114: 2489–2496.
59 Coenen, E.A., Raimondi, S.C., Harbott, J., et al. Prognostic significance of additional cytogenetic aberrations in 733 de novo pediatric 11q23/MLL-rearranged AML patients: results of an international study. *Blood Rev* 2011; 117: 7102–7111.
60 Hasle, H., Alonzo, T.A., Auvrignon, A., et al. Monosomy 7 and deletion 7q in children and adolescents with acute myeloid leukemia: an international retrospective study. *Blood Rev* 2007; 109: 4641–4647.
61 Breems, D.A., Van Putten, W.L., De Greef, G.E., et al. Monosomal karyotype in acute myeloid leukemia: a better indicator of poor prognosis than a complex karyotype. *J Clin Oncol* 2008; 26: 4791–4797.
62 Hollink, I.H., Zwaan, C.M., Zimmermann, M., et al. Favorable prognostic impact of NPM1 gene mutations in childhood acute myeloid leukemia, with emphasis on cytogenetically normal AML. *Leukemia* 2009; 23: 262–270.
63 Noronha, S.A., Farrar, J.E., Alonzo, T.A., et al. WT1 expression at diagnosis does not predict survival in pediatric AML: a report from the Children's Oncology Group. *Pediatr Blood Rev Cancer* 2009; 53: 1136–1139.
64 Ley, T.J., Ding, L., Walter, M.J., et al. DNMT3A mutations in acute myeloid leukemia. *N Engl J Med* 2010; 363: 2424–2433.
65 Mardis, E.R., Ding, L., Dooling, D.J., et al. Recurring mutations found by sequencing an acute myeloid leukemia genome. *N Engl J Med* 2009; 361: 1058–1066.
66 Figueroa, M.E., Abdel-Wahab, O., Lu, C., et al. Leukemic IDH1 and IDH2 mutations result in a hypermethylation phenotype, disrupt TET2 function and impair hematopoietic differentiation. *Cancer Cell* 2011; 18: 553–567.
67 Delhommeau, F., Dupont, S., Della Valle, V., et al. Mutation in TET2 in myeloid cancers. *N Engl J Med* 2009; 360: 2289–2301.
68 Figueroa, M.E., Lugthart, S., Li, Y., et al. DNA methylation signatures identify biologically distinct subtypes in acute myeloid leukemia. *Cancer Cell* 2010; 17: 13–27.
69 Valerio, D.G., Katsman-Kuipers, J.E., Jansen, J.H., et al. Mapping epigenetic regulator gene mutations in cytogenetically normal pediatric acute myeloid leukemia. *Haematologica* 2014; 99: e130–e132.

70 Damm, F., Bunke, T., Thol, F., et al. Prognostic implications and molecular associations of NADH dehydrogenase subunit 4 (ND4) mutations in acute myeloid leukemia. *Leukemia* 2012; 26: 289–295.
71 Hollink, I.H., Feng, Q., Danen-van Oorschot, A.A., et al. Low frequency of DNMT3A mutations in pediatric AML and the identification of the OCI-AML3 cell line as an in vitro model. *Leukemia* 2011; 26: 371–373.
72 Langemeijer, S.M., Jansen, J.H., Hooijer, J., et al. TET2 mutations in childhood leukemia. *Leukemia* 2011; 25: 189–192.
73 Ho, P.A., Alonzo, T.A., Kopecky, K.J., et al. Molecular alterations of the IDH1 gene in AML: a Children's Oncology Group and Southwest Oncology Group study. *Leukemia* 2010; 24: 909–913.
74 Yan, X.J., Xu, J., Gu, Z.H., et al. Exome sequencing identifies somatic mutations of DNA methyltransferase gene DNMT3A in acute monocytic leukemia. *Nat Genet* 2011; 43: 309–315.
75 Juhl-Christensen, C., Ommen, H.B., Aggerholm, A., et al. Genetic and epigenetic similarities and differences between childhood and adult AML. *Pediatr Blood Rev Cancer* 2012; 58: 525–531.
76 Stumpel, D.J., Schneider, P., van Roon, E.H., et al. Specific promoter methylation identifies different subgroups of MLL-rearranged infant acute lymphoblastic leukemia, influences clinical outcome and provides therapeutic options. *Blood Rev* 2009; 114: 5490–5498.
77 Lu, J., Getz, G., Miska, E.A., et al. MicroRNA expression profiles classify human cancers. *Nature* 2005; 435: 834–838.
78 Debernardi, S., Skoulakis, S., Molloy, G., Chaplin, T., Dixon-McIver, A., Young, B.D. MicroRNA miR-181a correlates with morphological sub-class of acute myeloid leukaemia and the expression of its target genes in global genome-wide analysis. *Leukemia* 2007; 21: 912–916.
79 Garzon, R., Volinia, S., Liu, C.G., et al. MicroRNA signatures associated with cytogenetics and prognosis in acute myeloid leukemia. *Blood Rev* 2008; 111: 3183–3189.
80 Jongen-Lavrencic, M., Sun, S.M., Dijkstra, M.K., Valk, P.J., Lowenberg, B. MicroRNA expression profiling in relation to the genetic heterogeneity of acute myeloid leukemia. *Blood Rev* 2008; 111: 5078–5085.
81 Danen-van Oorschot, A.A., Kuipers, J.E., Arentsen-Peters, S., et al. Differentially expressed miRNAs in cytogenetic and molecular subtypes of pediatric acute myeloid leukemia. *Pediatr Blood Rev Cancer* 2011; 58: 715–721.
82 Raghavan, M., Lillington, D.M., Skoulakis, S., et al. Genome-wide single nucleotide polymorphism analysis reveals frequent partial uniparental disomy due to somatic recombination in acute myeloid leukemias. *Cancer Res* 2005; 65: 375–378.
83 Balgobind, B.V., van Vlierberghe, P., van den Ouweland, A.M., et al. Leukemia-associated NF1 inactivation in patients with pediatric T-ALL and AML lacking evidence for neurofibromatosis. *Blood Rev* 2008; 111: 4322–4328.
84 Balgobind, B.V., van den Heuvel-Eibrink, M.M., de Menezes, R.X., et al. Evaluation of gene expression signatures predictive of cytogenetic and molecular subtypes of pediatric acute myeloid leukemia. *Haematologica* 2011; 96: 221–230.

85 Ross, M.E., Mahfouz, R., Onciu, M., et al. Gene expression profiling of pediatric acute myelogenous leukemia. *Blood Rev* 2004; 104: 3679–3687.
86 Valk, P.J., Verhaak, R.G., Beijen, M.A., et al. Prognostically useful gene-expression profiles in acute myeloid leukemia. *N Engl J Med* 2004; 350: 1617–1628.
87 Wouters, B.J., Jorda, M.A., Keeshan, K., et al. Distinct gene expression profiles of acute myeloid/T-lymphoid leukemia with silenced CEBPA and mutations in NOTCH1. *Blood Rev* 2007; 110: 3706–3714.
88 Balgobind, B.V., Zwaan, C.M., Reinhardt, D., et al. High BRE expression in pediatric MLL-rearranged AML is associated with favorable outcome. *Leukemia* 2010; 24: 2048–2055.
89 Kuipers, J.E., Coenen, E.A., Balgobind, B.V., et al. High IGSF4 expression in pediatric M5 acute myeloid leukemia with t(9;11)(p22;q23). *Blood Rev* 2011; 117: 928–935.
90 Shiba, N., Yoshida, K., Okuno, Y., et al. Whole exome sequencing reveals spectrum of gene mutations in pediatric AML. *ASH Annu Meet Abstr* 2012; 120: 124.
91 Ding, L., Ley, T.J., Larson, D.E., et al. Clonal evolution in relapsed acute myeloid leukaemia revealed by whole-genome sequencing. *Nature* 2012; 481: 506–510.
92 Meshinchi, S., Ries, R.E., Trevino, L.R., et al. Identification of novel somatic mutations, regions of recurrent loss of heterozygosity (LOH) and significant clonal evolution from diagnosis to relapse in childhood AML determined by exome capture sequencing – an NCI/COG Target AML study. *ASH Annu Meet Abstr* 2012; 120: 123.
93 Bachas, C., Schuurhuis, G.J., Hollink, I.H., et al. High-frequency type I/II mutational shifts between diagnosis and relapse are associated with outcome in pediatric AML: implications for personalized medicine. *Blood Rev* 2010; 116: 2752–2758.
94 Beekman, R., Valkhof, M.G., Sanders, M.A., et al. Sequential gain of mutations in severe congenital neutropenia progressing to acute myeloid leukemia. *Blood Rev* 2012; 119: 5071–5077.
95 Kohlmann, A., Martinelli, G., Hofmann W.-K., et al. The Interlaboratory Robustness of Next-Generation Sequencing (IRON) Study Phase II: Deep-sequencing analyses of hematological malignancies performed by an international network involving 26 laboratories. *ASH Annu Meet Abstr* 2012; 120: 1399.
96 Kohlmann, A., Weissmann, S., Schoeck, U., et al. First results of a 31-gene panel targeted to investigate myeloid malignancies by next-generation amplicon deep-sequencing. *ASH Annu Meet Abstr* 2012; 120: 883.
97 Zarrinkar, P.P., Gunawardane, R.N., Cramer, M.D., et al. AC220 is a uniquely potent and selective inhibitor of FLT3 for the treatment of acute myeloid leukemia (AML). *Blood Rev* 2009; 114: 2984–2992.
98 Inaba, H., Rubnitz, J.E., Coustan-Smith, E., et al. Clinical activity, pharmacokinetics and pharmacodynamics of sorafenib in pediatric acute myeloid leukemia. *Blood Rev* 2010; 116: abstract 1073.
99 Levis, M., Ravandi, F., Wang, E.S., et al. Results from a randomized trial of salvage chemotherapy followed by lestaurtinib for patients with FLT3 mutant AML in first relapse. *Blood Rev* 2011; 117: 3294–3301.

100 Smith, C.C., Wang, Q., Chin, C.-S., et al. Validation of ITD mutations in FLT3 as a therapeutic target in human acute myeloid leukaemia. *Nature* 2012; 485: 260–263.

101 Goemans, B.F., Zwaan, C., Miller, M., et al. Mutations in KIT and RAS are frequent events in pediatric core-binding factor acute myeloid leukemia. *Leukemia* 2005; 19: 1536–1542.

102 Burnett, A.K., Russell, N.H., Culligan, D., et al. The addition of the farnesyl transferase inhibitor, tipifarnib, to low dose cytarabine does not improve outcome for older patients with AML. *Br J Haematol* 2012; 158: 519–522.

103 Hartsink-Segers, S.A., Zwaan, C.M., Exalto, C., et al. Aurora kinases in childhood acute leukemia: the promise of aurora B as therapeutic target. *Leukemia* 2013; 27: 560–568.

104 Chen, L.S., Redkar, S., Taverna, P., Cortes, J.E., Gandhi, V. Mechanisms of cytotoxicity to Pim kinase inhibitor, SGI-1776, in acute myeloid leukemia. *Blood Rev* 2011; 118: 693–702.

105 Berg, T., Bug, G., Ottmann, O.G., Strebhardt, K. Polo-like kinases in AML. *Expert Opin Investig Drugs* 2012; 21: 1069–1074.

106 Grimwade, D., Hills, R.K., Moorman, A.V., et al. Refinement of cytogenetic classification in acute myeloid leukemia: determination of prognostic significance of rare recurring chromosomal abnormalities among 5876 younger adult patients treated in the United Kingdom Medical Research Council trials. *Blood Rev* 2010; 116: 354–365.

107 Mrozek, K., Heerema, N.A., Bloomfield, C.D. Cytogenetics in acute leukemia. *Blood Rev* 2004; 18: 115–136.

CHAPTER 5

Acute lymphoblastic leukaemia

Anna Andersson, Anthony V. Moorman, Christine J. Harrison and Charles Mulligan

Introduction

Acute lymphoblastic leukaemia (ALL) is characterized by an expansion of immature hematopoietic cells, so-called blasts, in the bone marrow and, frequently, peripheral blood. ALL is the most common childhood malignancy, with a peak incidence at around 2–5 years of age, with the total incidence being 3–4 cases per 100,000 each year. It is rarer among adults, with an incidence of around one case per 100,000 per year. Treatment improvements have increased the cure rate to more than 80% for children and about 40% for adults, even though there are differences among the specific ALL subtypes.[1,2] ALL comprises T-cell- (T-ALL) and B-cell precursor-ALL (BCP-ALL). The discovery of the Philadelphia chromosome in chronic myeloid leukaemia (CML) in 1960[3] paved the way for the characterization of leukaemia as a genetic disease. Since then, it has become evident that chromosomal translocations serve as hallmarks of leukaemia and that they are intimately associated with specific leukaemia subtypes and frequently with prognosis. Today, specific chromosomal abnormalities are used in combination with clinical information to diagnose and risk stratify patients to clinically relevant subgroups receiving different therapies.

Chromosomal translocations are the result of the exchange of genetic material between two chromosomes, resulting in either (i) the juxtaposition of an oncogene to the vicinity of the strong promoter elements of the immunoglobulin heavy genes (*IGH*) or the T-cell receptor genes (*TCR*), leading to aberrant expression of the oncogene, or, more commonly, (ii) the disruption of two genes and subsequent rejoining of their coding sequences resulting in the creation of a chimeric fusion gene. Typically, the genes rearranged by the latter mechanism are transcription factors, which serve as master regulators of normal haematopoiesis. Thus, these

The Genetic Basis of Haematological Cancers, First Edition. Edited by Sabrina Tosi and Alistair G. Reid.

rearrangements lead to alteration or disruption of the normal genetic programmes controlled by these transcription factors.[4] However, even though chromosomal translocations are a hallmark of leukaemia, it is well accepted that additional genetic hits are needed for overt leukaemia to occur. The evidence comes from experimental animal models where most leukaemia-associated fusion genes alone fail to cause leukaemia. In addition, it has been shown that paediatric leukaemia can arise *in utero* with no evidence of overt leukaemia until several years later, again indicating the need for additional genetic hits.[5,6] Recently, with the introduction of high-resolution single-nucleotide polymorphism (SNP) arrays and next-generation sequencing (NGS) technologies, it has become clear that in addition to chromosomal rearrangements, DNA copy-number alterations (CNA) such as sub-microscopic deletions and amplifications and sequence mutations are common associated genetic events. Similarly to genes targeted by chromosomal rearrangements in acute leukaemia, those targeted by CNA commonly play important roles in key cellular pathways, including the transcriptional regulation of lymphoid development, cell cycle regulation, tumour suppression, lymphoid signalling, regulation of apoptosis and epigenetic modifications.[7,8] However, even though whole-genome sequencing studies are starting to define the genomic landscape,[9] the total complement of genetic lesions in leukaemia remains to be determined.

Chromosomal aberrations in BCP-ALL

In BCP-ALL, genetic changes serve an important role in diagnosis, whilst providing important clinical information. In about 75% of these cases, significant specific chromosomal rearrangements occur, including high hyperdiploidy (51–65 chromosomes), the translocation t(12;21)(p13;q22) (encoding *ETV6-RUNX1*, also known as *TEL-AML1*), t(1;19)(p13;q22) (*TCF3-PBX1* or *E2A-PBX1*), hypodiploidy (≤46 chromosomes), rearrangements of *MLL* at 11q23, t(9;22)(q34;q11.1)/*BCR-ABL1*, rearrangements of the immunoglobulin heavy chain (*IGH*) and intrachromosomal amplification of chromosome 21 (iAMP21), as listed in Table 5.1. Their relative distributions are shown in Fig. 5.1. This chapter considers both childhood and adult ALL. Although it is tempting to think of them as two different entities from the genetic point of view, classification of these cytogenetic abnormalities indicates a variable distribution according to age, as shown in Figure 5.2 and indicated in Table 5.1.

Table 5.1 The most frequent established chromosomal abnormalities in BCP-ALL.

Abnormality type	Abnormality	Specific aberration	Molecular genetic features	Age group
Good risk	*ETV6-RUNX1* fusion	t(12;21)(p13;q22)	*ETV6*, *RUNX1*	Young children
	High hyperdiploidy	51–65 chromosomes	Whole chromosome gains, *FLT3*, *NRAS*, *KRAS*, *PTPN11*, *PAX5* mutations	Mostly children, some adults
Poor risk	Philadelphia chromosome	t(9;22)(q34;q11)	*BCR*, *ABL1*	Mainly adults, some children
	KMT2A (MLL) rearrangement	t(4;11)(q21;q23)	*MLL*, *AFF1*	Mainly infants, some children and adults
		t(6;11)(q27;q23)	*MLLT4*, *MLL*	All
		t(9;11)(p21;q23)	*MLLT3*, *MLL*	All
		t(10;11)(p12;q23)	*MLLT10*, *MLL*	All
		t(11;19)(q23;p13.3)	*MLLT1*, *MLL*	Mainly infants, some children and adults
		t(17;19)(q22;p13)	*TCF3*, *HLF*	All
	Near-haploidy	24–29 chromosomes	Frequent doubling of chromosome number. Ras pathway and *IKZF3* mutations	Children

(*continued*)

Table 5.1 (*continued*)

Abnormality type	Abnormality	Specific aberration	Molecular genetic features	Age group
	Low hypodiploidy	30–39 chromosomes	*IKZF2* alterations, deletion *CDKN2A/B*, *RB1*, germline *TP53* mutations	Adults
	IGH translocation	t(14;19)(q32;q13)	*CEBPA*, *IGH*	Older children and adults
		t(14;20)(q32;q13)	*CEBPB*, *IGH*	Older children and adults
		t(8;14)(q11;q32)	*CEBPD*, *IGH*	Older children and adults
		inv(14)(q11q32)	*CEBPE, IGH*	Older children and adults
		t(14;14)(q11;q32)	*CEBPE, IGH*	Older children and adults
		t(6;14)(p22;q32)	*ID4*, *IGH*	Older children and adults
		t(14;19)(q32;p13)	*EPOR*, *IGH*	
		t(X;14)(p22;q32)	*CRLF2*, *IGH*	
		t(Y;14)(p11;q32).	*CRLF2*, *IGH*	
	iAMP21	Grossly abnormal chromosome 21	?*RUNX1* and other genes on chromosome 21	Older children

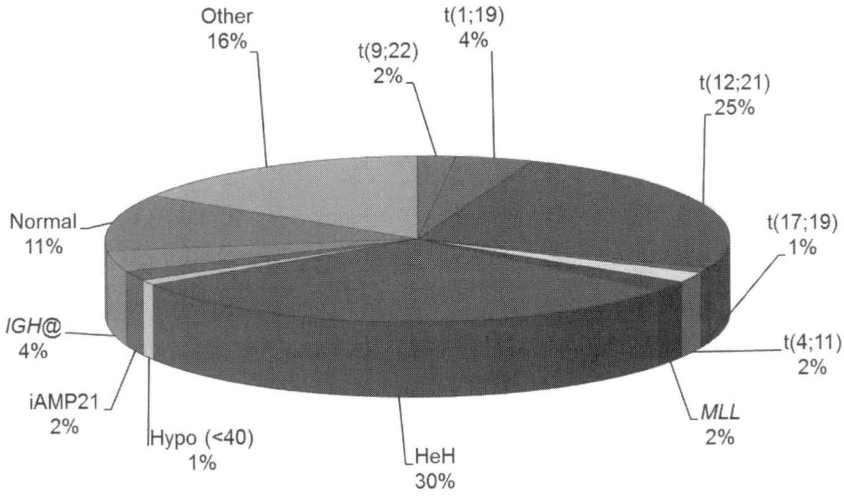

Figure 5.1 Distribution of the major established cytogenetic abnormalities in childhood BCP-ALL. t(9;22), t(9;22)(q34;q11); t(12;21), t(12;21)(p13;q22); t(17;19), t(17;19)(q22;p13); t(4;11), t(4;11)(q21;q23); *MLL*, other rearrangements involving the *MLL* gene; HeH, high hypodiploidy; Hypo (<40), hypodiploidy with less than 40 chromosomes; iAMP21, intrachromosomal amplification of chromosome 21; *IGH*, translocations involving *IGH*; Normal, normal karyotype; Other, other chromosomal abnormalities.

For example, the high prevalence of *MLL* rearrangements, particularly the translocation t(4;11)(q21;q23), in infants less than 1 year of age is evident. The dramatic decrease in high hyperdiploidy and *ETV6-RUNX1* fusion after the age of 10 years is mirrored for by an increase in the proportion of patients with the translocation t(9;22)(q34;q11) and *IGH* translocations into adulthood. iAMP21 occurs in older children and young adults.

In view of the association with prognosis, based on detailed analysis over a number of clinical treatment trials, cytogenetic subgroups can be grouped together according to their known risk group to produce simplified survival curves indicating good, intermediate and poor outcomes. The components of each risk group are indicated in Figure 5.3a for childhood and Figure 5.3b for adult ALL. Among the adults, two intermediate-risk groups have been identified with differing outcomes.

High hyperdiploidy

High hyperdiploidy (51–65 chromosomes) is characterized by a non-random gain of chromosomes, X, 4, 6, 8, 10, 14, 17 and 18,

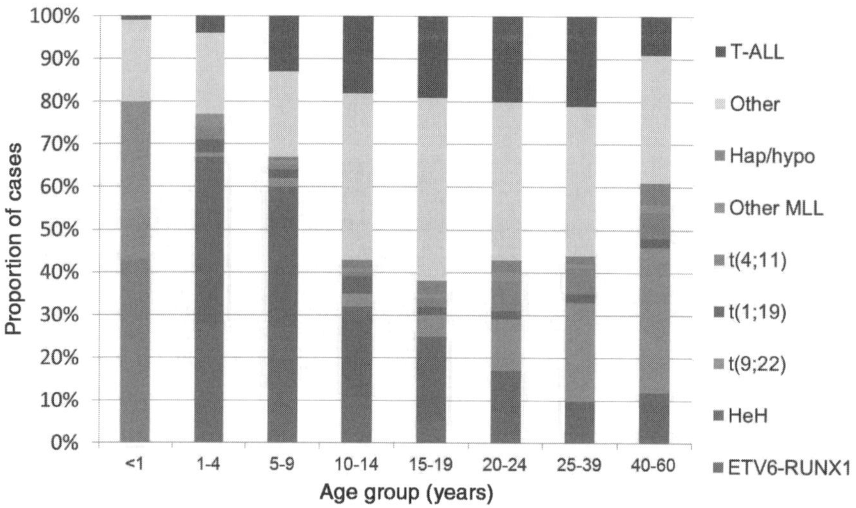

Figure 5.2 Distribution of the most common chromosomal abnormalities according to age. The abnormalities are colour coded according to the key on the right. Other, other chromosomal abnormalities; Hap/hypo, hypodiploidy with less than 40 chromosomes; Other *MLL*, other rearrangements involving the *MLL* gene; t(4;11), t(4;11)(q21;q23); t(1;19), t(1;19)(q23;p13); t(9;22), t(9;22)(q34;q11); HeH, high hypodiploidy; *ETV6-RUNX1*, fusion from t(12;21)(p13;q22). *(See plate section for color representation of this figure.)*

as shown in the karyogram in Figure 5.4a. It is strongly associated with paediatric BCP-ALL, found in ~30%, and rarely seen in adult ALL, ~10%, and even then found among younger adults.[10] There is a pronounced age peak at 2–4 years of age (median age 3.7 years). It is associated with a favourable prognosis in children, with an overall survival of ~90%. However, 15–20% of cases relapse, needing salvage therapy.[11,12] Attempts have been made to identify cytogenetic subtypes among high hyperdiploid ALL that correlate with outcome, with suggestions that gains of chromosomes 4, 10, 17 and 18 are associated with a favourable prognosis.[11,13] About 50% of cases contain structural abnormalities in addition to the chromosomal gains; most commonly they are unbalanced changes and include partial gains of the long arm of chromosome 1 (1q), deletions of 6q and isochromosome of 7q or 17q; balanced translocations are rarely seen.[14] Importantly, cells from high hyperdiploid cases are difficult to culture with normal cells outgrowing the abnormal cells, hence cytogenetic analysis often fails. In such cases, DNA indexing and/or fluorescence in situ hybridization (FISH) analyses are fundamental technologies to detect high hyperdiploidy accurately in clinical diagnosis.[15]

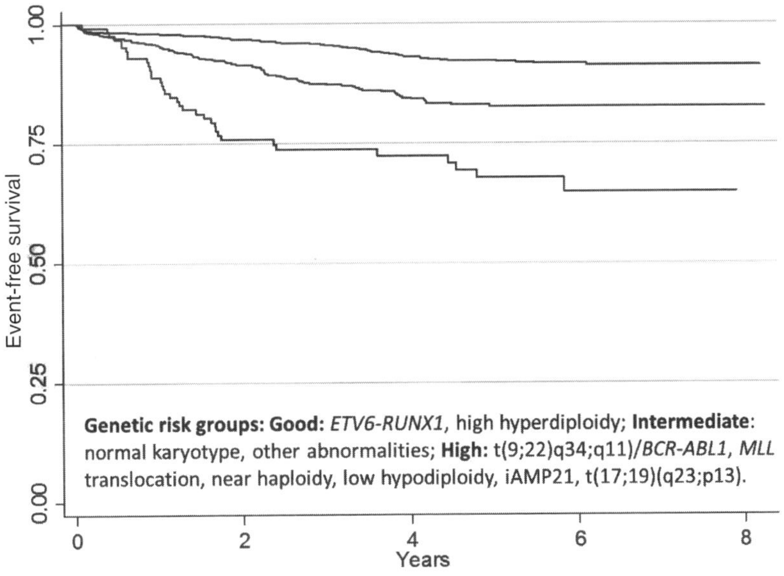

(a)

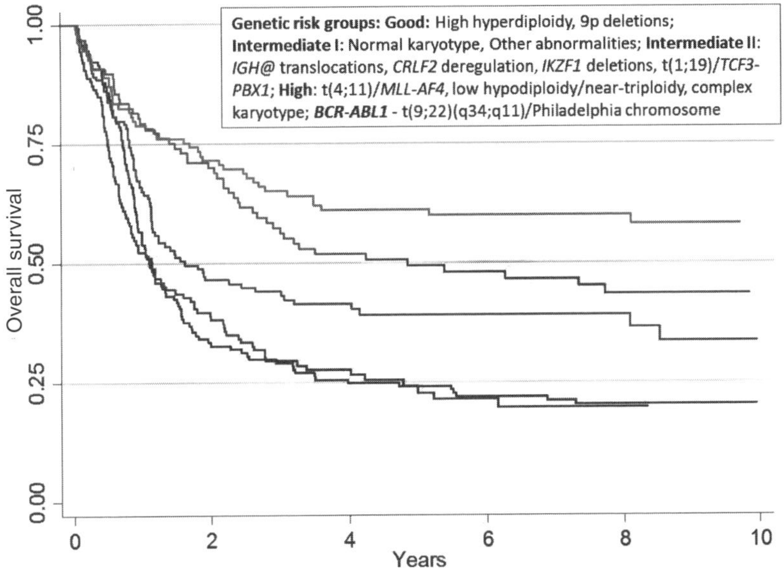

(b)

Figure 5.3 Kaplan–Meier survival curves. (a) Event-free survival of childhood BCP-ALL classified according to genetic risk group as indicated. Blue, good risk; red, intermediate risk; green, high risk. (b) Overall survival of adult BCP-ALL classified according to genetic risk group as indicated. Data taken from the childhood ALL trial, UK ALL2003 and UKALLXII/ECOG2993 adult trial. MRC UKALLXII[57,66]. Yellow, good risk, green, two intermediate risk groups; red, *BCR-ABL1* positive; blue, high risk. *(See plate section for color representation of this figure.)*

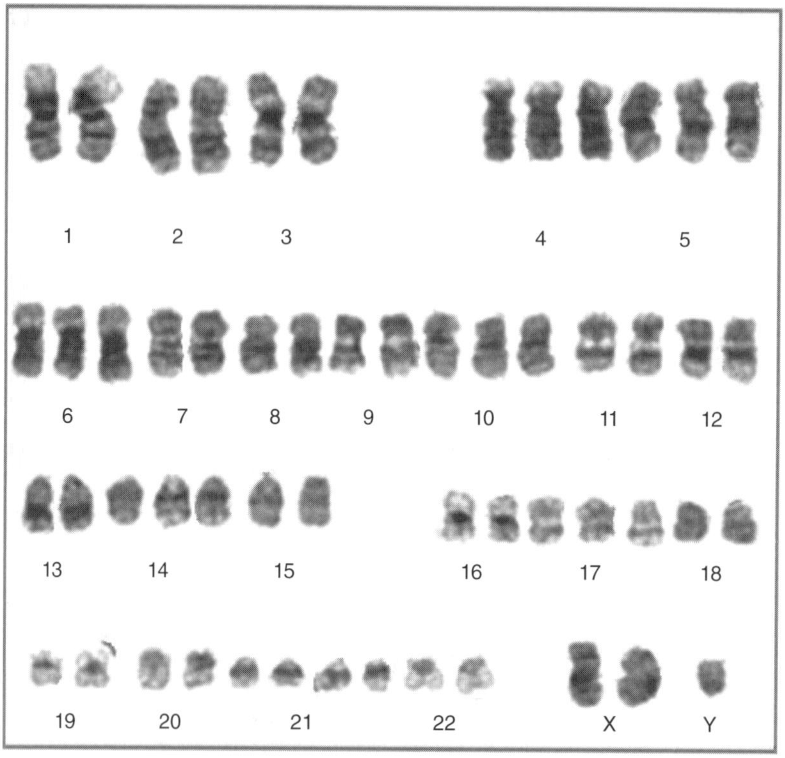

(a)

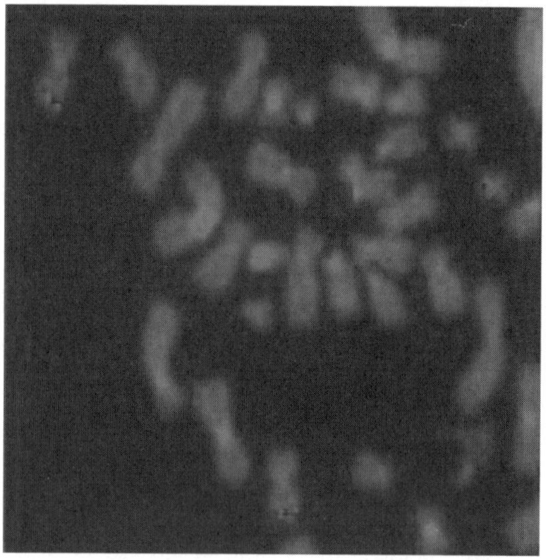

(b)

Figure 5.4 Common cytogenetic abnormalities found in BCP-ALL. (a) A high hyperdiploid karyogram with gains of chromosomes 4, 5, 6, 10, 14, 17, 21 (gain of 2 copies) and X. (b) FISH of the *MLL* rearrangement, t(11;19)(q23;p13.6)/MLL-MLLT1. The normal chromosome 11 shows the closely apposed red and green signals, the abnormal chromosome 11 shows the green signal (5′MLL) only with the red signal (3′MLL) translocated to the abnormal chromosome 19. (c) Karyogram showing the translocation t(9;22)(q34;q11). *(See plate section for color representation of this figure.)*

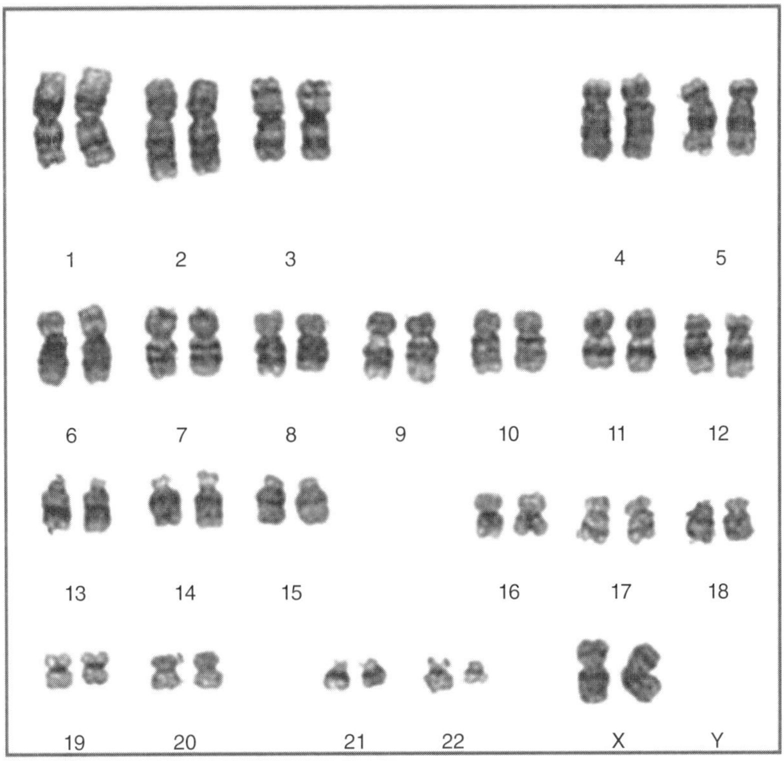

(c)

Figure 5.4 *(continued)*

Little is known about the molecular consequences of high hyper-diploidy, but a general dose effect of certain loci on the gained chromosomes has been suggested, in addition to imprinting,[16,17] which is selective expression of a gene dependent on the parent of origin. However, subsequent studies have been unable to support this suggestion.[18,19] Indeed, gene expression studies have shown that generally there is an increase in the expression of genes on the gained chromosomes, while some genes on the gained chromosomes display a substantially higher or lower expression. This observation suggests that alternative mechanisms, such as mutations or methylation, may result in deregulated gene expression.[20,21] This hypothesis is supported by recent genome-wide cytosine methylation profiling of ALL, showing that a substantial proportion of genes on triploid chromosomes that do not show increased expression are subject to methylation-induced silencing.[22] The molecular evidence behind the formation of high hyperdiploidy points to two major routes, with the most common (70%) being simultaneous gain of chromosomes in a single abnormal cell division and, less commonly (30%), initial tetraploidy with subsequent loss of chromosomes.[14]

Cooperating mutations in high hyperdiploid ALL include activating mutations in the receptor tyrosine kinase/RAS pathway (*FLT3* in 10–25%, *KRAS/NRAS* in 15–30% and *PTPN11* in 10–15%). These mutations appear to be mutually exclusive, suggesting that activation of the RAS pathway or kinase signalling are important cooperating events in this ALL subtype, with approximately one-third of such cases having activation of these pathways upon target gene resequencing.[23] Recently, a study of 16 cases of relapsed high hyperdiploid ALL identified mutations in the *CREB*-binding protein (*CREBBP*) in 63% of cases.[24] *CREBBP* mutations have also been found in 18% of relapsed high-risk ALL.[25] The increased frequency of *CREBBP* mutations in relapsed high hyperdiploid ALL suggest that *CREBBP* status could be used as a parameter to predict early relapse.

t(12;21)(p13;q22)/*ETV6-RUNX1*

The t(12;21)(p13;q22)/*ETV6-RUNX1* fusion is identified in about 25% of childhood BCP-ALL, yet is extremely rare in adults.[26–31] The translocation is usually cryptic on cytogenetic analysis and was first identified by FISH in 1994.[32] This translocation fuses the ETS variant 6 (*ETV6*, formerly *TEL*) gene at 12p13 to the runt-related transcription factor 1 (*RUNX1*, formerly *AML1*) gene at 21q22 to create the *ETV6-RUNX1* fusion gene.[33,34]

The presence of *ETV6-RUNX1* at diagnosis correlates with a good prognosis and recent data indicate that the overall survival can be further improved to 99% on contemporary risk-directed therapy.[35] The rearrangement commonly arises *in utero* but the prolonged latency to overt leukaemia and twin studies, together with screening of normal cord blood, have shown that the *ETV6-RUNX1* fusion gene is present at a 100-fold higher incidence than the corresponding risk of the leukaemia, suggesting that additional genetic events are needed for the development of this disease.[36–39] The most common secondary genetic event in *ETV6-RUNX1*-positive ALL is deletion of the other *ETV6* allele,[40,41] and also other sub-microscopic copy-number alterations.[7]

RUNX1 (or another RUNX family member, such as *RUNX2* or *RUNX3*) encodes the heterodimeric partner of the core-binding factor gene

(*CBFB*). Together they constitute the core binding factor (CBF) transcription factor complex, which is a master regulator of genes that are essential for haematopoiesis. The CBF complex regulates the expression of a large number of genes with pivotal roles in haematopoiesis, including IL3[42] and GM-CSF.[43] RUNX1 also functions as a transcriptional repressor[43] with the C-terminal part interacting with Groucho-related co-repressors.[44] Knock-out of *Runx1* or *Cbfb* causes central nervous system (CNS) haemorrhage, lack of fetal haematopoiesis and embryonic lethality.[45,46] *RUNX1* is a member of a family of transcription factors with homology to the *Drosophila* pair-rule gene, runt. It has a strong DNA binding domain (runt homology domain), which is retained in all fusion genes. Cloning of *ETV6-RUNX1* revealed that the N-terminal helix–loop–helix domain of *ETV6* is fused to almost the entire RUNX1 protein.[33,34] ETV6 is widely expressed and, when fused to *RUNX1*, the expression of RUNX1 will be driven by the promoter of *ETV6*. It is concluded that the fusion protein likely functions as a transcriptional repressor that alters the expression of *RUNX1* target genes.

t(1;19)(q23;p13)/*TCF3-PBX1*

The translocation t(1;19)(q23;p13) results in expression of the *TCF3-PBX1* fusion gene, present in about 6% of childhood and adult BCP-ALL, with a higher incidence in younger adults.[47–49] *TCF3-PBX1* ALL was originally considered to be a high-risk leukaemia that often presented with CNS involvement and an increased risk of relapse.[50] On modern intensive protocols, patients with *TCF3-PBX1* fusion are classified as standard risk, although it remains an independent risk factor for CNS relapse.[51,52] At the cytogenetic level, the translocation occurs either as a balanced t(1;19) or, more commonly, as an unbalanced der(19)t(1;19) with duplication of 1q distal to *PBX1*. Originally, it was suggested that the unbalanced form correlated with an improved survival,[53] although this was not confirmed in later studies.[52]

The t(1;19)(q23;p13) disrupts the basic-loop–helix transcription factor *TCF3* (*E2A*) at 19q22 and the homeobox-containing gene *PBX1* at 1p13. The *TCF3* gene encodes two protein products, E12 and E47. Both are required for normal B-cell development. *PBX1* is a transcription factor normally not expressed in lymphoid lineages, which can bind directly to *HOX* genes or to *MEIS1*.[54] The *TCF3-PBX1* fusion retains the transactivation domain of *TCF3*, but loses its DNA binding domain. This domain is replaced by the homeodomain of *PBX1*, which likely functions as a transcriptional activator. Thus, the fusion protein continues to bind to *HOX* genes, but not *MEIS1*. *TCF3-PBX1* results in the deregulated expression of *PBX/HOX* target genes, and also disruption of *TCF3*, which is critical for normal B-cell development.

t(17;19)(q22;p13)/*TCF3-HLF*

A variant of the t(1;19) translocation results in fusion of the *HLF* gene, located at 17q22, to *TCF3* as a result of the t(17;19)(q23;p13) translocation.[55] At the molecular level, two types of rearrangements give rise to chimeric oncoproteins, which comprise either exons 1–13 (type 1) or exons 1–12 (type 2) of *TCF3* and exon 4 of *HLF*. These two molecular subgroups strongly correlate with specific clinical features: type 1 with disseminated intravascular coagulation and type 2 with hypercalcaemia.[56] This translocation is very rare, with an estimated incidence of 0.1% in BCP-ALL.[57] Patients are older with a median age of 13 years (range 8–18 years) and a low white blood cell count ($>50 \times 10^9$/L). However, all known patients relapsed and died within 2 years of diagnosis.[50] Therefore, despite the rarity of this translocation, it is important that these patients are accurately identified.

Hypodiploidy

Hypodiploidy (<46 chromosomes) is a rare subtype of ALL, seen in about 5% of BCP-ALL. It can be divided into three genetic subtypes; near haploidy (23–29 chromosomes), low hypodiploidy (30–39 chromosomes) and high hypodiploidy (40–45 chromosomes). The poor outcome of hypodiploid ALL is mainly restricted to patients with <45 chromosomes.[58,59] It is common for leukaemic cells with 23–29 or 30–39 chromosomes to undergo doubling of their chromosome number by endoreduplication so that a cell population coexists with a modal number in the hyperdiploid or triploid range. Importantly, the hypodiploid clone may not always be evident at diagnosis if present in only a minor subclone. Hence interphase FISH and/or flow cytometry should be used in combination with cytogenetic analysis at diagnosis to ensure accurate detection of hypodiploid clones, which determines the prognosis.

Patients with near haploidy (23–29 chromosomes) tend to be younger, with a median age of 7 years. The prognosis is dismal, with a 3-year event-free survival of only 29%. The most frequent chromosomal additions to the haploid set are chromosome 21 and the sex chromosomes, chromosomes 14 and 18. Structural rearrangements are rare.

Low hypodiploid (30–39 chromosomes) patients tend to be older than those with near haploidy, the majority being 10 years or older with a median age of 15 years. Similarly to near haploidy, duplication of the low hypodiploid clone is common, while structural abnormalities are more frequent. The pattern of chromosomal gains is distinct from the near haploid cases with the exception of the gain of chromosomes 21, X or Y, 14 and 18, which are shared between the two groups. The most commonly gained chromosomes onto the haploid complement include chromosomes 1, 11, 19, 10, 22, 5, 6, 8, 2 and 12. Only chromosomes 7 and 17 have reported to always be monosomic. The prognosis of this ALL subtype is equally poor to that of near haploid ALL.

High hypodiploidy (42–45 chromosomes), accounts for the majority of hypodiploid cases, with the modal number of 45 being by far the most frequent. Among these patients, the most common loss is that of a sex chromosome, with loss of chromosomes 7, 9, 13 and 17 less frequently described. In a large MRC trial, this ALL subtype had an event-free survival of 66%.[58] However, in a more recent study, there was no difference in the poor outcome among childhood cases with 24–29, 30–39 or a subgroup of patients characterized by 40–43 chromosomes.[59]

Among the high hypodiploid group, the overall reduction in chromosome number was commonly due to unbalanced translocations in the form of dicentric chromosomes. Chromosome 9 was the most frequently involved chromosome, particularly in the formation of dic(9;20), followed by chromosomes 7 and 12. Most cases had a complex karyotype with no duplication of the original clone.

Until recently, little was known about the additional genetic alterations underlying the pathogenesis and poor prognosis of hypodiploid ALL. A recent study of over 120 hypodiploid ALL cases, using a range of molecular and NGS techniques, clearly demonstrated that near haploid and low hypodiploid ALL have distinct transcriptomic signatures, sub-microscopic DNA copy-number alterations and sequence mutations that differ from other BCP-ALL subtypes.[60] The majority of near haploid cases harbour mutations activating Ras signalling (*NF1* in 40% of cases, but also *NRAS, KRAS* and *PTPN11*) and inactivating deletions and mutations of the IKAROS family gene *IKZF3* (AIOLOS). Close to 100% of low hypodiploid cases have mutations of the tumour suppressor gene *TP53* and inactivating mutations of a third IKAROS family member, *IKZF2* (HELIOS). Hypodiploid cells from both near and low hypodiploid cases exhibit activation of Ras-Raf-MEK-ERK and phosphatidylinositol-3-OH kinase (PI3K) signalling that is sensitive to PI3K and PI3K/mTOR inhibitors, suggesting that PI3K inhibition may provide a novel therapeutic approach. An unexpected finding was that the *TP53* sequence mutations identified in low hypodiploid ALL were commonly present in matched non-tumour DNA, suggesting germline inheritance. This has been confirmed in a limited number of family studies, indicating that low hypodiploid ALL is a manifestation of Li-Fraumeni syndrome.[60,61] Additional deleterious germline mutations were identified in other hypodiploid ALL cases, including activating mutations of *NRAS* and *PTPN11*. Hence detailed analysis of the role of inherited mutations in the pathogenesis of ALL is of increasing interest.

11q23/KMT2A (MLL) gene rearrangements

Rearrangements of the mixed lineage leukaemia (*MLL*) gene at chromosome band 11q23 are common in acute leukaemia, in particular among infants, where >75% carry a rearrangement.[62,63] Although *MLL* rearrangements are seen in all age groups, and in both lymphoid and myeloid leukaemia, they are present in 3% of childhood ALL[64] and in 9% of adult leukaemia cases.[65,66] The prognosis of *MLL*-rearranged infant leukaemia is very poor, with an event-free survival of ~37%.[67,68] Among older children with ALL, the event-free survival is ~45%, with the t(4;11)(q21;q23) being associated with a worse prognosis.[69]

MLL is a promiscuous gene, with more than 100 fusion sites identified so far,[70] although five specific partners account for around 80% of the *MLL* fusions: t(4;11)(q21;q23)/*MLL-AFF1(AF4)*, t(9;11)(p22;q23)/*MLL-MLLT3(AF9)*, t(11;19)(q23;p13.3)/*MLL-MLLT1(ENL)*, t(10;11)(p12;q23)/*MLL-MLLT10(AF10)* and t(6;11)(q27;q23)/*MLL-MLLT4 (AF6)*. For accurate diagnosis of *MLL* rearrangements, a combination of genetic and molecular analyses is required, in addition to cytogenetics (Figure 5.4b).

Childhood leukaemias containing *MLL* rearrangements are characterized by an early leukaemic initiation (likely *in utero* for most childhood cases), both lymphoid and myeloid features, and a poor outcome, representing a unique entity. DNA sequence analysis of *MLL* revealed striking homology to the *Drosophila trithorax* (*trx*) gene, spanning ~100 kb. It consists of at least 37 exons and encodes two protein products, one major and one minor.[71,72] Chromosomal translocations disrupt *MLL* and create fusion genes with the N-terminal portion of *MLL* fused to a partner gene. The breakpoints cluster in an 8.3 kb region (exons 8–14) located N-terminal of the proteolytic cleavage sites, resulting in the loss of the SET domain.

MLL fusions are transcriptional activators leading to increased expression of *HOXA* cluster genes, in particular *HOXA9* and its cofactor, *MEIS1*, but also some miRNAs, such as miR-17–93 and miR-196b.[73–75] In addition, miR-150, was recently identified to be down-regulated in leukaemias containing an *MLL* rearrangement; miR-150 normally down-regulates the expression of *FLT3*, which is often over-expressed in *MLL*-rearranged leukaemia and known to cooperate with *MLL* in the leukaemogenic process.[76] Targeted therapies for *MLL*-transformed leukaemia is attractive owing to the aggressiveness of this disease. One promising target is the methyltransferase DOT1L, which interacts directly or indirectly with several of the *MLL* fusion partners. Selective killing of cells with *MLL* rearrangements upon exposure to EPZ004777, a potent inhibitor towards DOT1L, has been shown.[77] In addition, inhibitors directed towards the protein–protein interaction between MLL fusion proteins and menin show promising results, with reversal of the oncogenic activity of *MLL*-rearranged leukaemias.[78]

t(9;22)(q34;q11.1)/*BCR-ABL1*

The t(9;22)(q34;q11.1)/*BCR-ABL1* results in the formation of the Philadelphia chromosome (Ph), which is the hallmark of chronic myeloid leukaemia (CML). It is found in 25% of adult ALL and in about 3% of paediatric ALL, in which these children tend to be older (median age 8 years).[79] The Ph was first identified by Nowell and Hungerford in 1960,[3] and in 1973 Dr Janet Rowley discovered that the Philadelphia chromosome was the product of a reciprocal chromosomal translocation between chromosomes 9 and 22,[80] as shown in Figure 5.4c. More than 10 years later, in 1985, it was recognized that the t(9;22)(q34;q11) fused the human homolog of the Abelson murine leukaemia virus, *ABL1*, at 9q34 to a 5.8 kb region of chromosome 22, named the breakpoint cluster region (*BCR*) gene at 22q11.[81] Subsequently, this rearrangement was shown to result in the *BCR-ABL1* fusion gene, which encodes either a 210 kDa chimeric protein, a variant which is tightly associated with CML, or a smaller 190 kDa variant protein, which is more often associated with ALL.[82] Both proteins were shown to have altered tyrosine kinase activity.[83] The *BCR* gene encodes a serine/threonine kinase and the *ABL1* gene encodes a protein that is structurally similar to the Src family of kinases. The leukaemogenic properties of BCR-ABL1 are linked to increased activity of ABL1, also required for transformation of haematopoietic cells. Upon transformation, multiple signalling pathways become activated, including RAS/MAPK, STAT, PI-3 kinase, JNK/SAPK and NF-κB,[84] resulting in deregulation of apoptosis, differentiation and cell adhesion.

At the chromosomal level, about 60% of Philadelphia-positive ALL have additional secondary aberrations present at diagnosis, including gain of a second copy of the Ph, a hyperdiploid karyotype or −7/7q−.[10,85] Therefore, in these cases with a hyperdiploid karyotype, it is important to identify the Ph so that the patient is treated appropriately with tyrosine kinase inhibitor (TKI) therapy. Approximately 70% of Ph-positive ALL have deletions of the B-cell differentiation gene *IKZF1* (see below).[86] In a minority of cases, the t(9;22) is cryptic by cytogenetic analysis, although the *BCR-ABL1* fusion can be detected by polymerase chain reaction (PCR).[87]

The *BCR-ABL1* fusion is associated with poor prognosis in all age groups, a high incidence of CNS involvement at diagnosis, a high white blood cell count and early development of multidrug resistance.[88] In paediatric Ph-positive ALL, the 3-year event-free survival was about 25%.[89] However, with the recent introduction of the tyrosine kinase inhibitor imatinib mesylate, the historically poor outcome of *BCR-ABL1*-positive ALL has improved. In one study, where a combination of conventional intensive chemotherapy and imatinib was used, a significantly improved event-free survival of 80% was achieved in children.[90] In addition, by using second-generation TKIs, such as dasatinib and nilotinib, the suppression of the BCR-ABL1 kinase activity may become even more potent, potentially resulting in further improved outcomes.[91,92] Ph-positive ALL in adults is also associated with very poor prognosis and a high relapse rate.[93] However, the use of imatinib has also resulted in improved outcome for this age group.[79,94]

Intrachromosomal amplification of chromosome 21 (iAMP21)

iAMP21 is characterized by gain of at least three additional copies of a large region of chromosome 21 that always includes *RUNX1*.[95–98] The abnormal chromosome 21 has a complex structure of alternating gain

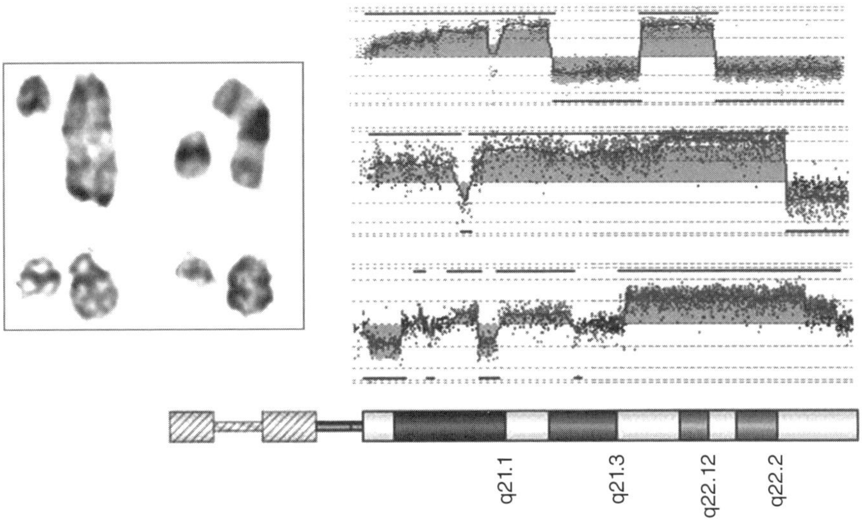

Figure 5.5 iAMP21 cytogenetics and array profiles. In the box on the left, four pairs of chromosomes 21 are shown from four different patients, indicating the variability of the iAMP21 chromosome. On the right are three array CGH profiles of chromosome 21 from three different iAMP21 patients showing the variability in copy number between patients.

and loss along the chromosome 21 long arm (Fig. 5.5). It is defined as a primary cytogenetic change, usually observed in patients lacking other key cytogenetic alterations, although rare cases of iAMP21 in association with *ETV6-RUNX1* and *BCR-ABL1* have been described.[99] Patients with iAMP21 have a high risk of relapse when treated on standard therapy, although outcome has been dramatically improved with intensive chemotherapy.[100,101] The nature of cooperating lesions and the role of iAMP21 in driving an aggressive leukaemia are currently poorly understood.

Complex karyotype

Few studies have classified ALL karyotypes according to their complexity. The definition used within adult ALL is the presence of five or more chromosomal abnormalities in the absence of an established translocation or ploidy group, which occurs in ~5% of adult BCP-ALL.[66] Although this subgroup does not appear to be associated with sex, white blood cell count or immunophenotype, there is an indication that the incidence increases with age among adults.[47] Most significantly, karyotype complexity in adult ALL is associated with increased risk of relapse and death, with an overall survival of 20–25%.[47,102] In UK adult ALL treatment trials, these patients are now treated as high risk.

Submicroscopic genetic alterations in BCP-ALL

Microarray-based profiling of DNA permits the identification of genomic copy-number alterations (CNA) (deletions and gains) at sub-kilobase resolution. Widely used platforms include single-nucleotide polymorphism (SNP) microarrays (e.g. Affymetrix and Illumina) that also permit interrogation of copy-neutral loss of heterozygosity (CN-LOH, also known as acquired uniparental disomy) and array-based comparative genomic hybridization (array-CGH) (e.g. Agilent and Roche-Nimblegen).[25] Several groups have reported SNP array and array-CGH profiling results in childhood ALL.[7,103–105] These studies have shown that, although ALL genomes typically harbour fewer structural alterations than many solid tumours, over 50 recurring deletions or gains have been identified, many of which involve a single gene or a few genes.

The genes most frequently involved in CNA encode proteins with key roles in lymphoid development (e.g. *PAX5, IZKF1, EBF1*), cell cycle regulation and tumour suppression (*CDKN2A/CDKN2B, RB1*), putative regulation of apoptosis (*BTG1*), lymphoid signalling, transcriptional regulation and co-activation (*ETV6, ERG*), regulation of chromatin structure and epigenetics,[7,104] and also deletion of the glucocorticoid receptor *NR3C1*. Their relative distribution is shown in Figure 5.6a. Sanger sequencing studies have identified recurring sequence mutations, which in B-lineage ALL most commonly affect lymphoid development (*PAX5* and less commonly, *IKZF1*), Ras signalling (*NRAS, KRAS* and *NF1*), cytokine receptor signalling (*IL7R, JAK2*) and tumour suppression (*TP53*) (Figure 5.6b).[106] Importantly, several genes are involved in multiple types of genetic aberrations, including CNA, translocations and sequence mutation (e.g. *PAX5*).

The nature and frequency of genetic lesions are subtype dependent. For example, *MLL*-rearranged leukaemia harbours very few additional structural or sequence alterations, whereas in contrast, the majority of non-*MLL* ALL harbour increased numbers of recurring sub-microscopic deletions, for example, at least 6–8 per case in *ETV6-RUNX1* and *BCR-ABL1* ALL.[7,86,107–109] In BCP-ALL, this is driven in part by the activity of the recombinase activating genes (RAG) that induce focal deletions resulting in a selective advantage of lymphoid progenitors.

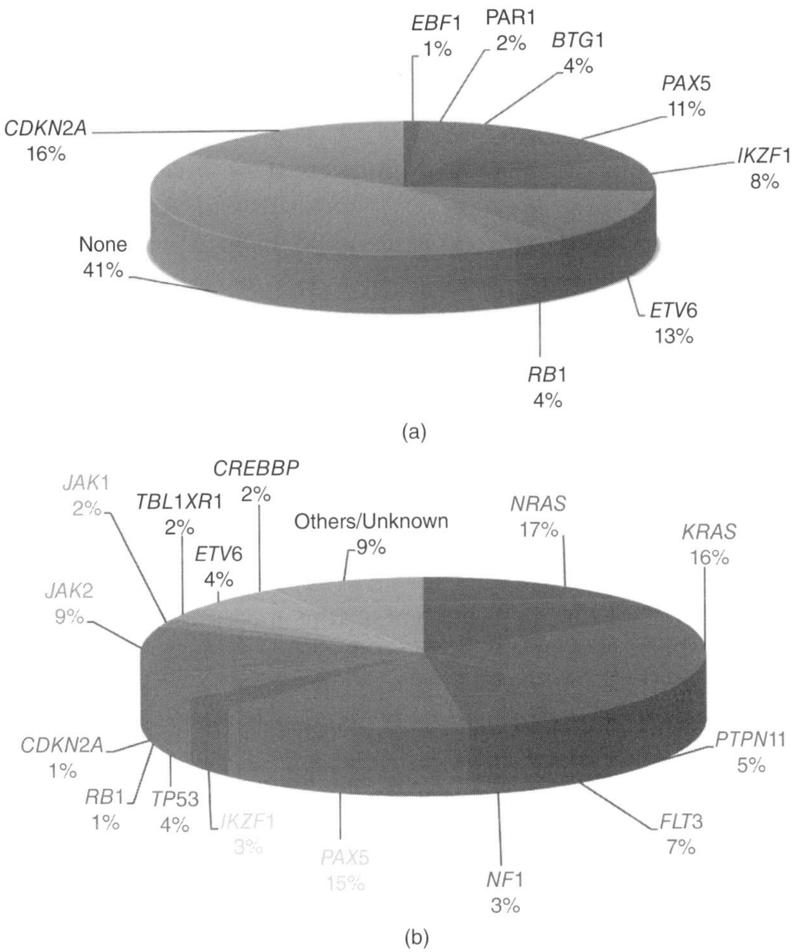

Figure 5.6 Distribution of copy number abnormalities and mutations in BCP-ALL. (a) Deletions of significant genes, PAR1, deletions within the pseudoautosomal region of the sex chromosomes. (b) Mutations in significant genes colour coded according to the signalling pathway to which they belong: purple, RAS signalling; blue, B-cell development genes; red, cell cycle control; green, JAK-STAT pathway, black, others. Deletions and mutations are not mutually exclusive. *(See plate section for color representation of this figure.)*

Emerging experimental data have shown that several of these alterations cooperate in leukaemogenesis.[110,111]

Alteration of transcription factors in BCP-ALL

Deletion, sequence mutation or rearrangement of genes encoding transcriptional regulators of lymphoid development is a hallmark of BCP-ALL. Alteration of *PAX5* (~35%), *IKZF1* (~15%) and *EBF1* (~5%)

are the most common alterations, with at least two-thirds of BCP-ALL harbouring one or more lesions in this pathway.[7,112] These alterations are usually loss of function or dominant negative lesions resulting in arrested lymphoid maturation, which is characteristic of leukaemic cells. Notably, although *PAX5* alterations are the most common genetic alteration in BCP-ALL, they are not associated with outcome.[112,113] In contrast, alteration of *IZKF1* (IKAROS) is a hallmark of two types of high-risk ALL: Ph-positive ALL[86,114,115] and *BCR-ABL1*-like (Ph-like) ALL.[116–118] *IKZF1* encodes IKAROS, the founder member of a family of zinc finger transcription factors required for the development of all lymphoid lineages.[119] *IKZF1* alterations include focal or large deletions that result in loss of expression of IKZF1. Focal deletions of coding exons 4–7 remove the N-terminal DNA-binding zinc fingers, leading to expression of a dominant negative isoform, IK6. *IKZF1* alterations are present in over 70% of *BCR-ABL1*-positive ALL, including *de novo* ALL and chronic myeloid leukaemia (CML) at progression to lymphoid blast crisis,[86] and are associated with a worse outcome in Ph positive ALL.[115]

CRLF2 rearrangements and Janus kinase mutations in ALL

The cytokine receptor *CRLF2* is rearranged or mutated in ~7% of childhood and adult BCP-ALL and 50% of Down syndrome ALL (DS-ALL).[120–123] *CRLF2* is located in the pseudoautosomal region of the sex chromosomes (PAR1) at Xp22.3/Yp11.3. It encodes cytokine receptor-like factor 2 (thymic stromal lymphopoietin receptor, TSLPR). With interleukin-7 receptor alpha, CRLF2 forms a heterodimeric receptor for the ligand TSLP (thymic stromal lymphopoietin). *CRLF2* is rearranged by translocation into the immunoglobulin heavy-chain locus (*IGH-CRLF2*) or by a focal deletion upstream of *CRLF2* that results in expression of *P2RY8-CRLF2*, which encodes full-length CRLF2. Both rearrangements result in aberrant over-expression of CRLF2 on the cell surface of leukaemic lymphoblasts that may be detected by flow cytometric immunophenotyping.[122] Less commonly, a CRLF2 p.Phe232Cys mutation results in receptor dimerization and over-expression.[120]

Approximately half of *CRLF2*-rearranged ALL harbour activating mutations of the Janus kinase genes, *JAK1* and *JAK2*,[122–124] otherwise uncommon in BCP-ALL. The JAK mutations are most often missense mutations at or near R683 in the pseudokinase domain of JAK2, distinct from the JAK2 V617F mutations that are a hallmark of myeloproliferative diseases. Less common are activating mutations in the kinase domain of *JAK1* and *JAK2*. The *JAK1/2* mutant alleles alone and in cooperation with CRLF2 over-expression are transforming *in vitro*, suggesting that these two lesions are central to lymphoid transformation.[125–127] *CRLF2*-rearranged leukaemic cells with deregulated CRLF2 exhibit activation of JAK-STAT and PI3K/mTOR pathways and are sensitive to JAK and mTOR inhibitors *in vitro* and *in vivo*.[128,129] An early phase trial of the JAK inhibitor ruxolitinib (ADVL1011) in relapsed and refractory childhood tumours, including cases with *CRLF2* rearrangements and/or JAK mutations, has recently been completed (clinicaltrials.gov identifier NCT01164163).

In non-DS ALL, *CRLF2* alterations and *JAK* mutations are associated with *IKZF1* deletion/mutation and poor outcome, particularly in cohorts of high-risk B-ALL.[130–133] Recent studies performed by the Children's Oncology Group (COG) have confirmed that *CRLF2* and *IKZF1* alterations are associated with inferior outcome in multiple cohorts and, notably, that elevated *CRLF2* expression in the absence of rearrangement is also an adverse prognostic feature.[134]

BCR-ABL1-like or Ph-like ALL

Recently, a new subgroup of BCP-ALL has been described characterized by an expression profile similar to *BCR-ABL1*-positive ALL, deletion of *IKZF1* and poor outcome, named *BCR-ABL1*-like or Ph-like ALL.[116,117,135] It is common, comprising up to 10–15% of childhood and up to one-third of BCP-ALL in adolescents and young adults. Approximately half of *BCR-ABL1*-like ALL harbour *CRLF2* rearrangements and concomitant JAK1/2 mutations. Recent transcriptome and whole-genome sequencing has shown that non-*CRLF2*-rearranged *BCR-ABL1*-like ALL harbour a diverse range of genomic alterations that activate cytokine receptors and tyrosine kinases, including *ABL1*, *ABL2*, *EPOR*, *JAK2* and *PDGFRB*.[118] These alterations are most commonly chromosomal rearrangements resulting in chimeric fusion genes deregulating cytokine receptors, for example, *IGH-EPOR* and tyrosine kinases, including *NUP214-ABL1*, *ETV6-ABL1*, *RANBP2-ABL1*, *RCSD1-ABL1*, *BCR-JAK2*, *PAX5-JAK2* and *STRN3-JAK2*, of which *EBF1-PDGFRB* is the most common (Fig. 5.7). In up to 20% of *BCR-ABL1*-like cases, alternative alterations activating

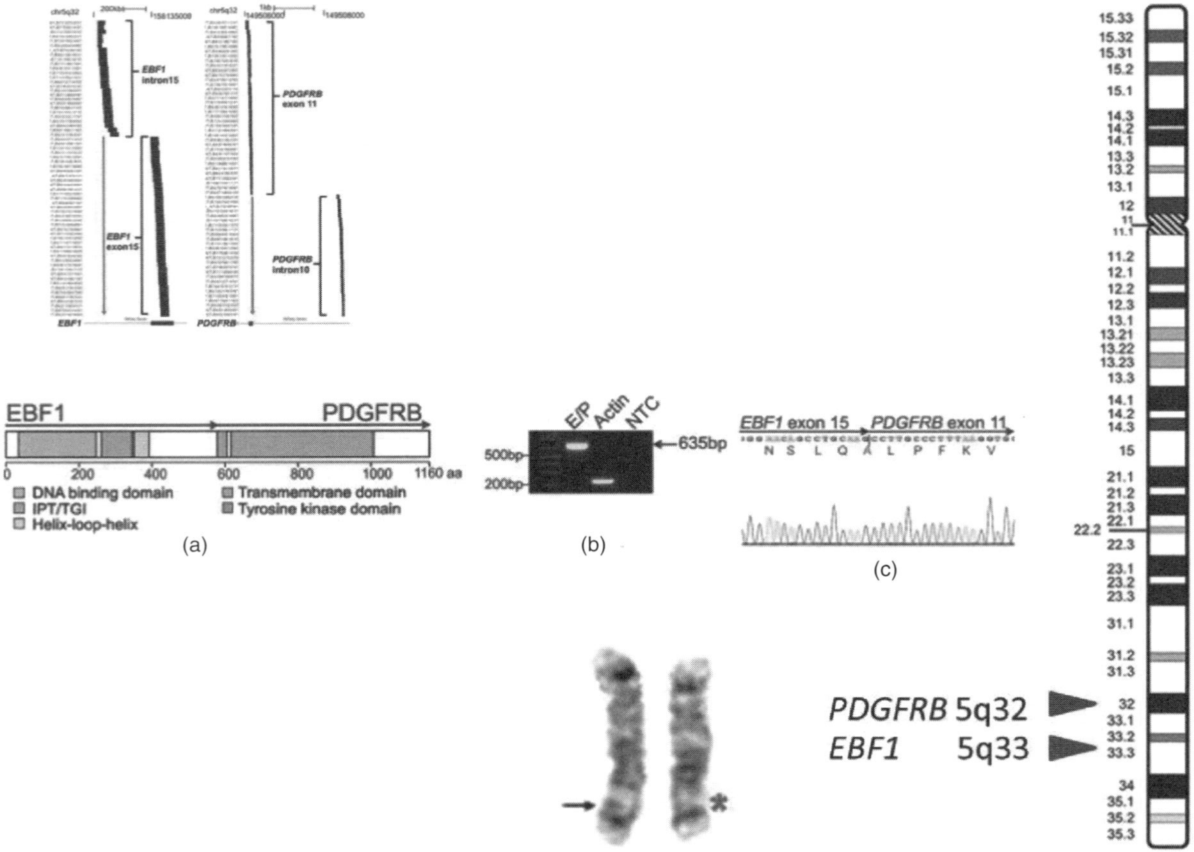

Figure 5.7 *EBF1-PDGFRB* fusion. Top, data from transcriptome sequencing. Middle: A, predicted domain structure; B, result from RT-PCR; C, result from Sanger sequencing. Bottom: a pair of chromosomes 5 with the asterisk (*) marking the deletion indicated by the arrows in the ideogram on the right.

kinase signalling occur, including activating mutations of *FLT3* and *IL7R*, in addition to focal deletions of *SH2B3*, also known as LNK, which constrains JAK signalling. These diverse genetic alterations activate a limited number of signalling pathways, notably *ABL1*, *PDGFRB* and JAK-STAT signalling. These rearrangements have been shown to activate signalling pathways in model cell lines. In addition, primary leukaemic cells and xenografts of *BCR-ABL1*-like ALL are highly sensitive to tyrosine kinase inhibitors (TKI) *in vivo*.[118,128] Anecdotal reports are emerging of responsiveness of refractory *BCR-ABL1*-like ALL to appropriate TKI therapy, for example *EBF1-PDGFRB* ALL to imatinib.[136,137] Thus it is predicted that the majority of *BCR-ABL1*-like ALL will be amenable to therapy with a limited number of imatinib-class TKI for *ABL1*, *ABL2* and *PDGFRB* rearrangements and JAK inhibitors, such as ruxolitinib, for alterations activating JAK-STAT signalling (*EPOR*, *IL7R*, *JAK2* and *SH2B3*). NGS studies of childhood and adult ALL are ongoing in order to identify comprehensively all kinase-activating alterations in *BCR-ABL1*-like ALL and to implement TKI therapy in clinical trials.

ERG-altered ALL

Alteration of the ETS-family transcription factor *ERG* (ETS-related gene) occurs exclusively in cases lacking known chromosomal rearrangements and it is a hallmark of a novel subtype of BCP-ALL with a distinct gene expression profile. *ERG* deletions involve an internal subset of exons resulting in loss of the central inhibitory and pointed domains, leading to expression of an aberrant C-terminal ERG fragment that retains the ETS and transactivation domains. It functions as a competitive inhibitor of wild-type ERG. Notably, despite the presence of *IKZF1* alterations in a proportion of *ERG*-deregulated cases, the outcome of this subtype of ALL is favourable.[132,138,139]

Genetic rearrangements in T-lineage ALL

T-ALL accounts for approximately 15% childhood and 25% adult ALL. It is characterized by an older age of onset, male sex predominance and inferior outcome in comparison with BCP-ALL.[140] Chromosomal abnormalities are evident on cytogenetic analysis in up to 70% of T-ALL cases and commonly involve one of the T-cell antigen receptor loci, including *TRA* and *TRD* at 14q11, *TRB* at 7q34 and *TRG* at 7p14. The

Table 5.2 Common genetic aberrations in T-ALL.

Type of aberration	Aberration	Molecular genetic features
Aberrant expression of transcription factors and related genes	t(1;7)(p34;q34)	*LCK, TRB*
	TAL1 deletion	*TAL1, STIL*
	t(6;7)(q23;q34	*MYB, TRB*
	t(7;9)(q34;q32)	*TAL2, TRB*
	t(7;9)(q34;q34.3)	*NOTCH1, TRB*
	t(7;11)(q34;p13)	*LMO1, TRB*
	t(7;11)(q34;p15)	*LMO2, TRB*
	t(7;12)(q34;p13.3)	*CCND2 , TRB*
	t(7;19)(q34;p13)	*LYL1, TRB*
	t(8;14)(q24;q11)	*MYC, TRA/D*
	t(11;14)(p13;q11)	*LMO1, TRA/TRD*
	t(11;14)(p15;q11)	*LMO2, TRA/TRD*
	t(12;14)(p13;q11)	*CCND2, TRA*
	inv(14)(q11q32)	*BCL11B, TRD*
	t(14;14)(q11;q32)	*BCL11B, TRD*
	NKX2–1 rearrangements	*NKX2–1*
	NKX2–2 rearrangements	*NKX2–2*
	MEF2C rearrangements	*MEF2C*
	t(14;21)(q11;q22)	*OLIG2, TRA*
Abnormalities of homeodomain genes	t(7;10)(q34;q24)	*TLX1, TRB*
	t(10;14)(q24;q11)	*TLX1, TRA/TRD*
	t(5;14)(q35;q32)	*TLX3, BCL11B*
Abnormalities of the *HOXA* cluster	inv(7)(p15q34)	*HOXA, TRB*
	t(7;7)(p15;q34)	*HOXA, TRB*
	t(7;14)(p15;q11)	*HOXA ,TRD*
	t(7;14)(p15;q32)	*HOXA, BCL11B*
Fusion transcripts	t(6;11)(q27;q23)	*MLLT4, MLL*
Copy-number changes	t(9;9)(q34;q34)	*NUP214, ABL1*
	t(9;14)(q34;q32)	*EML1, ABL1*
	t(10;11)(p12;q14)	*PICALM, MLLT10*
	MYB duplication	*MYB*
	del(9p)	*CDKN2A*
	del(18)(p11)	*PTPN2*
Mutations	*NOTCH1* mutations	*NOTCH1*
	FBXW7 mutations	*FBXW7*
	CNOT mutations	*CNOT*
	PFH6 mutations	*PFH6*

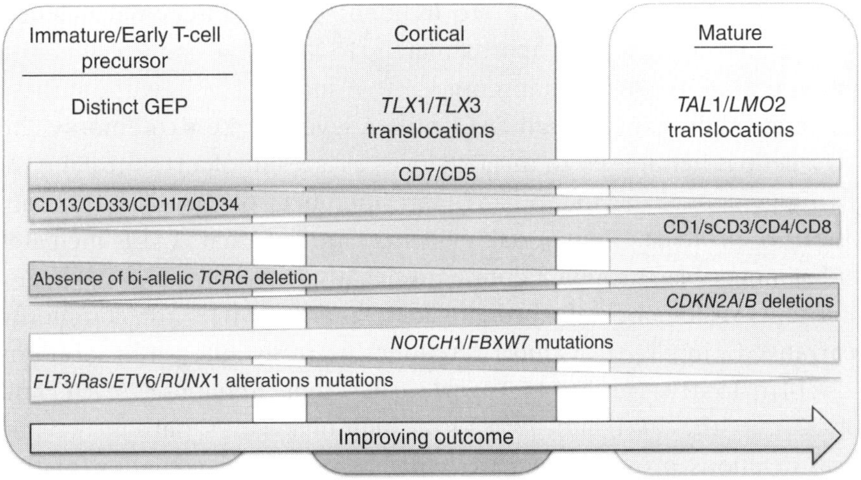

Figure 5.8 Cartoon illustrating the subtypes of T-ALL according to morphological and genetic type with their immunophenotypes, mutations and outcome indicated across the subtypes.

most common rearrangements are listed in Table 5.2. They occur in approximately one-third of T-ALL, but may be cryptic on cytogenetic analysis. Similar to rearrangements of *MYC*,[141] these rearrangements may arise from aberrant antigen receptor gene recombination errors in the normal recombination process, leading to the generation of functional antigen receptors.[142] Rearrangements in T-ALL commonly dysregulate transcription factors, including members of the bHLH family (*MYC*, *TAL1*, *TAL2*, *LYL1* and *BHLHB1*), genes encoding the LIM-only domain proteins (*LMO1* and *LMO2*) and homeodomain genes (*HOX11* and *HOX11L2*). In addition, T-ALL cases frequently harbour cryptic rearrangements of *ABL1*, activation mutations of *NOTCH1* and a spectrum of sub-microscopic genetic alterations commonly involving *CDKN2A*/*CDKN2B*, *PTEN* and *MYB*.[143] Essentially, T-ALL can be subdivided into three subtypes based on morphology, immunophenotype and genetics, as indicated in Fig. 5.8.[144]

TAL1/LMO2 rearranged T-ALL

T-ALL with rearrangements of *TAL1*/*LMO2* are classified as mature disease with a characteristic immunophenotype (Fig. 5.8). Alteration of *TAL1* (*SCL*, *TCL5*) at 1p32 is the most frequent transcription factor rearrangement in T-ALL. It arises from either the translocation t(1;14)(p32;q11), which occurs in 3% of cases and juxtaposes *TAL1* to the *TRA/TRD* locus, or the more frequent cryptic interstitial deletion at 1p32 that is present in approximately 15% of cases and results in a chimeric *SIL-TAL1* fusion transcript.[145,146] Additional cases without these rearrangements express high TAL1 mRNA levels.[147] Less commonly, the *TAL2* gene is juxtaposed to the *TRB* locus as a result of a t(7;9)(q34;q32) rearrangement.[148] TAL1 and LYL1 are members of the class II family of bHLH proteins. Functional evidence implies that TAL1 mediates leukaemogenesis through a dominant negative mechanism.[149]

The LIM-domain-only proteins LMO1 and LMO2 are commonly rearranged in T-ALL, most frequently from the translocations t(11;14)(p15;q11) and t(11;14)(p13;q11) that juxtapose *LMO1* and *LMO2* into the *TRA* and *TRD* loci. Additional cases harbour cryptic focal deletions proximal to *LMO2* that result in dysregulation of this locus.[7,150] Expression of *LMO1* and *LMO2* results in T-cell self-renewal and leukaemia when expressed in thymocytes.[151]

TLX1/TLX3 rearranged T-ALL

The cortical subtype of T-ALL is characterized by rearrangements of *TLX1* (*HOX11*) and *TLX3* (*HOX11L2*). More generally, the homeobox family of transcription factors comprises two classes of genes. Class I HOX genes are in four clusters (*HOXA*, *HOXB*, *HOXC* and *HOXD*) and class II genes are distributed throughout the genome. The HOX genes exert key roles in the regulation of haematopoiesis and leukaemogenesis.[152] The two HOX genes, *TLX1* and *TLX3*, are often rearranged in T-ALL. Approximately 7% of childhood T-ALL cases have ectopic expression of *TLX1* arising from the translocation t(10;14)(q24;q11) and the variant t(7;10)(q35;q24) that juxtapose *TLX1* to the *TRA* or *TRB* loci.[153,154] Additional cases lacking *TLX1* rearrangement exhibit over-expression of this gene.[147]

Approximately 20% of childhood T-ALL cases exhibit over-expression of *TLX3* (*HOX11L2*),[155,156] most commonly from the cryptic translocation t(5;14)(q35;q32) that juxtaposes *TLX3* to *BCL11B*,[157] a zinc finger protein expressed during T-cell ontogeny, recently identified as a target of deletion and somatic sequence mutation in T-ALL.[158] Several variant translocations have also been identified, including *BCL11B* to *NKX2-5*[159,160] and rearrangement of *CDK6* to *TLX3*.[161] Data regarding the prognostic importance of *TLX1* and *TLX3* in T-ALL are conflicting, which may be due in part to the presence of additional sub-microscopic genetic alterations and mutations in these cases.[162]

Recent exome sequencing has identified novel targets of mutation, including *CNOT3*, a member of transcriptional regulatory complex and ribosomal proteins.[163] To gain further insight into the male sex preponderance of T-ALL, Ferrando and colleagues performed targeted capture and sequencing of X chromosome genes. They identified sequence mutations and deletions of PHF6 in 16% and 38% of childhood and adult T-ALL, respectively.[164] *PHF6* alterations result in loss of *PHF6* expression and are associated with *TLX1/3* and *TAL1* rearranged ALL.[164] Although the role of PHF6 in leukaemogenesis is poorly understood, it may have complex and multifactorial roles as a tumour suppressor gene.

Early T-cell precursor ALL

Recently, a subtype of immature T-lineage ALL was described in which the leukaemic cells lack expression of mature/cortical thymic markers such as CD1a, CD8 and CD5 and exhibit aberrant expression of myeloid and stem cell markers. These cells exhibit a gene expression profile reminiscent of the murine early thymic (double-negative 1 stage) T-cell precursor that retains myeloid/macrophage differentiation capacity. These early T-cell precursor (ETP) ALL comprise an aggressive subtype with a dismal prognosis.[165,166] There are active efforts to identify the genetic basis of this subtype of ALL and recent reports have identified rearrangements of *MEF2C* in a proportion of them.[167]

WGS of tumour and matched non-tumour DNA of 12 ETP ALL cases and mutation recurrence testing of selected genes in 94 additional ETP and non-ETP T-ALL cases[168] unexpectedly showed marked diversity in the frequency and nature of genetic alterations. Several cases exhibited complex, multi-chromosomal structural alterations with the hallmarks of chromothripsis,[169] but no common genomic alteration was identified. However, three pathways were frequently mutated: haematopoietic development, cytokine receptor and Ras signalling, in addition to chromatin modification.[170–175] Loss-of-function alterations in genes encoding regulators of haematopoietic development are present in two-thirds of ETP T-ALL and most commonly involve *ETV6, GATA3, IKZF1* and *RUNX1*. It is notable that many of these genes are known targets of mutation and rearrangement in other subtypes of ALL and AML. Activating mutations in cytokine receptor and Ras signalling were also present in the majority of cases, including *NRAS, KRAS, FLT3, JAK1, JAK3* and *IL7R*, similar to those previously reported in other leukaemia subtypes. Activating mutations of IL7R, encoding the alpha chain of the interleukin 7 receptor, have also been reported.[175,176] These mutations are usually complex in-frame insertion mutations that introduce a cysteine into the transmembrane domain of IL7R, resulting in dimerization of the receptor and constitutive activation of JAK-STAT signalling in the absence of ligand. In cell lines and primary mouse bone marrow, the IL7R mutations induce cytokine-independent proliferation and activation of JAK-STAT signalling that is abrogated by JAK inhibitors such as ruxolitinib.[168] Although IL7R mutations are present in only a proportion of ETP ALL cases, evidence of JAK-STAT activation on phosphoflow cytometry or gene expression profiling is present in the majority of cases, suggesting that JAK inhibitors are a rational therapeutic strategy in this high-risk leukaemia.

An unexpected finding in ETP ALL was a high frequency of mutations of epigenetic regulators. Most common were mutations or deletions of genes encoding components of the polycomb repressor complex 2 (PRC2; *EZH2, SUZ12, EED*), which normally mediates histone 3 lysine 27 (H3K27) trimethylation. A range of deleterious mutations in the SET domain and elsewhere in EZH2 are observed that are predicted to be loss of function. EZH2 and PRC2 also interact with the histone methyltransferase DNMT3A, which is mutated in adult, but not childhood, ETP ALL.[177]

Other T-ALL genetic subtypes: *MLL* rearranged and *PICALM-MLLT10*

MLL is rearranged in about 5% of T-ALL cases, most commonly to *MLLT1 (ENL)*[178] and more frequently in adolescents. *MLL*-rearranged T-ALL represents a distinct biological entity with a transcriptional profile that differs from that in other *MLL*-rearranged cases.[147,179]

The translocation t(10;11)(p13;q14) may be cytogenetically cryptic and results in expression of the *PICALM-MLLT10 (CALM-AF10)* fusion.[180] It is observed in up to 10% of T-ALL. Notably, both partner genes are infrequently fused to *MLL* and, like *MLL*-rearranged ALL, *PICALM-MLLT10* cases exhibit upregulation of *HOX* genes and *MEIS1*, suggesting common oncogenic pathways. This rearrangement is typically seen in γδ T-ALL cases, in either immature or mature cells, and is associated with a poor outcome. Expression of the fusion in haematopoietic cells results in the development of leukaemia, which is often myeloid in phenotype.[181]

Additional recurring epigenetic target alterations include *SETD2*, encoding a histone 3 lysine 36 trimethylase, and the histone acetyltransferase and CREBBP homologue EP300 (p300). Additional new targets of mutation have been identified, including *DNM2*, *ECT2L* and *RELN*, and several specific somatic mutations that had previously been reported as germline mutations in inherited developmental disorders, notably those in the zinc finger domain of the haematopoietic transcription factor gene *GATA3*. As the mutational spectrum of ETP ALL is similar to that observed in myeloid leukaemias and the transcriptional profile of ETP ALL is similar to that of normal and malignant human haematopoietic stem cells and myeloid progenitors, but *not* the normal human early T-cell precursor,[168] 'early T-cell precursor' ALL is likely a misnomer and ETP ALL may be more appropriately considered to be part of a spectrum of immature leukaemias of variable and often ambiguous lineage.

Relapsed ALL

Several chromosomal alterations, such as *BCR-ABL1* and *MLL* rearrangement, are associated with a high risk of treatment failure. However, relapse occurs across the spectrum of ALL subtypes. It has long been recognized that ALL genomes are not static, but exhibit acquisition of chromosomal abnormalities over time.[182] There is therefore intense interest in genomic profiling of matched diagnosis and relapse samples to dissect the genetic basis of clonal heterogeneity in ALL and the relationship of such heterogeneity to risk of relapse. Although the primary chromosomal abnormality is usually retained between diagnosis and relapse, it has been shown that the majority of ALL show changes in the patterns of their secondary genomic alterations from diagnosis to relapse[183,184] and that many relapse-acquired lesions, including *IKZF1* and *CDKN2A*, are present at low levels at diagnosis.[184,185] Recurring mutations have been identified that influence drug sensitivity and risk of relapse. Mutations in the transcriptional coactivator and acetyl transferase CREBBP (CREB-binding protein or CBP) is a relapse-acquired lesion in up to 20% of relapsed ALL samples.[24,186] CREBBP acetylates both histone and non-histone targets and has a role in regulating the transcriptional response to glucocorticoid therapy. CREBBP has an important role in mediating the transcriptional response to glucocorticoids[187,188] and histone deacetylase inhibitors were active in steroid-resistant ALL cell lines.[186] Recently, two groups independently identified relapse-acquired mutations in the 5′ nucleotidase gene *NT5C2* that confer increased resistance to purine analogues.[189,190] Hence mutations that confer resistance to drugs commonly used to treat ALL represent a key mechanism of treatment failure and resistance.

Future directions

We have seen recently how the outcome of *BCR-ABL1*-positive ALL has been dramatically improved by treatment with TKI, reducing the requirement for bone marrow transplantation in Ph-positive adults. These studies have clearly shown that the application of novel agents in the appropriate biological arena to a suitable target can dramatically improve survival. Evolving studies are revealing other potential candidates with promise for future therapies. However, there remain many challenges ahead before these novel drugs become integrated into routine clinical practice. The discovery of germline mutations, for example, the high incidence of germline *TP53* mutations in patients with hypodiploid ALL,[60] has highlighted the role of genetic predisposition to certain subtypes of disease, which are clearly more widespread than previously envisaged. We should continue to search for novel targets that will surely emerge from the detailed analysis of accumulating data from state-of-the-art next-generation sequencing technologies in parallel with expression, proteomic and epigenetic studies. Total cure for ALL maybe achieved within the not too distant future.

References

1 Inaba, H., Greaves, M., Mullighan, C.G. Acute lymphoblastic leukaemia. *Lancet* 2013; 381: 1943–1955.

2 Pui, C.H., Robison, L.L., Look, A.T. Acute lymphoblastic leukaemia. *Lancet* 2008; 371: 1030–1043.

3 Nowell, P.C., Hungerford, D.A. A minute chromosome in human granulocytic leukemia. *Science* 1960; 132: 1497–1497.

4 Look, A.T. Oncogenic transcription factors in the human acute leukemias. *Science* 1997; 278: 1059–1064.

5 Ford, A.M., Ridge, S.A., Cabrera, M.E., et al. In utero rearrangements in the trithorax-related oncogene in infant leukaemias. *Nature* 1993; 363: 358–360.

6 Gale, K.B., Ford, A.M., Repp, R., et al. Backtracking leukemia to birth: identification of clonotypic gene fusion sequences in neonatal blood spots. *Proc Natl Acad Sci U S A* 1997; 94: 13950–13954.

7 Mullighan, C.G., Goorha, S., Radtke, I., et al. Genome-wide analysis of genetic alterations in acute lymphoblastic leukaemia. *Nature* 2007; 446: 758–764.

8 Mullighan, C.G. Molecular genetics of B-precursor acute lymphoblastic leukemia. *J Clin Invest* 2012; 122: 3407–3415.

9 Mullighan, C.G. Genome sequencing of lymphoid malignancies. *Blood* 2013; 122: 3899–3907.

10 Chilton, L., Buck, G., Harrison, C.J., et al. High hyperdiploidy among adolescents and adults with acute lymphoblastic leukaemia (ALL): cytogenetic features, clinical characteristics and outcome. *Leukemia* 2014; 28: 1511–1518.

11 Moorman, A.V., Richards, S.M., Martineau, M., et al. Outcome heterogeneity in childhood high-hyperdiploid acute lymphoblastic leukemia. *Blood* 2003; 102: 2756–2762.

12 Tallen, G., Ratei, R., Mann, G., et al. Long-term outcome in children with relapsed acute lymphoblastic leukemia after time-point and site-of-relapse stratification and intensified short-course multidrug chemotherapy: results of trial ALL-REZ BFM 90. *J Clin Oncol* 2010; 28: 2339–2347.

13 Sutcliffe, M.J., Shuster, J.J., Sather, H.N., et al. High concordance from independent studies by the Children's Cancer Group (CCG) and Pediatric Oncology Group (POG) associating favorable prognosis with combined trisomies 4, 10 and 17 in children with NCI Standard-Risk B-precursor Acute Lymphoblastic Leukemia: a Children's Oncology Group (COG) initiative. *Leukemia* 2005; 19: 734–740.

14 Paulsson, K., Johansson, B. High hyperdiploid childhood acute lymphoblastic leukemia. *Genes Chromosomes Cancer* 2009; 48: 637–660.

15 Harrison, C.J., Moorman, A.V., Barber, K.E., et al. Interphase molecular cytogenetic screening for chromosomal abnormalities of prognostic significance in childhood acute lymphoblastic leukaemia: a UK Cancer Cytogenetics Group study. *Br J Haematol* 2005; 129: 520–530.

16 Haas, O.A. Is genomic imprinting involved in the pathogenesis of hyperdiploid and haploid acute lymphoblastic leukemia of childhood? *Acta Genet Med Gemellol (Roma)* 1996; 45: 239–242.

17 Gruszka-Westwood, A.M., Horsley, S.W., Martinez-Ramirez, A., et al. Comparative expressed sequence hybridization studies of high-hyperdiploid childhood acute lymphoblastic leukemia. *Genes Chromosomes Cancer* 2004; 41: 191–202.

18 Paulsson, K., Panagopoulos, I., Knuutila, S., et al. Formation of trisomies and their parental origin in hyperdiploid childhood acute lymphoblastic leukemia. *Blood* 2003; 102: 3010–3015.

19 Paulsson, K., Morse, H., Fioretos, T., Behrendtz, M., Strombeck, B., Johansson, B. Evidence for a single-step mechanism in the origin of hyperdiploid childhood acute lymphoblastic leukemia. *Genes Chromosomes Cancer* 2005; 44: 113–122.

20 Andersson, A., Olofsson, T., Lindgren, D., et al. Molecular signatures in childhood acute leukemia and their correlations to expression patterns in normal hematopoietic subpopulations. *Proc Natl Acad Sci U S A* 2005; 102: 19069–19074.

21 Ross, M.E., Zhou, X., Song, G., et al. Classification of pediatric acute lymphoblastic leukemia by gene expression profiling. *Blood* 2003; 102: 2951–2959.

22 Figueroa, M.E., Chen, S.C., Andersson, A.K., et al. Integrated genetic and epigenetic analysis of childhood acute lymphoblastic leukemia. *J Clin Invest* 2013; 123: 3099–3111.

23 Paulsson, K., Horvat, A., Strombeck, B., et al. Mutations of FLT3, NRAS, KRAS and PTPN11 are frequent and possibly mutually exclusive in high hyperdiploid childhood acute lymphoblastic leukemia. *Genes Chromosomes Cancer* 2008; 47: 26–33.

24 Inthal, A., Zeitlhofer, P., Zeginigg, M., et al. CREBBP HAT domain mutations prevail in relapse cases of high hyperdiploid childhood acute lymphoblastic leukemia. *Leukemia* 2012; 26: 1797–1803.

25 Mullighan, C.G. Single nucleotide polymorphism microarray analysis of genetic alterations in cancer. *Methods Mol Biol* 2011; 730: 235–258.

26 Aguiar, R.C.T., Sohal, J., van Rhee, F., et al. TEL-AML1 fusion in acute lymphoblastic leukaemia of adults. *Br J Haematol* 1996; 95: 673–677.

27 Al-Obaidi, M.S.J., Martineau, M., Bennett, C.F., et al. ETV6/AML1 fusion by FISH in adult acute lymphoblastic leukemia. *Leukemia* 2002; 16: 669–674.

28 Golub, T., McLean, T., Stegmaier, K., et al. TEL-AML1: the most common gene rearrangement in childhood ALL. *Blood* 1995; 86: 2377.

29 Raynaud, S., Mauvieux, L., Cayuela, J.M., et al. TEL/AML1 fusion gene is a rare event in adult acute lymphoblastic leukemia. *Leukemia* 1996; 10: 1529–1530.

30 Romana, S.P., Poirel, H., Leconiat, M., et al. High-frequency of T(12–21) in childhood B-lineage acute lymphoblastic leukemia. *Blood* 1995; 86: 4263–4269.

31 Shurtleff, S.A., Buijs, A., Behm, F.G., et al. TEL/AML1 fusion resulting from a cryptic t(12;21) is the most common genetic lesion in pediatric ALL and defines a subgroup of patients with an excellent prognosis. *Leukemia* 1995; 9: 1985–1989.

32 Romana, S.P., Le Coniat, M., Berger, R. t(12;21): a new recurrent translocation in acute lymphoblastic leukemia. *Genes Chromosomes Cancer* 1994; 9: 186–191.

33 Golub, T.R., Barker, G.F., Bohlander, S.K., et al. Fusion of the TEL gene on 12p13 to the AML1 gene on 21q22 in acute lymphoblastic-leukemia. *Proc Natl Acad Sci U S A* 1995; 92: 4917–4921.

34 Romana, S.P., Mauchauffe, M., Leconiat, M., et al. The t(12;21) of acute lymphoblastic-leukemia results in a tel-AML1 gene fusion. *Blood* 1995; 85: 3662–3670.

35 Bhojwani, D., Pei, D., Sandlund, J.T., et al. ETV6-RUNX1-positive childhood acute lymphoblastic leukemia: improved outcome with contemporary therapy. *Leukemia* 2012; 26: 265–270.

36 Ford, A.M., Bennett, C.A., Price, C.M., Bruin, M.C.A., Van Wering, E.R., Greaves, M. Fetal origins of the TEL-AML1 fusion gene in identical twins with leukemia. *Proc Natl Acad Sci U S A* 1998; 95: 4584–4588.

37 Mori, H., Colman, S.M., Xiao, Z.J., et al. Chromosome translocations and covert leukemic clones are generated during normal fetal development. *Proc Natl Acad Sci U S A* 2002; 99: 8242–8247.

38 Wiemels, J.L., Cazzaniga, G., Daniotti, M., et al. Prenatal origin of acute lymphoblastic leukaemia in children. *Lancet* 1999; 354: 1499–1503.

39 Wiemels, J.L., Ford, A.M., Van Wering, E.R., Postma, A., Greaves, M. Protracted and variable latency of acute lymphoblastic leukemia after *TEL-AML1* gene fusion in utero. *Blood* 1999; 94: 1057–1062.

40 Raynaud, S., Cave, H., Baens, M., et al. The 12;21 translocation involving TEL and deletion of the other TEL allele: two frequently associated alterations found in childhood acute lymphoblastic leukemia. *Blood* 1996; 87: 2891–2899.

41 SennanaSendi, H., Dastugue, N., Talmant, P., et al. Deletions of the non-translocated allele of TEL (ETV6) seem to be constant in childhood B-lineage ALLs with a t(12;21). Results of a FISH study on 148 children

included in the 58881 therapeutic trial from the CLCG (EORTC). *Blood* 1996; 88: 274.

42 Shoemaker, S.G., Hromas, R., Kaushansky, K. Transcriptional regulation of interleukin 3 gene expression in T lymphocytes. *Proc Natl Acad Sci U S A* 1990; 87: 9650–9654.

43 Takahashi, A., Satake, M., Yamaguchiiwai, Y., et al. Positive and negative regulation of granulocyte-macrophage colony-stimulating factor promoter activity by AML1-related transcription factor, PEBP2. *Blood* 1995; 86: 607–616.

44 Imai, Y., Kurokawa, M., Tanaka, K., et al. TLE, the human homolog of Groucho, interacts with AML1 and acts as a repressor of AML1-induced transactivation. *Biochem Biophys Res Commun* 1998; 252: 582–589.

45 Okuda, T., van Deursen, J., Hiebert, S.W., Grosveld, G., Downing, J.R. AML1, the target of multiple chromosomal translocations in human leukemia, is essential for normal fetal liver hematopoiesis. *Cell* 1996; 84: 321–330.

46 Wang, Q., Stacy, T., Miller, J.D., et al. The CBFβ subunit is essential for CBFα2 (AML1) function in vivo. *Cell* 1996; 87: 697–708.

47 Moorman, A.V., Chilton, L., Wilkinson, J., Ensor, H.M., Bown, N., Proctor, S.J. A population-based cytogenetic study of adults with acute lymphoblastic leukemia. *Blood* 2010; 115: 206–214.

48 Privitera, E., Kamps, M.P., Hayashi, Y., et al. Different molecular consequences of the 1;19 chromosomal translocation in childhood B-cell precursor acute lymphoblastic leukemia. *Blood* 1992; 79: 1781–1788.

49 Raimondi, S.C., Behm, F.G., Roberson, P.K., et al. Cytogenetics of pre-B-cell acute lymphoblastic leukemia with emphasis on prognostic implications of the t(1;19). *J Clin Oncol*1990; 8: 1380–1388.

50 Hunger, S.P. Chromosomal translocations involving the E2A gene in acute lymphoblastic leukemia: clinical features and molecular pathogenesis. *Blood* 1996; 87: 1211–1224.

51 Jeha, S., Pei, D., Raimondi, S.C., et al. Increased risk for CNS relapse in pre-B cell leukemia with the t(1;19)/TCF3-PBX1. *Leukemia* 2009; 23: 1406–1409.

52 Pui, C.H., Campana, D., Pei, D., et al. Treating childhood acute lymphoblastic leukemia without cranial irradiation. *N Engl J Med* 2009; 360: 2730–2741.

53 Secker-Walker, L.M., Berger, R., Fenaux, P., et al. Prognostic significance of the balanced t(1;19) and unbalanced der(19)t(1;19) translocations in acute lymphoblastic leukemia. *Leukemia* 1992; 6: 363–369.

54 Shanmugam, K., Green, N.C., Rambaldi, I., Saragovi, H.U., Featherstone, M.S. PBX and MEIS as non-DNA-binding partners in trimeric complexes with HOX proteins. *Mol Cell Biol* 1999; 19: 7577–7588.

55 Inaba, T., Roberts, W.M., Shapiro, L.H., et al. Fusion of the leucine zipper gene HLF to the E2A gene in human acute B-lineage leukemia. *Science* 1992; 257: 531–534.

56 Hunger, S.P., Devaraj, P.E., Foroni, L., Secker-Walker, L.M., Cleary, M.L. Two types of genomic rearrangements create alternative E2A-HLF fusion proteins in t(17;19)-ALL. *Blood* 1994; 83: 2970–2977.

57 Moorman, A.V. The clinical relevance of chromosomal and genomic abnormalities in B-cell precursor acute lymphoblastic leukaemia. *Blood Rev* 2012; 26: 123–135.
58 Harrison, C.J., Moorman, A.V., Broadfield, Z.J., et al. Three distinct subgroups of hypodiploidy in acute lymphoblastic leukaemia. *Br J Haematol* 2004; 125: 552–559.
59 Nachman, J.B., Heerema, N.A., Sather, H., et al. Outcome of treatment in children with hypodiploid acute lymphoblastic leukemia. *Blood* 2007; 110: 1112–1115.
60 Holmfeldt, L., Wei, L., Diaz-Flores, E., et al. The genomic landscape of hypodiploid acute lymphoblastic leukemia. *Nat Genet* 2013; 45: 242–252.
61 Powell, B.C., Jiang, L., Muzny, D.M., et al. Identification of TP53 as an acute lymphocytic leukemia susceptibility gene through exome sequencing. *Pediatr Blood Cancer* 2013; 60: E1–E3.
62 Pui, C.H., Behm, F.G., Downing, J.R., et al. 11q23/MLL rearrangement confers a poor prognosis in infants with acute lymphoblastic leukemia. *J Clin Oncol* 1994; 12: 909–915.
63 Rubnitz, J.E., Link, M.P., Shuster, J.J., et al. Frequency and prognostic significance of Hrx rearrangements in infant acute lymphoblastic leukemia – a Pediatric Oncology Group study. *Blood* 1994; 84: 570–573.
64 Moorman, A.V., Ensor, H.M., Richards, S.M., et al. Prognostic effect of chromosomal abnormalities in childhood B-cell precursor acute lymphoblastic leukaemia: results from the UK Medical Research Council ALL97/99 randomised trial. *Lancet Oncol* 2010; 11: 429–438.
65 Mancini, M., Scappaticci, D., Cimino, G., et al. A comprehensive genetic classification of adult acute lymphoblastic leukemia (ALL): analysis of the GIMEMA 0496 protocol. *Blood* 2005; 105: 3434–3441.
66 Moorman, A.V., Harrison, C.J., Buck, G.A., et al. Karyotype is an independent prognostic factor in adult acute lymphoblastic leukemia (ALL): analysis of cytogenetic data from patients treated on the Medical Research Council (MRC) UKALLXII/Eastern Cooperative Oncology Group (ECOG) 2993 trial. *Blood* 2007; 109: 3189–3197.
67 Mann, G., Attarbaschi, A., Schrappe, M., et al. Improved outcome with hematopoietic stem cell transplantation in a poor prognostic subgroup of infants with mixed-lineage-leukemia (MLL)-rearranged acute lymphoblastic leukemia: results from the Interfant-99 study. *Blood* 2010; 116: 2644–2650.
68 Pieters, R., Schrappe, M., De Lorenzo, P., et al. A treatment protocol for infants younger than 1 year with acute lymphoblastic leukaemia (Interfant-99): an observational study and a multicentre randomised trial. *Lancet* 2007; 370: 240–250.
69 Pui, C.H., Chessells, J.M., Camitta, B., et al. Clinical heterogeneity in childhood acute lymphoblastic leukemia with 11q23 rearrangements. *Leukemia* 2003; 17: 700–706.
70 Meyer, C., Kowarz, E., Hofmann, J., et al. New insights to the MLL recombinome of acute leukemias. *Leukemia* 2009; 23: 1490–1499.

71 Djabali, M., Selleri, L., Parry, P., Bower, M., Young, B.D., Evans, G.A. A trithorax-like gene is interrupted by chromosome 11q23 translocations in acute leukaemias. *Nat Genet* 1992; 2: 113–118.
72 Nilson, I., Lochner, K., Siegler, G., et al. Exon/intron structure of the human ALL-1 (MLL) gene involved in translocations to chromosomal region 11q23 and acute leukaemias. *Br J Haematol* 1996; 93: 966–972.
73 Armstrong, S.A., Staunton, J.E., Silverman, L.B., et al. MLL translocations specify a distinct gene expression profile that distinguishes a unique leukemia. *Nat Genet* 2002; 30: 41–47.
74 Mi, S., Li, Z., Chen, P., et al. Aberrant over-expression and function of the miR-17–92 cluster in MLL-rearranged acute leukemia. *Proc Natl Acad Sci U S A* 2010; 107: 3710–3715.
75 Popovic, R., Riesbeck, L.E., Velu, C.S., et al. Regulation of mir-196b by MLL and its over-expression by MLL fusions contributes to immortalization. *Blood* 2009; 113: 3314–3322.
76 Jiang, X., Huang, H., Li, Z., et al. Blockade of miR-150 maturation by MLL-fusion/MYC/LIN-28 is required for MLL-associated leukemia. *Cancer Cell* 2012; 22: 524–535.
77 Daigle, S.R., Olhava, E.J., Therkelsen, C.A., et al. Selective killing of mixed lineage leukemia cells by a potent small-molecule DOT1L inhibitor. *Cancer Cell* 2011; 20: 53–65.
78 Grembecka, J., He, S., Shi, A., et al. Menin-MLL inhibitors reverse oncogenic activity of MLL fusion proteins in leukemia. *Nat Chem Biol* 2012; 8: 277–284.
79 Liu-Dumlao, T., Kantarjian, H., Thomas, D.A., O'Brien, S., Ravandi, F. Philadelphia-positive acute lymphoblastic leukemia: current treatment options. *Curr Oncol Rep* 2012; 14: 387–394.
80 Rowley, J.D. A new consistent chromosomal abnormality in chronic myelogenous leukaemia identified by quinacrine fluorescence and Giemsa staining (Letter). *Nature* 1973; 243: 290–293.
81 Groffen, J., Stephenson, J.R., Heisterkamp, N., Deklein, A., Bartram, C.R., Grosveld, G. Philadelphia chromosomal breakpoints are clustered within a limited region, Bcr, on chromosome-22. *Cell* 1984; 36: 93–99.
82 Heisterkamp, N., Stam, K., Groffen, J., de Klein, A., Grosveld, G. Structural organization of the bcr gene and its role in the Ph′ translocation. *Nature* 1985; 315: 758–761.
83 McLaughlin, J., Chianese, E., Witte, O.N. Alternative forms of the BCR-ABL oncogene have quantitatively different potencies for stimulation of immature lymphoid cells. *Mol Cell Biol* 1989; 9: 1866–1874.
84 Sattler, M., Griffin, J.D. Molecular mechanisms of transformation by the BCR-ABL oncogene. *Semin Hematol* 2003; 40: 4–10.
85 Heerema, N.A., Harbott, J., Galimberti, S., et al. Secondary cytogenetic aberrations in childhood Philadelphia chromosome positive acute lymphoblastic leukemia are nonrandom and may be associated with outcome. *Leukemia* 2004; 18: 693–702.
86 Mullighan, C.G., Miller, C.B., Radtke, I., et al. BCR-ABL1 lymphoblastic leukaemia is characterized by the deletion of Ikaros. *Nature* 2008; 453: 110–114.

87 Van Rhee, F., Kasprzyk, A., Jamil, A., et al. Detection of the BCR-ABL gene by reverse transcription/polymerase chain reaction and fluorescence in situ hybridization in a patient with Philadelphia chromosome negative acute lymphoblastic leukaemia. *Br J Haematol* 1995; 90: 225–228.
88 Crist, W., Carroll, A., Shuster, J., et al. Philadelphia-chromosome positive childhood acute lymphoblastic leukemia – clinical and cytogenetic characteristics and treatment outcome – a Pediatric Oncology Group study. *Blood* 1990; 76: 489–494.
89 Arico, M., Valsecchi, M.G., Camitta, B., et al. Outcome of treatment in children with Philadelphia chromosome-positive acute lymphoblastic leukemia. *N Engl J Med 32000*; 42: 998–1006.
90 Schultz, K.R., Bowman, W.P., Aledo, A., et al. Improved early event-free survival with imatinib in Philadelphia chromosome-positive acute lymphoblastic leukemia: a Children's Oncology Group study. *J Clin Oncol* 2009; 27: 5175–5181.
91 Kantarjian, H., Giles, F., Wunderle, L., et al. Nilotinib in imatinib-resistant CML and Philadelphia chromosome-positive ALL. *N Engl J Med* 2006; 354: 2542–2551.
92 Ottmann, O., Dombret, H., Martinelli, G., et al. Dasatinib induces rapid hematologic and cytogenetic responses in adult patients with Philadelphia chromosome-positive acute lymphoblastic leukemia with resistance or intolerance to imatinib: interim results of a phase 2 study. *Blood* 2007; 110: 2309–2315.
93 Secker-Walker, L.M., Craig, J.M., Hawkins, J.M., Hoffbrand, A.V. Philadelphia positive acute lymphoblastic leukemia in adults: age distribution, BCR breakpoint and prognostic significance. *Leukemia* 1991; 5: 196–199.
94 Fielding, A.K., Rowe, J.M., Buck, G., et al. UKALLXII/ECOG2993: addition of imatinib to a standard treatment regimen enhances long-term outcomes in Philadelphia positive acute lymphoblastic leukemia. *Blood* 2014; 123: 843–850.
95 Harewood, L., Robinson, H., Harris, R., et al. Amplification of AML1 on a duplicated chromosome 21 in acute lymphoblastic leukemia: a study of 20 cases. *Leukemia* 2003; 17: 547–553.
96 Moorman, A.V., Richards, S.M., Robinson, H.M., et al. Prognosis of children with acute lymphoblastic leukemia (ALL) and intrachromosomal amplification of chromosome 21 (iAMP21). *Blood* 2007; 109: 2327–2330.
97 Robinson, H.M., Harrison, C.J., Moorman, A.V., Chudoba, I., Strefford, J.C. Intrachromosomal amplification of chromosome 21 (iAMP21) may arise from a breakage–fusion–bridge cycle. *Genes Chromosomes Cancer* 2007; 46: 318–326.
98 Strefford, J.C., van Delft, F.W., Robinson, H.M., et al. Complex genomic alterations and gene expression in acute lymphoblastic leukemia with intrachromosomal amplification of chromosome 21. *Proc Natl Acad Sci U S A* 2006; 103: 8167–8172.
99 Harrison, C.J., Moorman, A.V., Schwab, C.J., et al. An international study of intrachromosomal amplification of chromosome 21 (iAMP21): cytogenetic characterization and outcome. *Leukemia* 2014; 28: 1015–1021.
100 Heerema, N.A., Carroll, A.J., Devidas, M., et al. Intrachromosomal amplification of chromosome 21 is associated with inferior outcomes in children with acute

lymphoblastic leukemia treated in contemporary standard-risk children's oncology group studies: a report from the Children's Oncology Group. *J Clin Oncol* 2013; 31: 3397–3402.

101 Moorman, A.V., Robinson, H., Schwab, C., et al. Risk-directed treatment intensification significantly reduces the risk of relapse among children and adolescents with acute lymphoblastic leukemia and intrachromosomal amplification of chromosome 21: a comparison of the MRC ALL97/99 and UKALL2003 trials. *J Clin Oncol* 2013; 31: 3389–3396.

102 Granada, I., Sancho, J.-M., Oriol, A., et al. The prognostic significance of complex karyotype in Philadelphia chromosome-negative (Ph–) acute lymphoblastic leukemia (ALL) in adults is related with risk group. *ASH Annu Meet Abstr* 2007; 110: 3501.

103 Kawamata, N., Ogawa, S., Zimmermann, M., et al. Molecular allelokaryotyping of pediatric acute lymphoblastic leukemias by high-resolution single nucleotide polymorphism oligonucleotide genomic microarray. *Blood* 2008; 111: 776–784.

104 Kuiper, R.P., Schoenmakers, E.F., van Reijmersdal, S.V., et al. High-resolution genomic profiling of childhood ALL reveals novel recurrent genetic lesions affecting pathways involved in lymphocyte differentiation and cell cycle progression. *Leukemia* 2007; 21: 1258–1266.

105 Mullighan, C.G., Downing, J.R. Global genomic characterization of acute lymphoblastic leukemia. *Semin Hematol* 2009; 46: 3–15.

106 Zhang, J., Mullighan, C.G., Harvey, R.C., et al. Key pathways are frequently mutated in high-risk childhood acute lymphoblastic leukemia: a report from the Children's Oncology Group. *Blood* 2011; 118: 3080–3087.

107 Dobbins, S.E., Sherborne, A.L., Ma, Y.P., et al. The silent mutational landscape of infant MLL-AF4 pro-B acute lymphoblastic leukemia. *Genes Chromosomes Cancer* 2013; 52: 954–960.

108 Parker, H., An, Q., Barber, K., et al. The complex genomic profile of ETV6-RUNX1 positive acute lymphoblastic leukemia highlights a recurrent deletion of TBL1XR1. *Genes Chromosomes Cancer* 2008;47: 1118–1125.

109 Schwab, C.J., Chilton, L., Morrison, H., et al. Genes commonly deleted in childhood B-cell precursor acute lymphoblastic leukemia: association with cytogenetics and clinical features. *Haematologica* 2013; 98: 1081–1088.

110 Heltemes-Harris, L.M., Willette, M.J., Ramsey, L.B., et al. Ebf1 or Pax5 haploinsufficiency synergizes with STAT5 activation to initiate acute lymphoblastic leukemia. *J Exp Med* 2011; 208: 1135–1149.

111 Virely, C., Moulin, S., Cobaleda, C., et al. Haploinsufficiency of the IKZF1 (IKAROS) tumor suppressor gene cooperates with BCR-ABL in a transgenic model of acute lymphoblastic leukemia. *Leukemia* 2010; 24: 1200–1204.

112 Mullighan, C.G., Su, X., Zhang, J., et al. Deletion of IKZF1 and prognosis in acute lymphoblastic leukemia. *N Engl J Med* 2009; 360: 470–480.

113 Iacobucci, I., Lonetti, A., Paoloni, F., et al. The PAX5 gene is frequently rearranged in BCR-ABL1-positive acute lymphoblastic leukemia but is not associated with outcome. A report on behalf of the GIMEMA Acute Leukemia Working Party. *Haematologica* 2010; 95: 1683–1690.

114 Iacobucci, I., Storlazzi, C.T., Cilloni, D., et al. Identification and molecular characterization of recurrent genomic deletions on 7p12 in the IKZF1 gene in a large cohort of BCR-ABL1-positive acute lymphoblastic leukemia patients: on behalf of Gruppo Italiano Malattie Ematologiche dell'Adulto Acute Leukemia Working Party (GIMEMA AL WP). *Blood* 2009; 114: 2159–2167.
115 Martinelli, G., Iacobucci, I., Storlazzi, C.T., et al. IKZF1 (Ikaros) deletions in BCR-ABL1-positive acute lymphoblastic leukemia are associated with short disease-free survival and high rate of cumulative incidence of relapse: a GIMEMA AL WP report. *J Clin Oncol* 2009; 27: 5202–5207.
116 Den Boer, M.L., van Slegtenhorst, M., De Menezes, R.X., et al. A subtype of childhood acute lymphoblastic leukaemia with poor treatment outcome: a genome-wide classification study. *Lancet Oncol* 2009; 10: 125–134.
117 Mullighan, C.G., Su, X., Zhang, J., et al. Deletion of IKZF1 and prognosis in acute lymphoblastic leukemia. *N Engl J Med* 2009; 360: 470–480.
118 Roberts, K.G., Morin, R.D., Zhang, J., et al. Genetic alterations activating kinase and cytokine receptor signaling in high-risk acute lymphoblastic leukemia. *Cancer Cell* 2012; 22: 153–166.
119 Georgopoulos, K., Bigby, M., Wang, J.H., et al. The Ikaros gene is required for the development of all lymphoid lineages. *Cell* 1994; 79: 143–156.
120 Chapiro, E., Russell, L., Lainey, E., et al. Activating mutation in the TSLPR gene in B-cell precursor lymphoblastic leukemia. *Leukemia* 2010; 24: 642–645.
121 Moorman, A.V., Schwab, C., Ensor, H.M., et al. IGH@ translocations, CRLF2 deregulation and micro-deletions in adolescents and adults with acute lymphoblastic leukemia (ALL). *J Clin Oncol* 2012; 30: 3100–3108.
122 Mullighan, C.G., Collins-Underwood, J.R., Phillips, L.A., et al. Rearrangement of CRLF2 in B-progenitor- and Down syndrome-associated acute lymphoblastic leukemia. *Nat Genet* 2009; 41: 1243–1246.
123 Russell, L.J., Capasso, M., Vater, I., et al. Deregulated expression of cytokine receptor gene, CRLF2, is involved in lymphoid transformation in B-cell precursor acute lymphoblastic leukemia. *Blood* 2009; 114: 2688–2698.
124 Hertzberg, L., Vendramini, E., Ganmore, I., et al. Down syndrome acute lymphoblastic leukemia: a highly heterogeneous disease in which aberrant expression of CRLF2 is associated with mutated JAK2: a report from the iBFM Study Group. *Blood* 2010; 115: 1006–1017.
125 Bercovich, D., Ganmore, I., Scott, L.M., et al. Mutations of JAK2 in acute lymphoblastic leukaemias associated with Down's syndrome. *Lancet* 2008; 372: 1484–1492.
126 Kearney, L., Gonzalez De Castro, D., Yeung, J., et al. A specific JAK2 mutation (JAK2R683) and multiple gene deletions in Down syndrome acute lymphoblastic leukaemia. *Blood* 2008; 113: 646–648.
127 Mullighan, C.G., Zhang, J., Harvey, R.C., et al. JAK mutations in high-risk childhood acute lymphoblastic leukemia. *Proc Natl Acad Sci U S A* 2009; 106: 9414–9418.
128 Maude, S.L., Tasian, S.K., Vincent, T., et al. Targeting JAK1/2 and mTOR in murine xenograft models of Ph-like acute lymphoblastic leukemia. *Blood* 2012; 120: 3510–3518.

129 Tasian, S.K., Doral, M.Y., Borowitz, M.J., et al. Aberrant STAT5 and PI3K/mTOR pathway signaling occurs in human CRLF2-rearranged B-precursor acute lymphoblastic leukemia. *Blood* 2012; 120: 833–842.

130 Cario, G., Zimmermann, M., Romey, R., et al. Presence of the P2RY8-CRLF2 rearrangement is associated with a poor prognosis in non-high-risk precursor B-cell acute lymphoblastic leukemia in children treated according to the ALL-BFM 2000 protocol. *Blood* 2010; 115: 5393–5397.

131 Harvey, R.C., Mullighan, C.G., Chen, I.M., et al. Rearrangement of CRLF2 is associated with mutation of JAK kinases, alteration of IKZF1, Hispanic/Latino ethnicity and a poor outcome in pediatric B-progenitor acute lymphoblastic leukemia. *Blood* 2010; 115: 5312–5321.

132 Harvey, R.C., Mullighan, C.G., Wang, X., et al. Identification of novel cluster groups in pediatric high-risk B-precursor acute lymphoblastic leukemia with gene expression profiling: correlation with genome-wide DNA copy number alterations, clinical characteristics and outcome. *Blood* 2010; 116: 4874–4884.

133 Ensor, H.M., Schwab, C., Russell, L.J., et al. Demographic, clinical and outcome features of children with acute lymphoblastic leukemia and CRLF2 deregulation: results from the MRC ALL97 clinical trial. *Blood* 2011; 117: 2129–2136.

134 Chen, I.M., Harvey, R.C., Mullighan, C.G., et al. Outcome modeling with CRLF2, IKZF1, JAK and minimal residual disease in pediatric acute lymphoblastic leukemia: a Children's Oncology Group study. *Blood* 2012; 119: 3512–3522.

135 Loh, M.L., Zhang, J., Harvey, R.C., et al. Tyrosine kinome sequencing of pediatric acute lymphoblastic leukemia: a report from the Children's Oncology Group TARGET Project. *Blood* 2013; 121: 485–488.

136 Lengline, L., Beldjord, K., Dombret, H., Soulier, J., Boissel, N., Clappier, E. Successful tyrosine kinase inhibitor therapy in a refractory B-cell precursor acute lymphoblastic leukemia with EBF1-PDGFRB fusion. *Haematologica* 2013; 98: e146–e148.

137 Weston, B.W., Hayden, M.A., Roberts, K.G., et al. Tyrosine kinase inhibitor therapy induces remission in a patient with refractory EBF1-PDGFRB-positive acute lymphoblastic leukemia. *J Clin Oncol* 2013; 31: e413–e416.

138 Clappier, E., Auclerc, M.F., Rapion, J., et al. An intragenic ERG deletion (ERG) is a marker of an oncogenic subtype of B-cell precursor acute lymphoblastic leukemia with a favorable outcome despite frequent IKZF1 deletions. *Leukemia* 2014; 28: 70–77.

139 Zaliova, M., Zimmermanova, O., Dorge, P., et al. ERG deletion is associated with CD2 and attenuates the negative impact of IKZF1 deletion in childhood acute lymphoblastic leukemia. *Leukemia* 2014; 28: 182–185.

140 Aifantis, I., Raetz, E., Buonamici, S. Molecular pathogenesis of T-cell leukaemia and lymphoma. *Nat Rev Immunol* 2008; 8: 380–390.

141 Kuppers, R., Dalla-Favera, R. Mechanisms of chromosomal translocations in B cell lymphomas. *Oncogene* 2001; 20: 5580–5594.

142 Nickoloff, J.A., De Haro, L.P., Wray, J., Hromas, R. Mechanisms of leukemia translocations. *Curr Opin Hematol* 2008; 15: 338–345.

143 De Keersmaecker, K., Marynen, P., Cools, J. Genetic insights in the pathogenesis of T-cell acute lymphoblastic leukemia. *Haematologica* 2005; 90: 1116–1127.
144 Meijerink, J.P. Genetic rearrangements in relation to immunophenotype and outcome in T-cell acute lymphoblastic leukaemia. *Best Pract Res Clin Haematol* 2010; 23: 307–318.
145 Bash, R.O., Hall, S., Timmons, C.F., et al. Does activation of the TAL1 gene occur in a majority of patients with T-cell acute lymphoblastic leukemia?A Pediatric Oncology Group study. *Blood* 1995; 86: 666–676.
146 Bernard, O., Lecointe, N., Jonveaux, P., et al. Two site-specific deletions and t(1;14) translocation restricted to human T-cell acute leukemias disrupt the 5′ part of the tal-1 gene. *Oncogene* 1991; 6: 1477–1488.
147 Ferrando, A.A., Neuberg, D.S., Staunton, J., et al. Gene expression signatures define novel oncogenic pathways in T cell acute lymphoblastic leukemia. *Cancer Cell* 2002; 1: 75–87.
148 Xia, Y., Brown, L., Yang, C.Y., et al. TAL2, a helix–loop–helix gene activated by the (7;9)(q34;q32) translocation in human T-cell leukemia. *Proc Natl Acad Sci U S A* 1991; 88: 11416–11420.
149 Begley, C.G., Green, A.R. The SCL gene: from case report to critical hematopoietic regulator. *Blood* 1999; 93: 2760–2770.
150 Van Vlierberghe, P., van Grotel, M., Beverloo, H.B., et al. The cryptic chromosomal deletion del(11)(p12p13) as a new activation mechanism of LMO2 in pediatric T-cell acute lymphoblastic leukemia. *Blood* 2006; 108: 3520–3529.
151 McCormack, M.P., Young, L.F., Vasudevan, S., et al. The Lmo2 oncogene initiates leukemia in mice by inducing thymocyte self-renewal. *Science* 2010; 327: 879–883.
152 Argiropoulos, B., Humphries, R.K. Hox genes in hematopoiesis and leukemogenesis. *Oncogene* 2007; 26: 6766–6776.
153 Hatano, M., Roberts, C.W., Minden, M., Crist, W.M., Korsmeyer, S.J. Deregulation of a homeobox gene, HOX11, by the t(10;14) in T cell leukemia. *Science* 1991; 253: 79–82.
154 Lu, M., Gong, Z.Y., Shen, W.F., Ho, A.D. The tcl-3 proto-oncogene altered by chromosomal translocation in T-cell leukemia codes for a homeobox protein. *EMBO J* 1991; 10: 2905–2910.
155 Berger, R., Dastugue, N., Busson, M., et al. t(5;14)/HOX11L2-positive T-cell acute lymphoblastic leukemia. A collaborative study of the Groupe Français de Cytogénétique Hématologique (GFCH). *Leukemia* 2003; 17: 1851–1857.
156 Cave, H., Suciu, S., Preudhomme, C., et al. Clinical significance of HOX11L2 expression linked to t(5;14)(q35;q32), of HOX11 expression and of SIL-TAL fusion in childhood T-cell malignancies: results of EORTC studies 58881 and 58951. *Blood* 2004; 103: 442–450.
157 Bernard, O.A., Busson-LeConiat, M., Ballerini, P., et al. A new recurrent and specific cryptic translocation, t(5;14)(q35;q32), is associated with expression of the Hox11L2 gene in T acute lymphoblastic leukemia. *Leukemia* 2001; 15: 1495–1504.
158 De Keersmaecker, K., Real, P.J., Gatta, G.D., et al. The TLX1 oncogene drives aneuploidy in T cell transformation. *Nat Med* 2010; 16: 1321–1327.

159 Nagel, S., Kaufmann, M., Drexler, H.G., MacLeod, R.A. The cardiac homeobox gene NKX2–5 is deregulated by juxtaposition with BCL11B in pediatric T-ALL cell lines via a novel t(5;14)(q35.1;q32.2). *Cancer Res* 2003; 63: 5329–5334.
160 Nagel, S., Scherr, M., Kel, A., et al. Activation of TLX3 and NKX2–5 in t(5;14)(q35;q32) T-cell acute lymphoblastic leukemia by remote 3′-BCL11B enhancers and coregulation by PU.1 and HMGA1. *Cancer Res* 2007; 67: 1461–1471.
161 Su, X.Y., Busson, M., Della Valle, V., et al. Various types of rearrangements target TLX3 locus in T-cell acute lymphoblastic leukemia. *Genes Chromosomes Cancer* 2004; 41: 243–249.
162 Van Vlierberghe, P., Homminga, I., Zuurbier, L., et al. Cooperative genetic defects in TLX3 rearranged pediatric T-ALL. *Leukemia* 2008; 22: 762–770.
163 De Keersmaecker, K., Atak, Z.K., Li, N., et al. Exome sequencing identifies mutation in CNOT3 and ribosomal genes RPL5 and RPL10 in T-cell acute lymphoblastic leukemia. *Nat Genet* 2013; 45: 186–190.
164 Van Vlierberghe, P., Palomero, T., Khiabanian, H., et al. PHF6 mutations in T-cell acute lymphoblastic leukemia. *Nat Genet* 2010; 42: 338–342.
165 Coustan-Smith, E., Mullighan, C.G., Onciu, M., et al. Early T-cell precursor leukaemia: a subtype of very high-risk acute lymphoblastic leukaemia. *Lancet Oncology* 2009; 10: 147–156.
166 Inukai, T., Kiyokawa, N., Campana, D., et al. Clinical significance of early T-cell precursor acute lymphoblastic leukaemia: results of the Tokyo Children's Cancer Study Group Study L99–15. *Br J Haematol* 2012; 156: 358–365.
167 Homminga, I., Pieters, R., Langerak, A.W., et al. Integrated transcript and genome analyses reveal NKX2-1 and MEF2C as potential oncogenes in T cell acute lymphoblastic leukemia. *Cancer Cell* 2011; 19: 484–497.
168 Zhang, J., Ding, L., Holmfeldt, L., et al. The genetic basis of early T-cell precursor acute lymphoblastic leukaemia. *Nature* 2012; 481: 157–163.
169 Korbel, J.O., Campbell, P.J. Criteria for inference of chromothripsis in cancer genomes. *Cell* 2013; 152: 1226–1236.
170 Della Gatta, G., Palomero, T., Perez-Garcia, A., et al. Reverse engineering of TLX oncogenic transcriptional networks identifies RUNX1 as tumor suppressor in T-ALL. *Nat Med* 2012; 18: 436–440.
171 Neumann, M., Coskun, E., Fransecky, L., et al. FLT3 mutations in early T-cell precursor ALL characterize a stem cell like leukemia and imply the clinical use of tyrosine kinase inhibitors. *PLoS One* 2013; 8: e53190.
172 Neumann, M., Heesch, S., Gokbuget, N., et al. Clinical and molecular characterization of early T-cell precursor leukemia: a high-risk subgroup in adult T-ALL with a high frequency of FLT3 mutations. *Blood Cancer J* 2012; 2: e55.
173 Ntziachristos, P., Tsirigos, A., Vlierberghe, P.V., et al. Genetic inactivation of the polycomb repressive complex 2 in T cell acute lymphoblastic leukemia. *Nat Med* 2012; 18: 298–303.
174 Van Vlierberghe, P., Ambesi-Impiombato, A., Perez-Garcia, A., et al. ETV6 mutations in early immature human T cell leukemias. *J Exp Med* 2011; 208: 2571–2579.

175 Zenatti, P.P., Ribeiro, D., Li, W., et al. Oncogenic IL7R gain-of-function mutations in childhood T-cell acute lymphoblastic leukemia. *Nat Genet* 2011; 43: 932–939.
176 Shochat, C., Tal, N., Bandapalli, O.R., et al. Gain-of-function mutations in interleukin-7 receptor-α (IL7R) in childhood acute lymphoblastic leukemias. *J Exp Med* 2011; 208: 901–908.
177 Neumann, M., Heesch, S., Schlee, C., et al. Whole-exome sequencing in adult ETP-ALL reveals a high rate of DNMT3A mutations. *Blood* 2013; 121: 4749–4752.
178 Hayette, S., Tigaud, I., Maguer-Satta, V., et al. Recurrent involvement of the MLL gene in adult T-lineage acute lymphoblastic leukemia. *Blood* 2002; 99: 4647–4649.
179 Ferrando, A.A., Armstrong, S.A., Neuberg, D.S., et al. Gene expression signatures in MLL-rearranged T-lineage and B-precursor acute leukemias: dominance of HOX dysregulation. *Blood* 2003; 102: 262–268.
180 Asnafi, V., Radford-Weiss, I., Dastugue, N., et al. CALM-AF10 is a common fusion transcript in T-ALL and is specific to the TCRγδ lineage. *Blood* 2003; 102: 1000–1006.
181 Caudell, D., Zhang, Z., Chung, Y.J., Aplan, P.D. Expression of a CALM-AF10 fusion gene leads to Hoxa cluster over-expression and acute leukemia in transgenic mice. *Cancer Res* 2007; 67: 8022–8031.
182 Raimondi, S.C., Pui, C.H., Head, D.R., Rivera, G.K., Behm, F.G. Cytogenetically different leukemic clones at relapse of childhood acute lymphoblastic leukemia. *Blood* 1993; 82: 576–580.
183 Kawamata, N., Ogawa, S., Seeger, K., et al. Molecular allelokaryotyping of relapsed pediatric acute lymphoblastic leukemia. *Int J Oncol* 2009; 34: 1603–1612.
184 Yang, J.J., Bhojwani, D., Yang, W., et al. Genome-wide copy number profiling reveals molecular evolution from diagnosis to relapse in childhood acute lymphoblastic leukemia. *Blood* 2008; 112: 4178–4183.
185 Mullighan, C.G., Phillips, L.A., Su, X., et al. Genomic analysis of the clonal origins of relapsed acute lymphoblastic leukemia. *Science* 2008; 322: 1377–1380.
186 Mullighan, C.G., Zhang, J., Kasper, L.H., et al. CREBBP mutations in relapsed acute lymphoblastic leukaemia. *Nature* 2011; 471: 235–239.
187 Kino, T., Nordeen, S.K., Chrousos, G.P. Conditional modulation of glucocorticoid receptor activities by CREB-binding protein (CBP) and p300. *J Steroid Biochem Mol Biol* 1999; 70: 15–25.
188 Lambert, J.R., Nordeen, S.K. CBP recruitment and histone acetylation in differential gene induction by glucocorticoids and progestins. *Mol Endocrinol* 2003; 17: 1085–1094.
189 Meyer, J.A., Wang, J., Hogan, L.E., et al. Relapse-specific mutations in NT5C2 in childhood acute lymphoblastic leukemia. *Nat Genet* 2013; 45: 290–294.
190 Tzoneva, G., Perez-Garcia, A., Carpenter, Z., et al. Activating mutations in the NT5C2 nucleotidase gene drive chemotherapy resistance in relapsed ALL. *Nat Med* 2013; 19: 368–371.

CHAPTER 6

The genetics of mature B-cell malignancies

Jonathan C. Strefford, Jude Fitzgibbon, Matthew J.J. Rose-Zerilli and Csaba Bödör

Introduction

Mature B-cell neoplasms are defined by extreme biological and clinical heterogeneity. The clinical presentation includes leukaemic disease and tumours affecting the lymphatic system, frequently, but not exclusively within the lymph nodes or spleen. Patients with these malignancies have highly variable clinical courses, from a rather benign natural history, such as subtypes of chronic lymphocytic leukaemia (CLL) and follicular lymphoma (FL), to aggressive and often therapy refractory diseases, such as diffuse large B-cell lymphoma (DLBCL) and sub-types of mantle cell lymphoma (MCL).

This clinical heterogeneity is underpinned by considerable biological diversity, which contributes to disease pathogenesis driven by a variety of molecular and cellular mechanisms. It is now clear that much of this clinical heterogeneity originates from variability at the genetic level, exemplified by the presence of recurrent somatically acquired genetic lesions, many of which have been functionally linked to disease pathophysiology. Initially, chromosomal staining allowed the identification of an abnormal derivative chromosome 14 in patients with Burkitt lymphoma, which chromosome banding ultimately resolved as the marker chromosome that is the result of the t(8;14)(q24;q32) translocation.[1] This was one of the first recurrent translocations identified in mature B-cell malignancies and paved the way for considerable research into the presence of chromosomal abnormalities in these diseases, so that there are now karyotypes of several thousand patients publicly available,[2] and many chromosomal lesions have been characterized at the molecular level. The development of chromosomal banding allowed

The Genetic Basis of Haematological Cancers, First Edition. Edited by Sabrina Tosi and Alistair G. Reid.

a low-resolution analysis of the genome in these patients and further technological advances have continued to increase the resolution at which the genome can be studied. In the 1980s, this development began with molecular cytogenetic techniques such as fluorescence *in situ* hybridization (FISH). In the 1990s, microarray-based approaches, such as array-based comparative genomic hybridization (aCGH) and single nucleotide polymorphism (SNP) arrays, continued constantly to refine our picture of the genomic defects present in these patients. The 21st century heralded the publication of the human genome project and the development of next-generation sequencing (NGS) technologies, which have continued to deepen our understanding of the molecular pathogenesis of these diseases, improving patient management, and will ultimately fully catalogue all the biologically relevant genomic lesions in these conditions, helping to provide curative strategies for these diseases. This chapter focuses on recent advances in the genetics of two types of mature B-cell malignancy, chronic lymphocytic leukaemia (CLL) and the germinal centre lymphomas, follicular (FL) and diffuse-large B-cell lymphoma (DLBCL).

Chronic lymphocytic leukaemia

CLL is the most common form of leukaemia in adults.[3] CLL and its nodal variant, small lymphocytic lymphoma (SLL), are characterized by a clonal expansion of mature $CD5^+$ lymphocytes that arise in the bone marrow and infiltrate lymphoid tissue such as the lymph nodes and the spleen (Fig. 6.1).[4] Although CLL was historically considered to be the result of the accumulation of long-lived, but resting, lymphocytes, evidence now points to the presence of a substantial pool of proliferating CLL cells.[5] Most cases of CLL are preceded by monoclonal B-cell lymphocytosis (MBL), an indolent condition defined by clonal B-cell expansion of less than 5000 B-cells in the peripheral blood.[6] Clinically, CLL is characterized by considerable heterogeneity. Some CLL patients can survive for many years, without symptoms or the need for treatment, whereas others have rapidly fatal disease despite aggressive therapy. A small proportion of CLL patients will transform to Richter syndrome (RS), an aggressive form of lymphoma. In recognition of this clinical variability, several clinical staging systems were devised.[7,8] Although these systems remain the cornerstone on which a clinical management decision is built, they fail to predict the disease course in all patients, particularly in those diagnosed

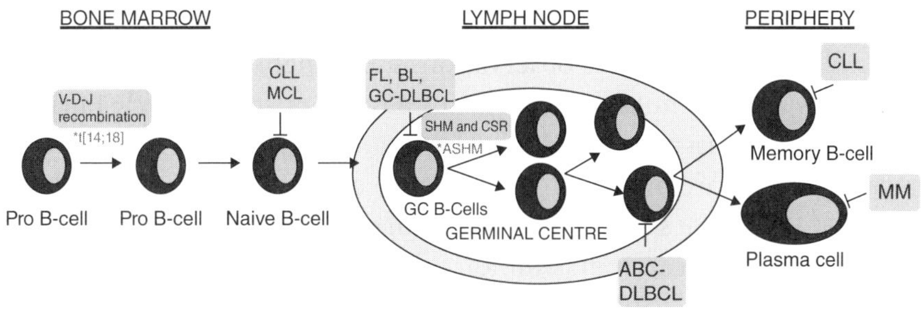

Figure 6.1 Cellular origin of mature B-cell malignancies. B-cell lymphomas arise at different stages of normal B-cell differentiation and can be assigned to their normal B-cell counterparts. Most lymphomas derive from germinal centre (GC) or post-GC B-cells, indicating the critical role of GCs in the pathogenesis of B-cell lymphomas. Abbreviations: CLL, chronic lymphocytic leukaemia; MCL, mantle cell lymphoma; FL, follicular lymphoma; BL, Burkitt's lymphoma; GC-DLBCL, germinal centre-type diffuse large B-cell lymphoma; SHM, somatic hypermutation; CSR, class switch recombination; ASHM, aberrant somatic hypermutation; ABC-DLBCL, activated B-cell type diffuse large B-cell lymphoma; MM, multiple myeloma.

in the early stages of the disease. Therefore, significant research efforts have been focused on the identification of biomarkers that accurately predict an aggressive clinical course. Although it is now clear that CLL is as diverse at the molecular and cellular level as it is clinically, there are features of the CLL cells that can be exploited in the clinical setting to identify patients destined to progress and require treatment. Of the biological processes that contribute to clinical behaviour in CLL, few are more fundamentally important to the process of pathogenesis than the genetics of the B-cell receptor (BCR) and the presence of somatically acquired chromosomal aberrations and gene mutations.

Immunoglobulin heavy-chain variable region gene mutational status

Extensive genetic analysis of the BCR indicates that 60% of CLL patients harbour evidence of somatic hypermutations within the immunoglobulin heavy-chain variable (*IGHV*) genes (mutated CLL or M-CLL).[9] In contrast, 40% of patients lack evidence of *IGHV* mutations (unmutated CLL or U-CLL) and have a germline *IGHV* sequence. As the process of somatic hypermutation occurs in the germinal centres (GC), the presence of these disease subsets provides valuable insight into the cellular origin of CLL. This suggests that M-CLL cases are derived from GC-experienced B-cells, whereas U-CLL patients have undergone differentiation independent of the GC.[10] Importantly, this observation is also highly relevant clinically, as patients with unmutated *IGHV* genes have poor survival.[11,12] Furthermore, CLL exhibits remarkable bias in the *IGHV* gene repertoire, where particular genes are over-represented, such as *IGHV3-34, -1-69, -3-21* and *-3-23*.[13,14] BCRs that share structural features are termed 'stereotypic' and their presence suggests that common antigens or autoantigens are recognized by CLL cells.[15] It has also been suggested that certain stereotypic BCRs have distinct clinical features, such as the inferior overall survival independently associated with stereotypic *IGHV3-21* usage.[16] Studies of the *IGHV* genes have greatly enhanced our understanding of CLL and our ability to define prognostic subgroups. Furthermore, an area of active research is the development of therapeutic molecules that target pathways known to be deregulated in M- or U-CLL cases.

Chromosomal banding and interphase molecular cytogenetics

Owing to the failure of CLL cells to undergo *in vitro* cell division without biological stimulation, early metaphase studies were largely unsuccessful. However, the introduction of polyclonal B-cell mitogens allowed the successful production of metaphase preparations from CLL cells[17] and led to the discovery of the first recurrent cytogenetic abnormality, trisomy 12, in CLL in 1980,[18,19] which was followed in 1987 by the presence of the interstitial deletion of 13q14.[20] With the use of more effective mitogens, initially with tetradecanoylphorbol acetate (TPA),[21] then CD40 ligand,[22] and more recently with CpG oligonucleotides combined with interleukins,[23,24] metaphases can be obtained in the majority of CLL cases and clonal aberrations are observed in 80%.[25] These earlier studies demonstrated the clinical importance of chromosomal alterations,[17,21,26] and later studies revealed that, surprisingly, reciprocal translocations occur in as many as 20% of cases. Some of these translocations are recurrent, involving the immunoglobulin gene locus at 14q32 and a number of partner genes, most commonly *BCL2* (18q22) and *BCL3* (19q13) and more infrequently *BCL11A* (2p15), *CCND3* (6p21) and *c-MYC* (8q24).[27–32] However, translocations remain rare in patients with CLL, a disease actually characterized by copy number changes rather than translocations, particularly the presence of deletion events, such as those targeting 11q, 13q and 17p.

The development of FISH overcame the necessity for dividing cells, by the application of centromeric or locus-specific probes to interphase cells. With these FISH probes, recurring chromosomal abnormalities could be detected with increased resolution compared with G-banded metaphase analysis. Twenty-five years of karyotypic analysis came to fruition with the development of a panel of FISH probes specific for regions of recurrent deletion and duplication in CLL. The application of these FISH probes to a cohort of 325 CLL patients resulted in a seminal study demonstrating the clinical importance of chromosomal abnormalities,[33] which is now standard practice in the clinical management of CLL. This study established a hierarchical prognostic model, based on the presence of five chromosomal categories, 17p deletion, 11q deletion, trisomy 12, the presence of no recurrent chromosomal lesion and 13q deletion as the sole abnormality, with 17p deletion and 13q deletion being the markers associated with the worst and best prognosis, respectively.[33] Similar data can be obtained using either multiplex ligation-dependent probe amplification (MLPA) or quantitative polymerase chain reaction (PCR), but a large-scale comparison of these methodologies has not so far been performed.[34,35]

Copy number alterations

Deletions of 13q14

Deletions of the long arm of chromosome 13 are the most frequent chromosomal aberration detected in patients with CLL, occurring in 60–80% of cases depending on the method of detection[33] (Table 6.1 and Figure 6.2a). These deletions can be present as monoallelic or biallelic deletions and studies of sequential samples from the same patients suggest that the monoallelic deletion is likely to be an early event, whereas the second deletion occurs later on in pathogenesis. These deletions can be variable both in genomic location and in gene content, and breakpoints can encompass many megabases (Mb) proximal and distal to the minimally deleted region (MDR). The identification of the target genes within this MDR has been a challenge. Mutation analysis of protein-coding genes revealed no inactivating mutations. In normal individuals, genes on 13q14, within and close to the MDR, are expressed, at least in some individuals, from a single allele only. The mechanism of allelic silencing and its importance as a method of gene silencing in

Table 6.1 Recurrent copy-number changes in CLL.

Gene name	Prevalence (%)	Principal candidate genes	Other candidate genes	References
del(13q)	60–80	*miR-15a/16-1, DLEU2*	*RB1, DLEU7*	20, 41, 43
del(11q)	10–20	*ATM*	*BIRC3, MRE11, H2AFX*	56
del(17p)	5–50	*TP53*	–	195
Trisomy 12	10–15	Unknown	–	18
del(6q)	5	Unknown	–	80
dup(2p)	5–28	*REL, BCL11A, XPO1*	*MYCN*	78, 82
dup(8q)	5	*CMYC*	–	69
del(15q)	4	*MGA*	–	69, 196

cases of CLL with either no detectable loss or heterozygous loss of 13q14 remain uncertain.[36,37]

A pivotal study showed that 13q14 deletions are associated with down-regulation of two microRNAs, miR-15a and miR-16-1, located within the MDR, although mutations within these genes are exceptionally rare in CLL.[38] This study provided the first tangible link between recurrent chromosomal alterations and microRNAs in cancer. Arguably the most important study into the pathogenic role of this locus was the generation of two mouse models deleted for the entire MDR or the miR-15a/miR-16-1 cluster.[39] In this model, the miR-15a/16-1 cluster regulates the transition from G0–G1 to S phase, through control of *CCND1, CCND2, CCNE1, CDK4* and *CDK6* expression. Both mouse models developed clonal lymphoproliferative disorders, including $CD5^+$ MBL, CLL and non-Hodgkin lymphoma (NHL). However, the frequency of these disorders was significantly higher in MDR deleted mice, suggesting that although loss of the miRNA cluster is sufficient for leukaemogenesis, additional components of the 13q14 deletion contribute to pathogenesis. It is now clear that even genes beyond the MDR can influence the prevalence of lymphoproliferative disorders in this model.[40]

A pathogenic role for genes outside the MDR is further supported by recent copy number profiling studies.[41–43] In spite of considerable breakpoint heterogeneity among 13q deletions, two breakpoint clusters (BCs) have been identified proximal and distal to the MDR that may be prone to breakage in CLL B-cells.[41] In the light of these BCs, it is clear that 13q deletions can be grouped thus: small deletions confined to a 2 Mb region where breakpoints often occur in the two BCs identified, which,

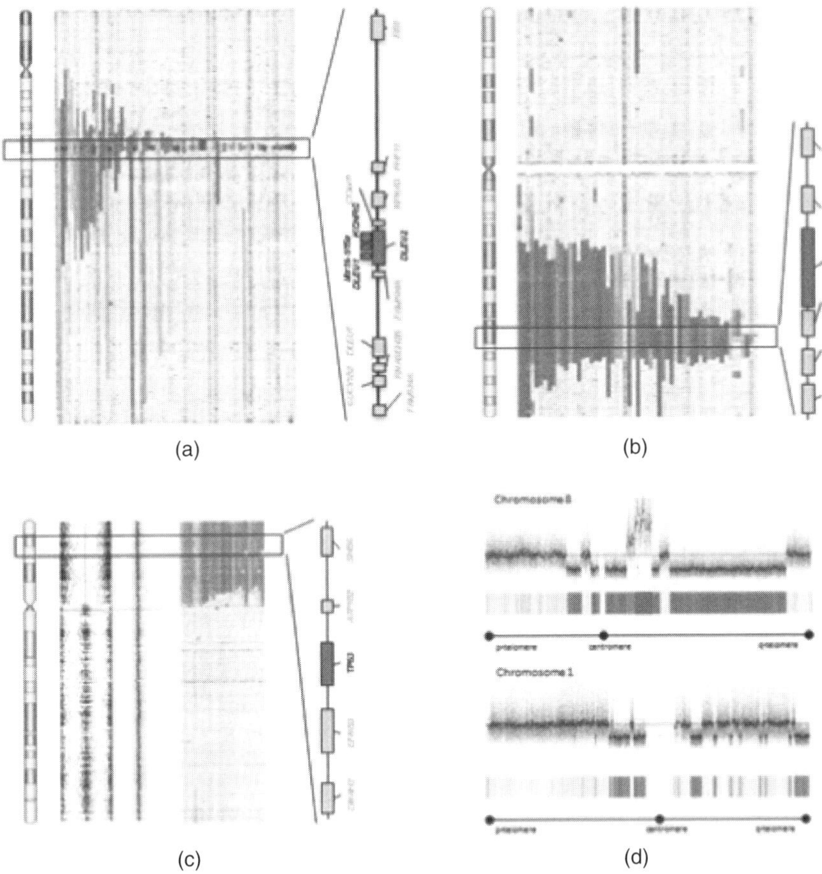

Figure 6.2 Copy-number changes in patients with CLL. (a), (b) and (c) show copy-number deletions of chromosomes 13, 11 and 17, respectively. Each chromosome runs vertically from p telomere (top), through the centromere to the q telomere (bottom) An idiogram is shown on the left, with a copy number heatmap of multiple patients, where blue, white and red show deletions, normal copy number and duplications, respectively. The location of each MDR is highlighted with an expanded view of the genes within these regions. For chromosome 17 an allelic ratio (left) and copy number (right) profile is also shown for a *TP53* mutated patient with copy number neutral LOH. (d) Two examples of chromothrypsis targeting chromosomes 8 (top) and 1 (bottom). The chromosome is positioned horizontally, running from p telomere (left) to q telomere (right). For each example, a copy-number profile and a heat map are shown. *(See plate section for color representation of this figure.)*

in addition to the genes within the MDR, include *FLJ31945, FAM10A4, BCMS* and *DLEU7*; larger deletions extended beyond this region in either a centromeric and/or telomeric direction, encompassing a large number of additional genes. Rather than being the result of consistent BCs, these larger deletions displayed highly heterogeneous breakpoints, suggesting that these deletions may be mechanistically dissimilar.[41] Although experimentally challenging, it has been possible to propose putative target genes proximal to the MDR.[41,43] Recently, research into 13q14 deletions has shifted to the gene *DLEU7*, which is often under-expressed in CLL cells,[44] providing preliminary evidence that genes outside the MDR may interact with the miR-15a/16-1 cluster to promote leukaemogenesis.

Deletions of 13q14 are associated with a favourable prognosis when identified as sole abnormalities using FISH analysis,[33] and monoallelic versus biallelic deletions carry no independent prognostic significance. There is evidence that both the number of cells harbouring a 13q deletion and the genomic size of the 13q deletion may have significant clinical implications. First, it has been suggested that patients with a higher clonal population of del(13q) cells exhibited reduced overall survival (OS) and treatment-free survival (TFS).[45] Second, larger 13q deletions that extend beyond the two BCs flanking the MDR are associated with progressive CLL,[41] elevated genomic complexity,[41,43] advanced clinical stage[42] and reduced OS and TFS.[43] Although these observations requires further validation, additional prognostic information could be provided in the clinical setting using FISH, MLPA and aCGH to distinguish between informative deletion sizes.

Trisomy 12

Trisomy 12 is the presence of three copies of chromosome 12 and occurs in ~10% of patients.[33] (Table 6.1). It is principally an early event detectable at diagnosis and acquisition of trisomy 12 during the disease course is exceptionally rare. The role of trisomy 12 in the molecular pathogenesis of CLL remains unclear and no firm candidate genes have been identified to date. Structural aberrations of chromosome 12 do occur and result in partial trisomy of 12q13–22. *MDM2* has been implicated as it resides within this part of chromosome 12 and is over-expressed in patients with trisomy 12.[46] Several other associations are worthy of note: (i) there is a strong association with both atypical lymphocyte morphology and immunophenotype;[47] (ii) trisomy 12 is frequently accompanied by additional trisomies, particularly of chromosomes 18 and 19;[48] (iii) *NOTCH1* mutations are significantly associated with the presence of trisomy 12;[49–51] and (iv) trisomy 12 patients exhibited a heightened responsiveness to hedgehog signalling inhibition.[52]

Clinically, trisomy 12 in isolation is associated with favourable response rates after treatment with chemo-immunotherapy and with longer survival times compared with cases with del(13q) or no FISH abnormality.[53] This benign disease may be due to the rare presence of *TP53* deletions or mutations[54] or the elevated surface expression of CD20 observed in these patients.[55]

Deletions of 11q24 and mutations of *ATM*

Interstitial deletions of the long arm of chromosome 11 [termed del(11q)][56] are present in ~10% of patients at diagnosis.[33] and in ~20% of patients requiring treatment (Table 6.1). The presence of del(11q) is significantly associated with a progressive phenotype and a requirement for treatment. Remission is shorter for del(11q) patients after initial treatment with chemotherapy,[57] although improved durations have been observed in del(11q) patients treatment with chemo-immunotherapy.[53] Deletions of 11q are variable in both location and gene content, but always encompass the *ATM* gene, a key pathogenic target of these aberrations (Figure 6.2b). Other genes within the del(11q) MDR have been studied, but no inactivating mutations have been identified in these genes.[58] Other 11q genes have been implicated in the pathogenesis of CLL and several are discussed later in this section. However, ongoing whole-genome/exome sequencing initiatives have failed to identify recurrent genes on 11q targeted by somatic mutations.

ATM remains the only gene on 11q that is mutated at relatively high frequencies. Mutations in *ATM* occur in the presence and absence of del(11q); with frequencies of ~10 and 25% in non-del(11q) and del(11q) patients, respectively.[59,60] The mutations are usually missense, nonsense or frameshift in nature and are likely to affect ATM protein function to a lesser or greater extent. Historically, the analysis of *ATM* mutations is challenging as the gene is large, with no recurrent mutational hotspot. Furthermore, the type of DNA variation that resides in the gene makes it difficult to identify bona fide somatic, functionally relevant mutations. In spite of these technical difficulties, progress on the clinical utility of *ATM* mutations has been made. Most notable is recent data emerging from the UK LRF CLL4 treatment trial, where the presence of *ATM* mutations was assessed in 224 patients.[61] In this study, the authors were able to subdivide del(11q) patients into two prognostically distinct groups based on the presence of an *ATM* mutation on the undeleted allele; patients with an *ATM* mutation and deletion exhibited survival similar to that of patients with mutations and/or deletion of *TP53* after initial treatment with chemotherapy. Interestingly, the same study showed that *ATM* mutations in isolation [without concomitant del(11q)] are not associated with reduced survival, suggesting that the poor outcome associated with del(11q) may be at least partially a result of the deregulation of other 11q genes.

Small deletions just containing *ATM* or genes within the MDR are rare, occurring in only ~5% of del(11q) patients. The majority of patients actually have larger deletions that result in the loss of several hundred genes. Therefore, as is the situation with other recurrent CNAs, the view that deletions only target genes within the MDR is a gross simplification of the pathogenic consequences of deletion events, which are likely to target a large numbers of genes and pathways. With the facts that *ATM* mutations occur in only the minority of del(11q) cases and that ATM protein function is apparently normal in del(11q) cases without *ATM* mutations, the implication is that other genes on 11q are important in the pathogenesis of del(11q) CLL, likely through haploinsufficiency, epigenetic deregulation or somatic mutation. Given the association between del(11q) and elevated genomic complexity, an attractive mechanism is the compound deletion of other genes involved in DNA damage response, such as deletion of *MRE11A* and *H2AFX*, both of which are located on 11q and deleted in a proportion of del(11q) CLL.[62] There are also 11q genes involved in other key pathways known to be deregulated in CLL, such as the NF-κB pathway. One such gene, *BIRC3*, a negative regulator of NF-κB, has recently been shown to be disrupted at the genomic levels in del(11q) patients either by deletion and/or mutation.[63] *BIRC3* disruption is absent in MBL and rare at CLL (4%) diagnosis. However, the frequency is significantly higher in fludarabine-refractory CLL (24%) and it has been suggested that *BIRC3* disruption represents an independent marker of poor outcome comparable to *TP53*. Consistent with the role of *BIRC3* as a negative regulator of NF-κB, functional studies have revealed the presence of constitutive non-canonical NF-κB activation in fludarabine-refractory CLL patients with molecular lesions of *BIRC3*.[63] Del(11q), *ATM* mutation and *BIRC3* disruption are not mutually exclusive and each aberration has been associated with reduced survival, hence a comprehensive study is required to demonstrate which molecular defect is most clinically informative.

Deletions of 17p13 and mutations of *TP53*

With the application of FISH, deletions of the short arm of chromosome 17 [termed del(17p)] are rare at diagnosis, accounting for <5% of CLL patients[33] (Table 6.1). However, the incidence increases in patients at first-line treatment indication and may account for as many as 50% of patients who ultimately become refractory to chemotherapy. Deletions are invariably large with centromeric breakpoints, are the result of several structural abnormalities such as deletions, unbalanced translocations and isochromosomes, and include the *TP53* gene at 17p13 (Figure 6.2c). Small, focal deletions of *TP53* do occur, albeit at a much lower frequency than whole-arm deletions. This genomic architecture is striking and dissimilar to the distribution of 17p deletions in other human malignancies, where deletions are far more heterogeneous in both location and size. This suggests that the true biological consequence of 17p deletion may involve the deregulation of other genes, but additional work is required to demonstrate this. The majority of 17p deletions are clonally dominant; they occur in the majority of cells when assessed by FISH, suggesting that this lesion provides a strong growth or survival advantage. Deletions of 17p are associated with dismal outcome, particularly after treatment with chemotherapy.[57] Patients are often refractory to chemotherapy and poor outcome is not overcome by combination chemo-immunotherapy with rituximab.[53] Patients with 17p deletions could benefit from treatments that kill CLL cells in a p53-independent manner and trials are currently under way, including such agents as alemtuzumab,[64] lenalidomide[65] and novel small molecules such as BCR signalling inhibitors.[66]

In addition to deletions, 17p can also be targeted by acquired uniparental disomy [aUPDs, also known as copy neutral loss of heterozygosity (cnLOH)], so that the region retains a normal copy number resulting from duplication of a single parental chromosome after a somatic recombination event.[67] These events result in the duplication of a *TP53* mutation.[68,69] Around 3–5% of CLL patients acquire a *TP53* mutation without loss of the other allele and these patients also exhibit poor survival.[70–73] Differences in the incidence of *TP53* loss and/or mutation among cohorts are a reflection of the differences in patient populations and methods for screening for *TP53* alterations. Most *TP53* mutations are missense and located within the DNA binding domain of p53 encoded by exons 5–8. Six hotspots are mutated in ~20% of patients. It is likely that chemotherapy selects pre-existing small *TP53* mutated clones, as the mutational spectrum is similar in both previously untreated and treated patients.[70] *In vitro* functional studies, in which double-stranded DNA breaks are induced in leukaemic cells and the expression of p53 and its downstream targets such as p21 and miR34a are measured, are also able to detect *TP53* abnormalities in CLL.[74] More recent studies suggest that primary abnormalities of p21 and miR34a expression in patients with no detectable *TP53* abnormality may also cause p53 dysfunction and can be associated with poor clinical outcome.[75,76] It is important to note that a subset of early-stage CLL patients with *TP53* abnormalities do exhibit a stable disease course. In contrast to the majority of patients with *TP53* deregulation, these stable cases have mutated *IGHV* genes.[77]

Other copy number alterations in CLL

A plethora of novel copy number changes have emerged from the literature,[78,79] but there is very little firm knowledge available regarding many of them. Those that have been most widely reported are deletions and duplications of 6q and 2p, respectively. Deletions of 6q can be detected in ~5% of the patients with CLL.[33] With FISH and array studies, multiple non-overlapping MDRs have been identified and the evidence supporting any candidate genes is not compelling.[69,80] The literature is equally uncertain regarding the clinical significance of 6q loss, but it has been associated with atypical lymphocyte morphology extensive lymphadenopathy but not chemoresistance.[81] Duplication of chromosome 2p has been consistently reported as a genomic abnormality in CLL studies using CGH and SNP arrays.[82] Overall occurrence has been reported in 5% of early-stage CLL patients, rising to 28% in Stage B and C disease. Conventional cytogenetics shows a variety of mechanisms resulting in 2p duplication, including the presence of dicentric chromosomes and unbalanced translocations.[83] However, 2p duplication rarely occurs in isolation and is often associated with adverse genetic abnormalities: del(11q) and del(17p) and unmutated *IGHV* genes. The two recurrently duplicated regions most commonly reported include the genes *REL*, *BCL11A*, *XPO1* and *MYCN*.[78,82] Expression of *MYCN* has been shown to be elevated in the presence of dup(2p), whereas *XPO1* gene mutations have been reported in CLL cases lacking these duplication events.[84]

Several other regions are targeted by recurrent copy number changes and although their importance is supported by several publications, the incidence is low. Duplications of 8q24 have been reported that often include the *c-MYC* locus, but it is currently unclear if this is the target gene. Deletion of the *MGA* gene at 15q15.1 has been observed by two studies and accounts for ~4% of CLL cases. Rare *MGA* mutations have been identified in CLL patients[69,84] and although its pathogenic importance is currently unknown, its role as a transcriptional repressor of cell proliferation and apoptosis suggests that it is worthy of further study.[85]

Genome complexity and chromothripsis

Although it is difficult to ascertain the true definition of genomic complexity,[86] recent array-based studies have confirmed and extended previous cytogenetic data showing that a subset of CLL patients have complex genomic profiles.[69,87–89] From these data, it is clear that although highly complex genomes do exist in CLL, the average genome contains fewer CNAs (0–3 CNAs per patient) than other solid tumours and mature B-cell malignancies. Genomic complexity is more common in advanced than in early-stage disease and sequential cytogenetic and FISH studies confirm that additional genomic abnormalities may be acquired during the course of the disease. Genomic complexity is frequently associated with and may be preceded by loss and/or mutation of the *TP53* and *ATM* genes, confirming their importance in DNA repair.[62] It has also been suggested that larger 13q deletions are associated with elevated complexity and although *RB1* may be a putative gene in the region, this is far from certain.[41,42] Recent data show a remarkable association between very short telomeres, telomere fusion events and genomic complexity in CLL.[90] A likely scenario is that defects in DNA damage checkpoints, such as those targeting *TP53* or *ATM*, allow telomeres to shorten below the length at which apoptosis or senescence is normally triggered and enable uncapped telomeres to fuse, resulting in genomic instability. This is supported by the association between extreme telomere erosion in *ATM*-mutated and 11q-deleted CLL.[91] Genomic complexity in CLL is not exclusively associated with *TP53* or *ATM* abnormalities, suggesting that either other DNA repair defects or other mechanisms of instability are important in these cases. Importantly, these cases with genomic complexity in the absence of *TP53/ATM* involvement appear to be clinically relevant.

Genomic complexity identified by both karyotypic and SNP array analysis is associated with reduced overall survival and it is likely that both approaches are identifying an overlapping subgroup of patients.[25,87] In a single cohort, genomic complexity was able to predict short overall survival, independent of a number of established biomarkers and clinical features.[87]

Although genomic complexity may result from the gradual accumulation of DNA damage over a protracted period, it is now clear that high levels of genomic complexity can be acquired quickly and even during a single mitotic cell division. Approximately 3% of human cancers show characteristics of a catastrophic mutational process termed chromothripsis[92] (Figure 6.2d). This process was first identified in a patient with CLL and involves genome shattering that occurs during a single mitotic cycle and results in a characteristic pattern of oscillating DNA copy number changes along a random single chromosome or a few random chromosomes.[92] This restricted genomic localization may be the result of a specific mechanism, such as the erosion of a single telomere to such an extent that the chromosome unwinds and breaks up. It has also been suggested that this restriction may be the result of a physically isolated chromosome in a micronucleus that becomes damaged and is finally reintegrated into the genome.[93] It is notable that recent data from the CLL8 trial support an increased frequency of chromothripsis in patients with mutations in DNA damage response genes such as *TP53* and *ATM*.[69,94] This may suggest that a compromised DNA damage response is critical in permitting the formation of chromothripsis or in the creation of an environment for the genomic damage to be tolerated. Recent data have suggested that in addition to the clustering of structural variants, multiple base-pair mutations can also be acquired in a single mitotic explosion, called kataegis.[95] This process drives cytosine-specific mutagenesis in regions flanking sites of genomic rearrangement and can result in up to 20 base-pair substitutions occurring rapidly. This mechanism has not so far been identified in patients with CLL, but would be a relevant mechanism for driving the accumulation of pathogenically relevant genetic instability in response to the tumour microenvironment.

The presence of chromothripsis and kataegis in cancer patients implies that multiple cancer genes can be disrupted in a single step, providing a leap forward in the malignant potential of a cancer clone. In both CLL and multiple myeloma, samples with evidence of chromothripsis were associated with reduced survival, suggesting that large-scale genomic disruption may render the leukaemic cells more malignant. SNP6 analysis of the German CLL8 trial identified genomic features consistent with chromothripsis in ~5% of patients,[69] with a strong enrichment in patients with an unmutated *IGHV* status (74%) and high-risk genomic aberrations (79%). Although these patients exhibited both inferior OS and PFS, further studies are required to identify chromothripsis as an independent marker of reduced survival.

Novel mutations in patients with CLL

For the first time, advances in sequencing technology have provided the opportunity to search the entire cancer genome for sequence alterations with base-pair resolution. Critically, these approaches generate billions of independent sequence reads in parallel, each derived from a single molecule of DNA, thereby providing a random sample of DNA molecules from a tumour sample. Using these approaches, it has been possible to construct comprehensive maps of the somatic mutations that occur in human cancer, in essence through the comparison of genomic sequence from a patient's cancer cells compared with the inherited variation in their germline DNA. In CLL, the application of high-throughput sequencing is transforming our understanding of the genetic lesions that underlie disease pathogenesis and has facilitated the identification of biologically and clinically relevant sequence alterations.

Initial studies focusing on small discovery cohorts, followed by targeted re-sequencing of recurrent variants in larger cohorts, have identified clinically significant mutations in a number of genes not targeted by copy-number alterations.[84,96] In addition, more recent studies using exome sequencing of extended patient series have also been reported, amassing nearly 300 published CLL exomes or genomes.[97–99] These early studies have not identified a principal driver mutation in most patients, and it is unlikely that such a mutation is present at the genomic level. However, these studies have identified recurrent mutations at lower frequencies and the genomic context in which they exist. It is now clear that the CLL genome harbours a relatively small number of somatic mutations, certainly lower than solid tumours such as lung and pancreatic cancer and at least comparable to other types of leukaemia. Studies have detected less than one mutation per megabase (Mb) of genomic DNA with between two and 76 non-synonymous mutations per patient.[97,98] By investigating the base-pair substitutions in CLL cases and the sequence context in which they occur, it has been proposed that the pattern of mutations in *IGHV* mutated and unmutated cases is consistent with somatic hypermutation in the *IGHV* mutated subgroup, where the mutations are potentially introduced by error-prone polymerase η.[84] Interestingly, it is also clear that novel mutations showed striking associations with standard prognostic markers, suggesting that particular combinations of genetic lesions may act in concert to drive leukaemogenesis.

NOTCH1

Recent high-throughput sequencing efforts have confirmed the importance of Notch signalling in CLL, a process that is important for a variety of developmental and physiological processes.[84,96] Biological data initially demonstrated the importance of Notch signalling in cell survival and apoptosis resistance in CLL cells.[100] The same group has now demonstrated that the use of Notch1 signalling inhibitors, such as γ-secretase, accelerated B-CLL cell apoptosis by proteasome inhibition and enhancing endoplasmic reticulum stress.[101]

The link between Notch signalling and mutations within *NOTCH1* was first established in a small series of 43 patients, where Sanger sequencing detected two patients (4.6%) with a heterozygous frameshift of 2 bp (ΔCT7544–7545, P2515Rfs*4) within the PEST domain of *NOTCH1*.[102] This variant creates a premature stop-codon and a lack of a C-terminal domain containing a PEST sequence, and finally results in the accumulation of an active Notch1 isoform in CLL cells.[84] With high-throughput sequencing studies, it is now evident that mutations cluster across exon 34 of *NOTCH1*.[84,96–98] Several subsequent studies have expanded our understanding of the prevalence of *NOTCH1* mutations and their clinical significance. Initial estimates of the frequency of *NOTCH1* mutations in discovery CLL were as high as 12%, but these cohorts were derived from untreated and relapsed CLL. At diagnosis, the frequency is much lower, between 6 and 10%,[103,104] but the frequency varies hugely based on the stage of diseases analysed. In MBL patients, the incidence is ~3%,[105] whereas >20% of patients with alkylating agent or purine analogue-refractory disease harbour a mutation.[96,103] Initially, it was suggested that patients with a mutant *NOTCH1* have survival comparable to those harbouring *TP53* abnormalities;[103] however, it is now clear that these mutations identify outcome similar to 11q deleted CLL.[50,106]

SF3B1

SF3B1 is a critical component of the RNA splicing machinery that achieves successful transcription and guarantees the functional diversity of protein species using alternative spicing. Mutations in RNA splicing genes were initially discovered in myeloid malignancies,[107] with a strong association between *SF3B1* gene mutations and cases of the myelodysplastic syndrome with increased ringed sideroblasts.[108] Subsequently, *SF3B1* mutations have also been found in CLL but not in other chronic B-cell lymphoproliferative disorders.[97,98,109] Mutations were documented in 5–17% of patients and were associated with advanced-stage, fludarabine-refractory disease in cases with no *TP53* abnormality and 11q23 deletions and with short TFT and OS independent of other prognostic variables.[97,98,109]

Consistent with such an essential role in the control of gene expression, the SF3B1 amino acid sequence shows great phylogenetic conservation, especially in the amino acids targeted by somatic mutations in CLL. The SF3B1 protein contains two well-defined domains: the N-terminal hydrophilic region which contains a number of protein-binding motifs and is known to interact directly with other spliceosome components, and a C-terminal region, which consists of 22 non-identical HEAT domains, the precise role of which is currently unclear. *SF3B1* mutations are principally within exons 14–16, which encode the HEAT3, HEAT4 and HEAT5 domains.[50,97,98,109] Computational modelling of the C-terminal domain of SF3B1 provides several suggestions: (i) most mutations (codon 623–701) are spatially close to one another and occur on the inner surface of the *SF3B1* structure that might be defined as a binding interface; and (ii) the remaining mutations (codon 741–894) occur within different domains, such as the putative external loop domain, and may therefore be functionally distinct.[98]

The analysis of the functional consequences of *SF3B1* mutations is an area of active research. To date, the following observations have been made: (i) using targeted and global experimental approaches, it is clear that *SF3B1* mutations do result in aberrant splicing;[97,98] (ii) more in-depth analysis with RNA sequencing (RNA-Seq) suggests that *SF3B1* mutations do not impair the general function of the protein, but alter certain specific activities;[98] and (iii) only a relatively small number of candidate genes are aberrantly spliced, but include known cancer genes. In *SF3B1* mutated patients, a highly expressed truncated *FOXP1* transcript, *FOXP1w*, was identified that lacks two putative PEST domain sequences involved in protein degradation.[98]

Other genes

Other recurrently mutated genes are emerging from ongoing high-throughput sequencing projects, although at low frequencies.[84,97,99] Furthermore, there is currently little information regarding their frequency in validation cohorts and their functional consequences. The mutational landscape of CLL is outlined in Table 6.2. Although data in this area are actively evolving, it is clear that other pathways are being identified targeted by somatic mutations in CLL. In addition to the predicted involvement of DNA damage response and cell-cycle control genes, additional genes are mutated involved in B-cell differentiation, Notch signalling, inflammatory pathways, telomere maintenance, epigenetic regulation, RNA splicing and processing and B-cell receptor signalling.[97–99] Although data are currently limited, several gene mutations, such as those targeting *BRAF* and *FBXW7*, may lead to the development of targeted treatments, for example, the use of BRAF inhibitors.[110]

Novel genetic mutations in clinical practice

Preliminary data support the clinical importance of several of these new mutated cancer genes, most notable disruption of *BIRC3* and mutations in *NOTCH1* and *SF3B1*. If the clinical significance of these lesions can be validated in large retrospective and prospective cohorts, the true importance of these biomarkers, in the context of established clinical and biological features, can be realized. With validation, these lesions may have utility as biomarkers with benefits that may include the ability to predict the natural history of patients presenting with early asymptomatic disease, predicting the outcome but also the choice of treatment, and may represent therapeutic targets. Two studies have contributed significantly to the goal of validating the prognostic significance of these new lesions. Recently, a large study developed a prognostic algorithm through the integration of gene mutations and chromosomal abnormalities.[106] By the parallel analysis of >1200 CLL patients using FISH and sequence analysis, they define four risk classifications: (1) high-risk patients with either *TP53* defects and/or *BIRC3* disruption; (2) intermediate-risk patients, harbouring *NOTCH1* and/or *SF3B1* mutations and/or del(11q); (3) low-risk patients, harbouring trisomy 12 or a

Table 6.2 Prevalence of recurrent mutations in CLL.

Gene name	Gene nomenclature	Frequency (%)	References
Tumour protein p53	*TP53*	5–20	197
Ataxia telangiectasia mutated	*ATM*	9–14	59
Notch 1	*NOTCH1*	12	96, 98, 102
Splicing factor 3b, subunit 1, 155 kDa	*SF3B1*	10	97, 98
Chromodomain helicase DNA-binding protein 2	*CHD2*	5	98
Low-density lipoprotein receptor-related protein 1B	*LRP1B*	5	98
Protection of telomeres 1	*POT1*	5	98
F-box and WD repeat domain containing 7, E3 ubiquitin protein ligase	*FBXW7*	4	97
Zinc finger, MYM-type 3	*ZMYM3*	4	97
DEAD (Asp–Glu–Ala–Asp) box polypeptide 3, X-linked	*DDX3X*	3	97, 99
Myeloid differentiation primary response gene 88	*MYD88*	3	84, 98
Mitogen-activated protein kinase 1	*MAPK1*	3	97
Histone cluster 1, H1e	*HIST1H1E*	3	99
BCL6 corepressor	*BCOR*	3	99
Receptor (TNFRSF)-interacting serine–threonine kinase 1	*RIPK1*	3	99
SAM domain and HD domain 1	*SAMHD1*	3	99
Sucrase–isomaltase (α-glucosidase)	*SI*	3	198
Exportin 1	*XPO1*	2.5	84
Kelch-like family member 6	*KLHL6*	2	84
V-raf murine sarcoma viral oncogene homologue B1	*BRAF*	2	199
V-Ki-ras2 Kirsten rat sarcoma viral oncogene homologue	*KRAS*	2	99
Mediator complex subunit 12	*MED12*	2	99
Inositol triphosphate 3-kinase B	*ITPKB*	2	99
Early growth response 2	*EGR2*	1	99
Interferon regulatory factor 4	*IRF4*	1.5	200
Baculoviral IAP repeat containing 3	*BIRC3*	1.5	63
Neuroblastoma RAS viral (v-ras) oncogene homologue	*NRAS*	1–3	99, 199
Retinoblastoma 1	*RB1*	2*	43

*Frequency in those patients with large 13q deletions which include the *RB1* locus. Defined as 'Type II' deletions.[42]

normal profile; and (4) a very low-risk group with del(13q), whose survival did not differ from that of a matched general population. This model adds significantly to the model based on the presence of aberrations detected by FISH,[33] due to the co-existence of poor-risk gene mutations in low-risk groups defined purely based on FISH.

Given the highly heterogeneous natural history of CLL and the often serendipitous date of initial diagnosis, it is important to confirm the prognostic relevance of novel biomarkers in the context of randomized clinical trials. The second study performed just such an analysis, assessing *NOTCH1* and *SF3B1* mutations in a cohort of 494 patients treated within the randomized phase III UK LRF CLL4 trial that compared chlorambucil and fludarabine with and without cyclophosphamide in previously untreated patients (Fig. 6.3).[57] This study showed that although *TP53* alterations remained the most informative marker of dismal survival in the UK LRF CLL4 cohort, *NOTCH1* and *SF3B1* mutations have added independent prognostic value. It will be critical to understand the biological role of these mutations and their clinical significance in the context of modern therapies, such as novel B-cell receptor signalling inhibitors and chemo-immunotherapy.

Although it has not so far been demonstrated, therapeutic targeting of *NOTCH1, SF3B1* and *BIRC3* alterations has clear promise. *NOTCH1* is a well-established therapeutic target in T-cell acute lymphoblastic leukaemia (T-ALL), and is possible that NF-κB pathway and spliceosome inhibitors could have activity against *BIRC3* and *SF3B1* defects, respectively. Other treatment avenues have emerged from investigations of the biological importance of the BCR and clearly show that this signalling cascade may be a target for therapy. Two such drugs are currently undergoing intensive clinical investigations, the Bruton tyrosine kinase (BTK) and phosphatidylinositol 3-kinase (PI3K) inhibitors.[66,111] It is evident that further understanding of the genetic lesions that contribute to CLL pathogenesis will continue to aid in the accurate risk-adapted stratification of CLL patients and identify molecular mechanisms that are amenable to therapeutic intervention.

Germinal centre lymphomas

The introduction of NGS technologies has also led to significant improvements in our understanding of the molecular pathogenesis of non-Hodgkin lymphomas, identifying a plethora of previously unappreciated mutation targets and oncogenic pathways that have contributed to the development and progression of these diseases. This part of the chapter summarizes the recent advances in the genetics of germinal centre (GC) lymphomas, with special emphasis on follicular

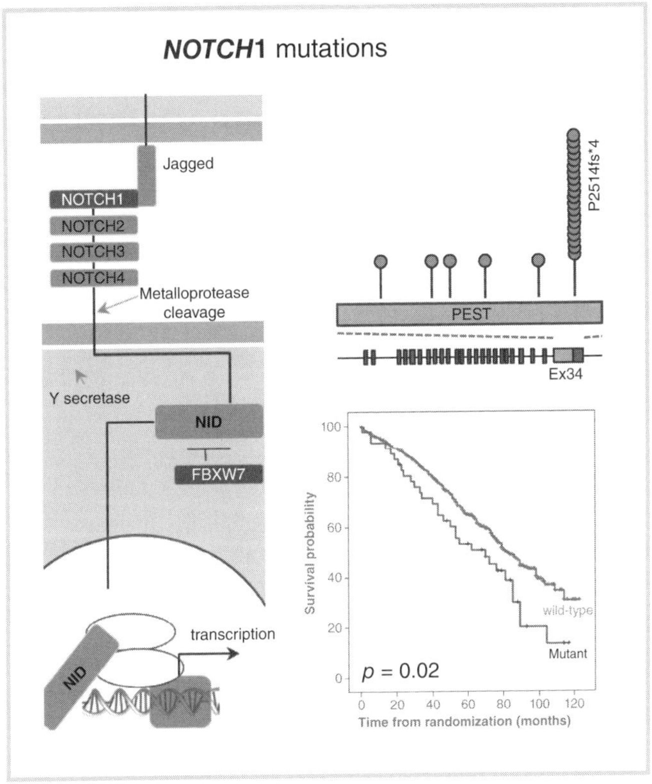

(a)

Figure 6.3 *NOTCH1* and *SF3B1* mutations in CLL. (a) The Notch signalling cascade and the distribution of mutations in *NOTCH1*. The *NOTCH1* gene contains 34 exons and encodes a protein with a C-terminal TAD-PEST domain, which is a hotspot for mutation in CLL. Part of exon 34 is magnified and the location of each mutation is shown. (b) Elements of the mRNA processing machinery and the distribution of mutation in *SF3B1*. The *SF3B1* gene contains 25 exons and encodes a protein with a C-terminal domain consisting of 22 HEAT domains. Exons 14, 15 and 16 are magnified and the locations of key hotspots are shown. Below each gene map is the overall survival data from the UK CLL4 trial. OS of CLL4 patients based on the mutational status of *NOTCH1* (left) and *SF3B1* (middle). The p values are derived from Kaplan–Meier analysis with a log-rank test and median survival times with 95% confidence. Adapted from Oscier et al.[50]

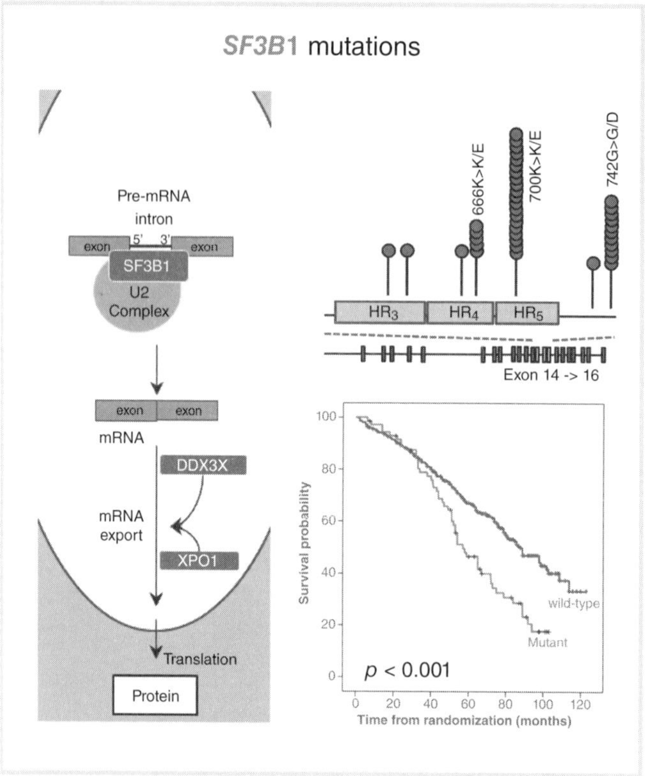

(b)

Figure 6.3 (*continued*)

lymphoma (FL) and diffuse large B-cell lymphoma (DLBCL), the two most common GC lymphoma entities (Fig. 6.1).

Follicular lymphoma

FL is the most common indolent lymphoma, accounting for ~25% of all B-cell non-Hodgkin lymphoma (NHL) cases.[112] It exhibits a heterogeneous clinical course with a median survival of 11–12 years, characterized by an initial good response to first-line therapy followed by multiple relapses and the need for recurrent therapeutic interventions.[113] In addition, a subset of patients (25–30%) undergo transformation to high-grade lymphoma (usually DLBCL), where the treatment options are limited, resulting in a much shortened OS.[114–116] Although the addition of rituximab to standard therapies resulted in a significant improvement in outcome measures, FL is considered incurable, with a proportion of patients becoming resistant to therapy.[117–119]

Genetic landscape of FL

Cytogenetics and copy number alterations

The molecular hallmark of the disease is the t(14;18)(q32;q21) translocation arising due to defective VDJ recombination of the *IGHV* genes in the bone marrow. It places the *BCL2* gene under control of the *IGHV* enhancer, leading to constitutive over-expression of the antiapoptotic BCL-2 protein.[120,121] This cellular survival advantage leads to accumulation of long-lived B-cells, which acquire additional secondary changes in the presence of the activation-induced cytidine deaminase (AID)-mediated genomic instability and from which the disease is thought to arise.[122] The t(14;18) translocation is present in more than 90% of the patients; however, it is also detected in the majority of healthy individuals, suggesting that it is not sufficient on its own to induce lymphomagenesis.[123–126]

In addition to the t(14;18) translocation, additional karyotypic events including somatically acquired CNAs and aUPD are universally found in FL and contribute to FL pathogenesis and transformation. Several groups reported recurrent CNAs based on aCGH and microarray studies with frequencies higher than 10%, including loss of 1p, 6q, 9p, 10q, 13q and 17p and also gains of 2p, 7, 8, 12q, 18q and X, combinations of which are present in almost all FL cases.[127–131]

The introduction of SNP microarrays allowed the identification of regions of aUPDs, with 1p, 6p, 12q and 16p being the most frequent (16–28%).[132–135] Some of these were reported to have prognostic relevance; in particular, aUPD on 1p36 correlated with shorter OS and aUPD on 16p was predictive of transformation and correlated with reduced PFS.[133] These areas can pinpoint mutations rendered homozygous over pre-existing monoallelic events.[132] Unlike other indolent B-cell malignancies such as CLL and multiple myeloma, where karyotype analysis can be used to risk stratify individual patients, this is not the case in FL, where profiling has largely been descriptive and pointing towards locations where tumour suppressor genes or oncogenes most likely reside. These investigations, in particular those analysing sequential biopsy samples, have provided insights into the evolution of FL and the existence of a B-cell progenitor pool of cells [called the common progenitor cell (CPC)], from which each episode of the disease is thought to arise, indicating the incurable nature of FL (Fig. 6.4).[131,132,136,137]

Although these karyotypic events in FL have been well documented, the identification of the target pathogenic loci have until recently remained obscure. With recurrent heterozygous deletions and aUPD in

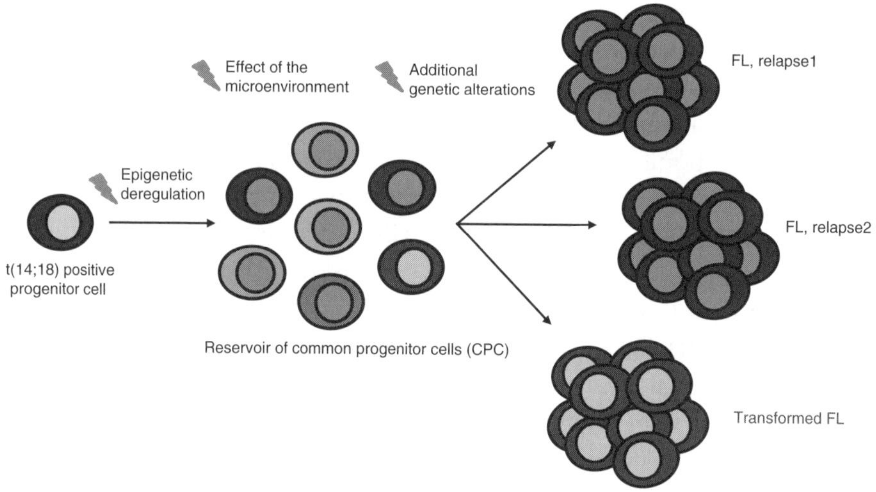

Figure 6.4 Model for FL pathogenesis and transformation. The t(14;18) positive progenitor cells acquire additional genetic and epigenetic alterations leading the development of an FL initiating pool of common progenitor cells (CPC), which give rise to each subsequent episode of the disease. Modified from Montoto and Fitzgibbon.[211]

more than 25% of FL patients, 1p36 represents one of the most frequent alterations in FL. In an attempt to identify the minimally deleted region at 1p36, *TNFRSF14* (tumour necrosis factor receptor superfamily 14), a cell-surface molecule expressed by haematopoietic cells was identified as the candidate tumour suppressor gene within this region and was shown to be mutated in 18.3% of the cases with mutations associated with inferior clinical outcome.[138] *TNFRSF14* mutations were also detected at a frequency of 46% in a second study, although the *TNFRSF14* variants were not linked with poor prognosis.[139] The majority of mutations in both studies (57%) resulted in the production of a truncated protein leading to decreased cell-surface expression of TNFRSF14. TNFRSF14 can act as both a receptor and a ligand and is able to transmit both stimulatory and inhibitory signals to T-cells and other components of the tumour microenvironment depending on its engagement with specific ligands.[140] Considering this complexity of signal transduction and interactions mediated by TNFRSF14, further studies are required to elucidate the role of these mutations in FL development.

The search for the candidate genes targeted by 6q deletions resulted in the identification of *TNFAIP3/A20*, a negative regulator of NF-κB signalling and ephrin receptor 7A (*EPHA7*) as tumour suppressors in

FL.[141,142] Interestingly, EPHA7 protein was absent in 72% of FL cases, which was attributed to deletion of one allele and suppression of the other allele via aberrant promoter hypermethylation.[141,143] The consequence of EPHA7 loss on FL development was demonstrated in a Bcl2 transgenic mouse model with *EPHA7* knockdown, that demonstrated a marked acceleration of disease onset and increase in penetrance. The authors revealed an intriguing tumour suppressor mechanism for EPHA7, where in normal B-cells the truncated form of EPHA7 ($EPHA^{TR}$) is secreted and binds to EPHA2, preventing its homodimerization and activation of the downstream signalling, while in FL in the absence of $EPHA^{TR}$, EPHA2 is able to dimerize and induce oncogenic signalling via ERK, STAT3 and SRC contributing to FL pathogenesis. The soluble tumour suppressor feature of EPHA7 also highlights its potential as a targeted therapeutic polypeptide.[144]

Epigenetic deregulation: aberrant DNA methylation

Epigenetic modifications such as DNA methylation are reversible changes associated with transcriptional regulation of gene expression. Aberrant cytosine methylation of the so-called CpG islands has been linked with silencing of various tumour suppressor genes in cancer.[145] This represents an attractive therapeutic target as the aberrant hypermethylation can be reversed using hypomethylating agents, approved for treatment of myelodysplastic syndromes (MDS) and currently under evaluation in clinical trials for FL.[146]

The initial methylation studies in FL focused on individual candidate tumour suppressor genes in small patient cohorts using non-quantitative methods.[147] These studies revealed hypermethylation of genes such as *DAPK1* and *SHP1* in 85–90% of the samples assayed.[148–151] Additionally, aberrant hypermethylation of cyclin-dependent kinase inhibitors *p15*, *p16* and *p57* was suggested to be associated with transformation of FL in a number of studies.[152–154]

With the introduction of newer array-based technologies, the focus of investigations has recently shifted to the global assessment of the FL methylome. One of the first array-based methylation profiling studies performed on 164 FL cases demonstrated widespread aberrant hypermethylation in FL, with 8% of the promoters displaying an increase in methylation compared with normal controls.[155] This provided a robust discriminator between the tumour and control normal samples, which is in line with observations from other studies.[156,157] The number of hypermethylated genes in FL was higher than in other B-cell lymphomas, e.g. CLL or MCL, indicating a pronounced hypermethylation pattern in FL.[157] However, the methylation profiles of sequential FL and transformed FL biopsies were conserved, suggesting that aberrant methylation represents an early event in FL pathogenesis and may not contribute significantly to transformation of FL.[158]

An intriguing finding of the methylation profiling studies in FL and other B-cell NHLs was the significant over-representation of stem cell targets of the Polycomb repressive complex 2 (PRC2) among the hypermethylated genes.[158–160] This was noteworthy since hypermethylation of PRC2 targets may 'lock in stem cell phenotypes', leading to aberrant clonal expansion.[160,161] The importance of this finding was established when the core catalytic subunit of PRC2, the histone methyltransferase enhancer of zeste homologue 2 (*EZH2*), was recently reported to harbour somatic gain of function mutations in ~7–22% of FLs and 14–22% of germinal centre-type DLBCLs in the first series of NGS experiments on these tumours.[162–164]

Epigenetic deregulation: recurrent mutations of epigenetic regulators in FL

The introduction of the NGS technologies has led to an unprecedented increase in the identification of novel mutational targets across a wide range of malignancies. In FL, these investigations have revealed a unique molecular portrait of the disease with recurrent mutations of various components and regulators of the epigenetic machinery in almost all cases of this malignancy (Table 6.3).[165,166] The specific pattern of the mutations leading to aberrant histone methylation and acetylation suggests that the coordinated repression of the transcriptome represents one of the critical events in FL pathogenesis (Fig. 6.5).

The first gene shown to be recurrently mutated in these studies was the histone methyltransferase *EZH2* with a mutation frequency of 7–22% of FL cases.[162–164] EZH2 serves as the enzymatic subunit of the PRC2 complex and catalyses the trimethylation of lysine 27 on histone H3 (H3K27me3), leading to transcriptional repression of the targeted loci.[167] This methyltransferase also plays a central role in early B-cell development by controlling immunoglobulin heavy-chain gene (*IGH@*) rearrangements, with the *EZH2* conditional knock-out mice demonstrating an impaired pro-B to pre-B-cell transition.[168] In germinal centre B-cells, EZH2 is highly expressed and binds to a specific set of target genes involved in differentiation, suppression of cell growth and proliferation, suggesting a GC-specific EZH2 regulatory programme in

Table 6.3 Recurrent mutations of epigenetic regulators in follicular lymphoma.

Gene symbol	Gene name	Function	Frequency (%)	Gain/loss of function
MLL2	Myeloid/lymphoid or mixed-lineage leukaemia 2	H3K4 MTase*	89	Loss
CREBBP	CREB-binding protein	HAT	33	Loss
EP300	E1A-binding protein p300	HAT	9	Loss
EZH2	Enhancer of zeste homologue 2	H3K27 MTase	7–22	Gain
MEF2B	Myocyte enhancer factor 2B	Enhances HAT* activity	13	Loss

*Abbreviations: MTase, methyltransferase; HAT, histone acetyltransferase.

normal and malignant B-cells.[169] Somatically acquired *EZH2* mutations lead to replacement of the critical tyrosine residue (Y646) within the catalytic SET domain of the protein, with the mutations acting through a unique gain of function mechanism, where the coordinated activity of the wild-type and the mutant enzyme leads to decreased mono- and dimethylation of H3K27 (H3K27me and H3K27me2) with a global increase of H3K27me3 levels.[170,171] This provided a compelling rationale for the development of EZH2-targeted therapies, and indeed the first selective EZH2 inhibitors have now been developed and are being evaluated in pre-clinical models. These early experiments have demonstrated a high selectivity against a panel of lymphoma cell lines carrying *EZH2* mutations.[172–174] These targeted epigenetic therapies are now gaining momentum and represent a promising new area of drug development with agents selectively targeting the aberrant patterning of chromatin marks. It will be of great interest to see the translation of these agents in clinical trials.

The sequencing experiments that identified *EZH2* served as the forerunner to the striking discovery of inactivating hetero- and homozygous mutations of the H3K4 histone methyltransferase *MLL2* in 90% of FL cases. This observation challenges the notion that the t(14;18) translocation is the essential primary genetic hit in FL pathogenesis.[165] Trimethylated H3K4 (H3K4me3) is associated with the presence of nonsense and indel mutations in the *MLL2* gene and results in the transcriptional activation of a series of developmental genes, and seems likely to behave as a tumour suppressor in FL.[175,176] In addition, the region on chromosome

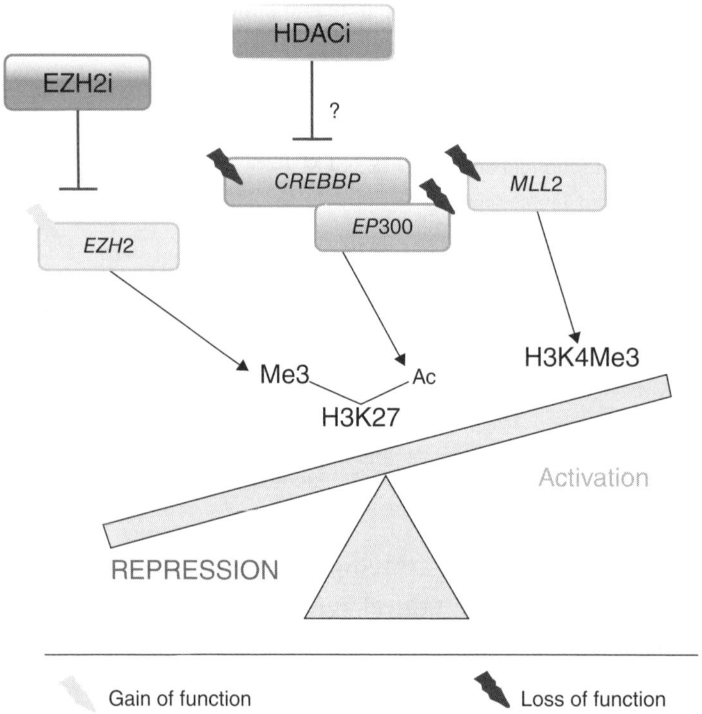

Figure 6.5 Epigenetic deregulation in follicular lymphoma. The specific pattern of mutations in various members of the epigenetic machinery, including gain of function mutations of the histone methyltransferases *EZH2* and *MLL2* and histone acetyltransferases *CREBBP* and *EP300*, indicates that global repression of the transcriptome plays a crucial role in FL pathogenesis. These changes represent attractive therapeutic targets for epigenetic therapeutic agents such as EZH2 inhibitors and histone deacetylase inhibitors (HDACi).

12 containing the *MLL2* gene is frequently the subject of aUPD, associated with homozygous gene mutation and complete *MLL2* inactivation. Although loss of H3K4 trimethylation via *MLL2* inactivation is the most frequent aberration in FL, suitable agents to modulate H3K4 modifications are not available at present.

The histone acetylation machinery is also deregulated due to the presence of somatically acquired mutations in FL with inactivating mutations of histone acetyltransferases (HAT) *CREBBP* (32.6%) and *EP300* (8.7%) and a regulator of histone acetylation, *MEF2B* (15.3%), leading to reduced acetylation the target genes.[165,166] Constitutive activation of BCL6 and a reduction in the tumour-suppressor activity of p53 were demonstrated as one of the consequences of these mutations.[166] Of note,

H3K27 is also the subject of acetylation by CREBBP and EP300. Similarly to *EZH2* gain-of-function mutations, loss-of-function mutations within *CREBBP* and *EP300* are likely to lead to reduced acetylation and, in turn, increased H3K27 trimethylation, which also points towards global repression of the gene expression in FL. Although speculative, the discovery of mutations targeting components of the acetylation machinery provides a rational basis for the re-evaluation of histone deacetylase inhibitors (HDACi) as a potential therapeutic strategy for patients stratified based on the presence of these mutations.

Diffuse large B-cell lymphoma

Diffuse large B-cell lymphoma (DLBCL) is the most common type of NHL, representing about 40% of cases, and is characterized by a highly variable clinical course.[177] Although addition of rituximab resulted in improvement in the outcome, approximately one-third of the patients remain incurable and will succumb to their disease. During the last decade, significant improvement in understanding of molecular pathogenesis of DLBCL has been achieved, primarily as a consequence of the development of novel technologies that have allowed global gene expression profiling, the documentation of chromosomal changes and, most recently, the introduction of whole-exome and -genome sequencing. As with CLL and FL, these innovations are not merely contributing to the molecular characterization of the disease but also permitting the development of treatment strategies based on their underlying genetic abnormalities.

Subtypes of DLBCL based on gene expression profiling

The seminal gene expression profiling (GEP) study by Alizadeh and colleagues[178] identified two major subgroups of DLBCL: germinal centre (GC) B-cell type and activated B-cell (ABC) type DLBCL. These subtypes are characterized by distinct clinical and biological features, the latter of which is a direct reflection of the fact that these DLBCL subtypes are derived from specific stages of B-cell differentiation.[178] Patients with germinal GC-like DLBCL had a significantly better overall survival than those with ABC-like DLBCL, providing a powerful predictor of outcome. Based on the most recent NGS studies, this dichotomy is also mirrored in the mutation spectra of DLBCL, with some alterations exclusively occurring in the GC subtype only and others are restricted to the ABC subtype. However, a considerable number of recurrent mutation events are shared between the two subtypes, suggesting some overlapping pathogenetic mechanisms.[166,179]

Novel mutation targets in DLBCL

Several groups investigated the genomic basis of DLBCL using whole-exome or -genome sequencing and reported a striking genetic heterogeneity with more than 300 recurrently mutated cancer genes identified.[179–181] The most frequently mutated genes and pathways include chromatin modification (*MLL2, CREBBP, EP300, MEF2B, AIRD1A*), NF-κB (*CARD11, MYB88, CD79B, TNFAIP3*) and the PI3 kinase signalling (*PIK3CD, PIK3R1, MTOR*) pathways. Compared with other malignancies, e.g. FL or CLL, DLBCL was characterized by a considerably higher mutation rate with a mean non-synonymous mutation rate of 3.2 mutations per megabase, indicating a profound genomic instability in this malignancy.[180]

Recurrent mutations in epigenetic regulators in DLBCL

Deregulation of histone methylation and acetylation supposedly leading to repression of the transcriptome appears to be a common mechanism linking FL and DLBCL as the mutations in methyltransferases *EZH2* and *MLL2* and acetyltransferases *CREBBP*, *EP300* and *MEF2B* arising in both diseases.[165] However, the pattern and frequency of these mutations in FL and DLBCL suggest mechanistic and biological distinctness; FL harbours multiple mutations in these epigenetic regulators, whereas these mutations tend to emerge in a mutually exclusive pattern in DLBCL.[165,166,179] Interestingly, the *EZH2* SET domain mutations are almost exclusively restricted to the GC subtype of the disease (14–22%), while the *MLL2* (32%), *CREBBP* (29%), *EP300* (10%) and *MEF2B* (11%) mutations can be detected in both GC and ABC subtypes of DLBCL. Similarly to the situation with FL, the first selective EZH2 inhibitors demonstrated encouraging potency in DLBCL cell lines and mouse xenograft models.[172,173] Of note, mutations of other epigenetic regulators histone 1 (H1) family proteins involved in chromatin compacting were also reported in DLBCL with two-thirds of patients harbouring mutation in H1 proteins.[180] The precise functional consequence of this observation is still open to speculation.

Mutations in components of B-cell receptor signalling in DLBCL

B-cell receptor (BCR) signalling provides survival signals for B-cells through activation of the PI3K/mTOR and NF-κB pathways and it was demonstrated that constitutive signal transduction via BCR is a critical

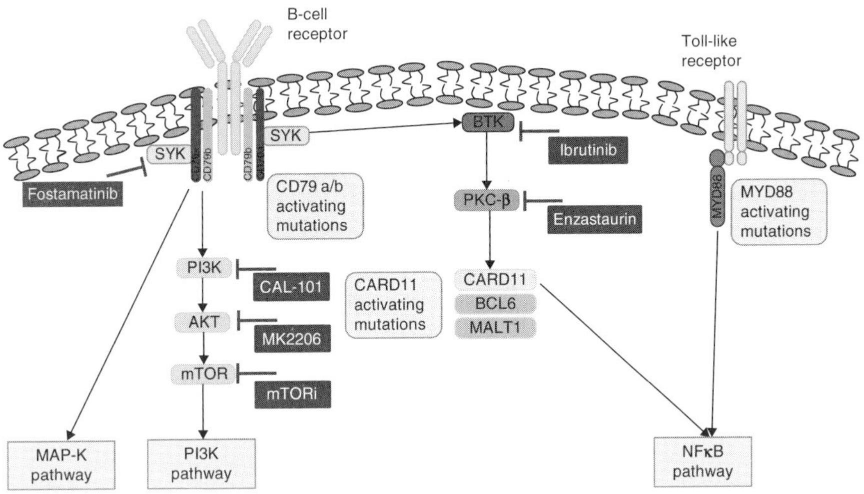

Figure 6.6 Oncogenic pathways in diffuse large B-cell lymphoma. Illustrated are the activated pathways via B-cell receptor and Toll-like receptor-mediated signalling. Activating mutations of *CD79a*, *CD79b*, *CARD11* and *MYD88* were identified in DLBCL, contributing the constitutive/chronic activation of these pathways. Several agents (highlighted in the dark-shaded boxes) targeting the effectors of these pathways are in clinical trials and have demonstrated promising results in DLBCL and other B-cell lymphoma entities.

factor for survival and proliferation of B-cell lymphomas (Fig. 6.6).[182] Early studies demonstrated that deregulation of BCR signalling and dependency on NF-κB activation seems to play a central role in the ABC DLBCLs, also highlighted by higher response rates to the proteasome inhibitor bortezomib.[183,184] Recent genomic studies provided some insights into the genetic background of altered BCR signalling and NF-κB activation in DLBCL. Lenz and colleagues reported activating mutations of the scaffold protein CARD11, leading to constitutive NF-κB activation in around 10% of the ABC DLBCL cases.[185] Interestingly, activating *CARD11* mutations were detected in a smaller subset of GC DLBCL cases (3%). Later, this repertoire has expanded with activating mutations of *CD79B* and *CD79A*, the proximal subunits of the BCR, being detected in about 20% of ABC DLBCLs and rarely in GC DLBCLs. The mutations targeted the immunoreceptor tyrosine-based activation motifs (ITAM) of *CD79A* and *CD79B*, leading to chronic active BCR signalling with subsequent activation of the NF-κB, PI3K and MAP-kinase pathways.[186] Notably, the same mutation that occurs in CLL, a recurrent mutation in *MYD88* (L265P), was found in 30% of ABC DLBCL. This mutation leads

Table 6.4 Next-generation sequencing studies in different B-cell non-Hodgkin lymphomas.

Entity	Mutated genes identified	References
Follicular lymphoma	*MLL2, EZH2, CREBBP, MEF2B, EP300*	165, 201, 202
Diffuse large B-cell lymphoma	*EZH2, MLL2, MEF2B, CREBBP, EP300, TNFAIP3, MYD88, Histone H1* genes	165, 201–204
Burkitt's lymphoma	*ID3, GNA13, RET, PIK3R1*	205, 206
Mantle cell lymphoma	*NOTCH1*	207
Splenic B-cell marginal zone lymphoma	*NOTCH2*	208, 209
Hairy-cell leukaemia	*BRAF*	210

to oncogenic activation of NF-κB and also JAK-mediated activation of STAT3.[180,187,188]

These findings have important clinical implications as the BCR signalling pathway offers a plethora of potential targets including SYK, BTK and PI3 kinases, which are ideal for therapeutic intervention using targeted therapies (Fig. 6.4), and indeed many of these agents demonstrate encouraging results in different B-cell lymphomas[189–191] (Table 6.4). The initial studies suggest that the mutation status of the aforementioned regulators of the BCR signalling pathway will likely predict response to these therapies, hence stratification of patients based on the mutation status of these genes will be important for achieving optimal outcomes with these novel targeted agents.[192]

Conclusions and future perspectives

Our knowledge of the molecular pathogenesis of these mature B-cell malignancies has expanded exponentially over the last 50 years. Over this time, significant technological advances have always preceded leaps forward in our understanding of these diseases. It began with the development of cytogenetic analysis that remains a relevant, albeit low-resolution whole-genome scanning approach. In the 1980s, the gap between molecular genetics and cytogenetics was bridged with the advent of FISH and other molecular cytogenetic approaches, which for the first time provided a truly integrated approach to genomic analysis. In the late 1990s and into the new millennium, developments in microarray technology have further bridged this gap, and developments in NGS are identifying relevant cancer genes and pathways with unbridled speed and precision. The application of NGS technology to unravelling the complexity of the cancer genome has been pioneered by the International Cancer Genome Consortium (ICGC; www.icgc.org) and the inclusion of many mature B-cell malignancies into this initiative promises that our understanding of these diseases will continue to expand. These ICGC programmes are ongoing, but have already transformed our perception of these neoplasms, but this next phase of cancer genomics is only just beginning. Significant research remains to be conducted that will not only catalogue the genomic architecture of these diseases, but will ultimately define the clinical utility and biological importance of these lesions in the context of disease diagnosis, natural history, prognostication and therapeutic response. Such approaches are listed below:

1 *Clinical trials.* Although a plethora of lesions have been identified in these diseases, it will be important to continue to validate their importance in the context of clinical treatment trials. These experiments may employ standard molecular techniques investigating candidate genes or NGS approaches to investigate gene panels or whole genomes/exomes. This will permit the following comparisons to be made: (i) to assess clinical utility of these novel mutations in homogeneous patient cohorts, where patient material is assessed at consistent time points during the natural history of the diseases; (ii) to assess the ability of these new lesions to predict therapeutic response, particularly after treatment with novel small molecules; and (iii) to identify the genetic component that contributes, or drives, therapeutic resistance by investigating sequential samples taken from the same patients preceding, during and after treatment.
2 *Large international studies.* It is likely that the genes mutated at high frequencies in these diseases have already been defined and due to this high frequency their clinical importance is more straightforward to ascertain. However, for less prevalent mutational events, particularly those where multiple genes are targeted in key biological pathways, it will be necessary to investigate very large patient cohorts or employ meta-analysis to collate large international studies.
3 *Functional validation.* There are a variety of genetic and bioinformatics approaches to predict the pathogenic importance of a given lesion, including recurrence and the likelihood that a lesion will compromise

protein function. However, to identify accurately so-called 'driver' mutations, functional analysis is required to prove causation. This analysis could include biochemical analysis, studies of the crystal structures of mutant proteins and the introduction of mutations into *in vitro* and *in vivo* model systems, with the assessment of appropriate biological readouts, such as cell death, proliferation, differentiation and signalling. The development of many of these systems is both technologically challenging and time consuming. However, where possible it will be important to consider improvements in the throughput of such approaches, so that the multitude of lesions emerging from NGS studies can be functionally validated on a time-scale that permits rapid clinical translation.

4 *Molecular diagnostics.* The translation of these genetic studies into real improvements in patient care will require focus on the following: (i) multiple, independent studies to validate the most informative panel of genes in the context of a given treatment modality; (ii) due to the protracted natural history of these diseases, it will be important to establish when these molecular diagnostic assays are to be performed; (iii) the most accurate technology will need to be optimized, which will require international agreement; and (iv) the realization of the full cost implication of these assays and the establishment of a funding strategy to support their use.

5 *Systems biology.* Our understanding of DNA sequence changes in these diseases is considerably more advanced than our understanding of epigenetic alterations or those at the transcriptome level. Exciting data are emerging, some of which are discussed in this chapter. Large studies of DNA methylation have been conducted in CLL that show considerable promise.[193] It has also been demonstrated that recurrent fusion transcripts do occur in CLL at the transcriptional level,[194] but further study and validation are required. Ultimately, the application of computational modelling and systems biology is required to integrate genomic, epigenetic, transcriptomic and proteomic data to provide a systematic analysis of the molecular lesions contributing to the pathogenesis of the B-cell neoplasms.

To conclude, the race continues between laboratories worldwide to uncover the secrets that underpin the pathophysiology of these malignancies. All parties involved in this race share great optimism that our increased knowledge will ultimately lead to improvements in the clinical management of patients with these diseases, which would include targeted therapies and, ultimately, a cure.

Acknowledgements

This work was funded by Leukaemia and Lymphoma Research, Cancer Research UK and the Kay Kendall Leukaemia Fund. C.B. is the recipient of a TAMOP 4.2.4.A/1 fellowship. The authors would like to thank Professor David Oscier for his help and guidance in the preparation of this chapter.

References

1 Zech, L., et al. Characteristic chromosomal abnormalities in biopsies and lymphoid-cell lines from patients with Burkitt and non-Burkitt lymphomas. *Int J Cancer* 1976; 17(1): 47–56.
2 Mitelman, F., Johansson, B., Mertens F. The impact of translocations and gene fusions on cancer causation. *Nat Rev Cancer* 2007; 7(4): 233–245.
3 Swerdlow, S.H., et al. *WHO Classification of Tumours of Haematopoietic and Lymphoid Tissues*, 4th edn. Lyon: World Health Organization, 2008.
4 Chiorazzi, N., Rai, K.R., Ferrarini, M. Chronic lymphocytic leukemia. *N Engl J Med* 2005; 352(8): 804–15.
5 Messmer, B.T., et al. In vivo measurements document the dynamic cellular kinetics of chronic lymphocytic leukemia B cells. *J Clin Invest* 2005; 115(3): 755–764.
6 Rawstron, A.C., et al. Monoclonal B-cell lymphocytosis and chronic lymphocytic leukemia. *N Engl J Med* 2008; 359(6): 575–583.
7 Rai, K.R., et al. Clinical staging of chronic lymphocytic leukemia. *Blood* 1975; 46(2): 219–234.
8 Binet, J.L., et al. A new prognostic classification of chronic lymphocytic leukemia derived from a multivariate survival analysis. *Cancer* 1981; 48(1): 198–206.
9 Hashimoto, S., et al. Somatic diversification and selection of immunoglobulin heavy and light chain variable region genes in IgG^+ $CD5^+$ chronic lymphocytic leukemia B cells. *J Exp Med* 1995; 181(4): 1507–1517.
10 Stevenson, F.K., et al. The occurrence and significance of V gene mutations in B cell-derived human malignancy. *Adv Cancer Res* 2001; 83: 81–166.
11 Hamblin, T.J., et al. Unmutated Ig V_H genes are associated with a more aggressive form of chronic lymphocytic leukemia. *Blood* 1999; 94(6): 1848–1854.
12 Damle, R.N., et al. Ig V gene mutation status and CD38 expression as novel prognostic indicators in chronic lymphocytic leukemia. *Blood* 1999; 94(6): 1840–1847.
13 Stamatopoulos, K., et al. Over 20% of patients with chronic lymphocytic leukemia carry stereotyped receptors: pathogenetic implications and clinical correlations. *Blood* 2007; 109(1): 259–270.
14 Ghia, E.M., et al. Use of *IGHV3–21* in chronic lymphocytic leukemia is associated with high-risk disease and reflects antigen-driven, post-germinal center leukemogenic selection. *Blood* 2008; 111(10): 5101–5108.

15 Murray, F., et al. Stereotyped patterns of somatic hypermutation in subsets of patients with chronic lymphocytic leukemia: implications for the role of antigen selection in leukemogenesis. *Blood* 2008; 111(3): 1524–1533.
16 Tobin, G., et al. Somatically mutated Ig V_H3–21 genes characterize a new subset of chronic lymphocytic leukemia. *Blood* 2002; 99(6): 2262–2264.
17 Gahrton, G., et al. Nonrandom chromosomal aberrations in chronic lymphocytic leukemia revealed by polyclonal B-cell-mitogen stimulation. *Blood* 1980; 56(4): 640–647.
18 Hurley, J.N., et al. Chromosome abnormalities of leukaemic B lymphocytes in chronic lymphocytic leukaemia. *Nature* 1980; 283(5742): 76–78.
19 Gahrton, G., et al. Extra chromosome 12 in chronic lymphocytic leukaemia. *Lancet* 1980; 2(8160): 146–147.
20 Fitchett, M., et al. Chromosome abnormalities involving band 13q14 in hematologic malignancies. *Cancer Genet Cytogenet* 1987; 24(1): 143–150.
21 Juliusson, G., et al. Prognostic subgroups in B-cell chronic lymphocytic leukemia defined by specific chromosomal abnormalities. *N Engl J Med* 1990; 323(11): 720–724.
22 Buhmann, R., et al. CD40L stimulation enhances the ability of conventional metaphase cytogenetics to detect chromosome aberrations in B-cell chronic lymphocytic leukaemia cells. *Br J Haematol* 2002; 118(4): 968–975.
23 Dicker, F., et al. Immunostimulatory oligonucleotide-induced metaphase cytogenetics detect chromosomal aberrations in 80% of CLL patients: a study of 132 CLL cases with correlation to FISH, IgVH status and CD38 expression. *Blood* 2006; 108(9): 3152–3160.
24 Put, N., et al. Improved detection of chromosomal abnormalities in chronic lymphocytic leukemia by conventional cytogenetics using CpG oligonucleotide and interleukin-2 stimulation: a Belgian multicentric study. *Genes Chromosomes Cancer* 2009; 48(10): 843–853.
25 Haferlach, C., et al. Comprehensive genetic characterization of CLL: a study on 506 cases analysed with chromosome banding analysis, interphase FISH, IgV_H status and immunophenotyping. *Leukemia* 2007; 21(12): 2442–2451.
26 Han, T., et al. Prognostic importance of cytogenetic abnormalities in patients with chronic lymphocytic leukemia. *N Engl J Med* 1984; 310(5): 288–292.
27 Martín-Subero, J.I., et al. A comprehensive genetic and histopathologic analysis identifies two subgroups of B-cell malignancies carrying a t(14;19)(q32;q13) or variant BCL3-translocation. *Leukemia* 2007; 21(7): 1532–1544.
28 Chapiro, E., et al. The most frequent t(14;19)(q32;q13)-positive B-cell malignancy corresponds to an aggressive subgroup of atypical chronic lymphocytic leukemia. *Leukemia* 2008; 22(11): 2123–2127.
29 Cavazzini, F., et al. Chromosome 14q32 translocations involving the immunoglobulin heavy chain locus in chronic lymphocytic leukaemia identify a disease subset with poor prognosis. *Br J Haematol* 2008; 142(2): 529–537.
30 Put, N., et al. Translocation t(14;18) is not associated with inferior outcome in chronic lymphocytic leukemia. *Leukemia* 2009; 23(6): 1201–1204.

31 Yin, C.C., et al. Chronic lymphocytic leukemia with t(2;14)(p16;q32) involves the *BCL11A* and *IgH* genes and is associated with atypical morphologic features and unmutated IgV_H genes. *Am J Clin Pathol* 2009; 131(5): 663–670.

32 Huh, Y.O., et al. MYC translocation in chronic lymphocytic leukaemia is associated with increased prolymphocytes and a poor prognosis. *Br J Haematol* 2008; 142(1): 36–44.

33 Döhner, H., et al. Genomic aberrations and survival in chronic lymphocytic leukemia. *N Engl J Med* 2000; 343(26): 1910–1916.

34 Bastard, C., et al. Comparison of a quantitative PCR method with FISH for the assessment of the four aneuploidies commonly evaluated in CLL patients. *Leukemia* 2007; 21(7): 1460–1463.

35 Buijs, A., Krijtenburg, P.J. Meijer, E. Detection of risk-identifying chromosomal abnormalities and genomic profiling by multiplex ligation-dependent probe amplification in chronic lymphocytic leukemia. *Haematologica* 2006; 91(10): 1434–1435.

36 Mertens, D., et al. Allelic silencing at the tumor-suppressor locus 13q14.3 suggests an epigenetic tumor-suppressor mechanism. *Proc Natl Acad Sci U S A* 2006; 103(20): 7741–7746.

37 Mertens, D., et al. Down-regulation of candidate tumor suppressor genes within chromosome band 13q14.3 is independent of the DNA methylation pattern in B-cell chronic lymphocytic leukemia. *Blood* 2002; 99(11): 4116–4121.

38 Calin, G.A., et al. Frequent deletions and down-regulation of micro-RNA genes *miR15* and *miR16* at 13q14 in chronic lymphocytic leukemia. *Proc Natl Acad Sci U S A* 2002; 99(24): 15524–15529.

39 Klein, U., et al. The *DLEU2/miR-15a/16-1* cluster controls B cell proliferation and its deletion leads to chronic lymphocytic leukemia. *Cancer Cell* 2010; 17(1): 28–40.

40 Lia, M., et al. Functional dissection of the chromosome 13q14 tumor-suppressor locus using transgenic mouse lines. *Blood* 2011; 119(13): 2981–2990.

41 Parker, H., et al. 13q deletion anatomy and disease progression in patients with chronic lymphocytic leukemia. *Leukemia* 2011; 25(3): 489–497.

42 Ouillette, P., et al. Integrated genomic profiling of chronic lymphocytic leukemia identifies subtypes of deletion 13q14. *Cancer Res* 2008; 68(4): 1012–1021.

43 Ouillette, P., et al. The prognostic significance of various 13q14 deletions in chronic lymphocytic leukemia. *Clin Cancer Res* 2011; 17(21): 6778–6790.

44 Palamarchuk, A., et al. 13q14 deletions in CLL involve cooperating tumor suppressors. *Blood* 2010; 115(19): 3916–3922.

45 Hernández, J.A., et al. A high number of losses in 13q14 chromosome band is associated with a worse outcome and biological differences in patients with B-cell chronic lymphoid leukemia. *Haematologica* 2009; 94(3): 364–371.

46 Dierlamm, J., et al. FISH identifies different types of duplications with 12q13–15 as the commonly involved segment in B-cell lymphoproliferative malignancies characterized by partial trisomy 12. *Genes Chromosomes Cancer* 1997; 20(2): 113–119.

47 Matutes, E., et al. Trisomy 12 defines a group of CLL with atypical morphology: correlation between cytogenetic, clinical and laboratory features in 544 patients. *Br J Haematol* 1996; 92(2): 382–388.
48 Sellmann, L., et al. Trisomy 19 is associated with trisomy 12 and mutated IGHV genes in B-chronic lymphocytic leukaemia. *Br J Haematol* 2007; 138(2): 217–220.
49 Balatti, V., et al. *NOTCH1* mutations in CLL associated with trisomy 12. *Blood* 2011; 119(2): 329–331.
50 Oscier, D.G., et al. The clinical significance of *NOTCH1* and *SF3B1* mutations in the UK LRF CLL4 trial. *Blood* 2013; 120(22): 4441–4443.
51 López, C., et al. Different distribution of *NOTCH1* mutations in chronic lymphocytic leukemia with isolated trisomy 12 or associated with other chromosomal alterations. *Genes Chromosomes Cancer* 2012; 51(9): 881–889.
52 Decker, S., et al. Trisomy 12 and elevated *GLI1* and *PTCH1* transcript levels are biomarkers for Hedgehog-inhibitor responsiveness in CLL. *Blood* 2012; 119(4): 997–1007.
53 Hallek, M., et al. Addition of rituximab to fludarabine and cyclophosphamide in patients with chronic lymphocytic leukaemia: a randomised, open-label, phase 3 trial. *Lancet* 2010; 376(9747): 1164–1174.
54 Zenz, T., et al. TP53 mutation profile in chronic lymphocytic leukemia: evidence for a disease specific profile from a comprehensive analysis of 268 mutations. *Leukemia* 2010; 24(12): 2072–2079.
55 Tam, C.S., et al. Chronic lymphocytic leukaemia CD20 expression is dependent on the genetic subtype: a study of quantitative flow cytometry and fluorescent in-situ hybridization in 510 patients. *Br J Haematol* 2008; 141(11): 36–40.
56 Döhner, H., et al. 11q deletions identify a new subset of B-cell chronic lymphocytic leukemia characterized by extensive nodal involvement and inferior prognosis. *Blood* 1997; 89(7): 2516–2522.
57 Catovsky, D., et al. Assessment of fludarabine plus cyclophosphamide for patients with chronic lymphocytic leukaemia (the LRF CLL4 Trial): a randomised controlled trial. *Lancet* 2007; 370(9583): 230–239.
58 Kalla, C., et al. Analysis of 11q22–q23 deletion target genes in B-cell chronic lymphocytic leukaemia: evidence for a pathogenic role of *NPAT, CUL5* and *PPP2R1B*. *Eur J Cancer* 2007; 43(8): 1328–1335.
59 Stankovic, T., et al. Inactivation of ataxia telangiectasia mutated gene in B-cell chronic lymphocytic leukaemia. *Lancet* 1999; 353(9146): 26–29.
60 Austen, B., et al. Mutation status of the residual ATM allele is an important determinant of the cellular response to chemotherapy and survival in patients with chronic lymphocytic leukemia containing an 11q deletion. *J Clin Oncol* 2007; 25(34): 5448–5457.
61 Skowronska, A., et al. Biallelic ATM inactivation significantly reduces survival in patients treated on the United Kingdom Leukemia Research Fund Chronic Lymphocytic Leukemia 44 trial. *J Clin Oncol* 2012; 30(36): 4524–4532.
62 Ouillette, P., et al. Aggressive chronic lymphocytic leukemia with elevated genomic complexity is associated with multiple gene defects in the response to DNA double-strand breaks. *Clin Cancer Res* 2010; 16(3): 835–847.

63 Rossi, D., et al. Alteration of BIRC3 and multiple other NF-κB pathway genes in splenic marginal zone lymphoma. *Blood* 2011; 118(18): 4930–4934.

64 Pettitt, A.R., et al. Alemtuzumab in combination with methylprednisolone is a highly effective induction regimen for patients with chronic lymphocytic leukemia and deletion of TP53: final results of the National Cancer Research Institute CLL206 trial. *J Clin Oncol* 2012; 30(14): 1647–1655.

65 Ferrajoli, A., et al. Lenalidomide induces complete and partial remissions in patients with relapsed and refractory chronic lymphocytic leukemia. *Blood* 2008; 111(11): 5291–5297.

66 Herman, S.E., et al. Bruton tyrosine kinase represents a promising therapeutic target for treatment of chronic lymphocytic leukemia and is effectively targeted by PCI-32765. *Blood* 2011; 117(23): 6287–6296.

67 Raghavan, M., et al. Genome-wide single nucleotide polymorphism analysis reveals frequent partial uniparental disomy due to somatic recombination in acute myeloid leukemias. *Cancer Res* 2005; 65(2): 375–378.

68 Saddler, C., et al. Comprehensive biomarker and genomic analysis identifies p53 status as the major determinant of response to MDM2 inhibitors in chronic lymphocytic leukemia. *Blood* 2008; 111(3): 1584–1593.

69 Edelmann, J., et al. High-resolution genomic profiling of chronic lymphocytic leukemia reveals new recurrent genomic alterations. *Blood* 2012; 120(24): 4783–4794.

70 Zenz, T., et al. Monoallelic TP53 inactivation is associated with poor prognosis in chronic lymphocytic leukemia: results from a detailed genetic characterization with long-term follow-up. *Blood* 2008; 112(8): 3322–3329.

71 Rossi, D., et al. The prognostic value of TP53 mutations in chronic lymphocytic leukemia is independent of Del17p13: implications for overall survival and chemorefractoriness. *Clin Cancer Res* 2009; 15(3): 995–1004.

72 Dicker, F., et al. The detection of TP53 mutations in chronic lymphocytic leukemia independently predicts rapid disease progression and is highly correlated with a complex aberrant karyotype. *Leukemia* 2009; 23(1): 117–124.

73 Gonzalez, D., et al. Mutational status of the TP53 gene as a predictor of response and survival in patients with chronic lymphocytic leukemia: results from the LRF CLL4 trial. *J Clin Oncol* 2011; 29(16): 2223–2229.

74 Pettitt, A.R., et al. p53 dysfunction in B-cell chronic lymphocytic leukemia: inactivation of ATM as an alternative to TP53 mutation. *Blood* 2001; 98(3): 814–822.

75 Zenz, T., et al. Detailed analysis of p53 pathway defects in fludarabine-refractory chronic lymphocytic leukemia (CLL): dissecting the contribution of 17p deletion, TP53 mutation, p53–p21 dysfunction and miR34a in a prospective clinical trial. *Blood* 2009; 114(13): 2589–2597.

76 Johnson, G.G., et al. A novel type of p53 pathway dysfunction in chronic lymphocytic leukemia resulting from two interacting single nucleotide polymorphisms within the p21 gene. *Cancer Res* 2009; 69(12): 5210–5217.

77 Best, O.G., et al. A novel functional assay using etoposide plus nutlin-3a detects and distinguishes between ATM and TP53 mutations in CLL. *Leukemia* 2008; 22(7): 1456–1459.

78 Schwaenen, C., et al. Automated array-based genomic profiling in chronic lymphocytic leukemia: development of a clinical tool and discovery of recurrent genomic alterations. *Proc Natl Acad Sci U S A* 2004; 101(4): 1039–1044.
79 Pfeifer, D., et al. Genome-wide analysis of DNA copy number changes and LOH in CLL using high-density SNP arrays. *Blood* 2007; 109(3): 1202–1210.
80 Stilgenbauer, S., et al. Incidence and clinical significance of 6q deletions in B cell chronic lymphocytic leukemia. *Leukemia* 1999; 13(9): 1331–1334.
81 Cuneo, A., et al. Chronic lymphocytic leukemia with 6q– shows distinct hematological features and intermediate prognosis. *Leukemia* 2004; 18(3): 476–483.
82 Chapiro, E., et al. Gain of the short arm of chromosome 2 (2p) is a frequent recurring chromosome aberration in untreated chronic lymphocytic leukemia (CLL) at advanced stages. *Leuk Res* 2010; 34(1): 63–68.
83 Jarosova, M., et al. Gain of chromosome 2p in chronic lymphocytic leukemia: significant heterogeneity and a new recurrent dicentric rearrangement. *Leuk Lymphoma* 2010; 51(2): 336–345.
84 Puente, X.S., et al. Whole-genome sequencing identifies recurrent mutations in chronic lymphocytic leukaemia. *Nature* 2011; 475(7354): 101–105.
85 Grandori, C., et al. The Myc/Max/Mad network and the transcriptional control of cell behavior. *Annu Rev Cell Dev Biol* 2000; 16: 653–699.
86 Gunnarsson, R., et al. Large but not small copy-number alterations correlate to high-risk genomic aberrations and survival in chronic lymphocytic leukemia: a high-resolution genomic screening of newly diagnosed patients. *Leukemia* 2010; 24(1): 211–215.
87 Ouillette, P., et al. Acquired genomic copy number aberrations and survival in chronic lymphocytic leukemia. *Blood* 2011; 118(11): 3051–3061.
88 Kujawski, L., et al. Genomic complexity identifies patients with aggressive chronic lymphocytic leukemia. *Blood* 2008; 112(5): 1993–2003.
89 Gunnarsson, R., et al. Screening for copy-number alterations and loss of heterozygosity in chronic lymphocytic leukemia – a comparative study of four differently designed, high resolution microarray platforms. *Genes Chromosomes Cancer* 2008; 47(8): 697–711.
90 Lin, T.T., et al. Telomere dysfunction and fusion during the progression of chronic lymphocytic leukemia: evidence for a telomere crisis. *Blood* 2010; 116(11): 1899–1907.
91 Britt-Compton, B., et al. Extreme telomere erosion in ATM-mutated and 11q-deleted CLL patients is independent of disease stage. *Leukemia* 2012; 26(4): 826–830.
92 Stephens, P.J., et al. Massive genomic rearrangement acquired in a single catastrophic event during cancer development. *Cell* 2011; 144(1): 27–40.
93 Crasta, K., et al. DNA breaks and chromosome pulverization from errors in mitosis. *Nature* 2012; 482(7383): 53–58.
94 Rausch, T., et al. Genome sequencing of pediatric medulloblastoma links catastrophic DNA rearrangements with TP53 mutations. *Cell* 2012; 148(1–2): 59–71.
95 Nik-Zainal, S., et al. Mutational processes molding the genomes of 21 breast cancers. *Cell* 2012; 149(5): 979–993.

96 Fabbri, G., et al. Analysis of the chronic lymphocytic leukemia coding genome: role of *NOTCH1* mutational activation. *J Exp Med* 2011; 208(7): 1389–1401.
97 Wang, L., et al. *SF3B1* and other novel cancer genes in chronic lymphocytic leukemia. *N Engl J Med* 2011; 365(26): 2497–2506.
98 Quesada, V., et al. Exome sequencing identifies recurrent mutations of the splicing factor *SF3B1* gene in chronic lymphocytic leukemia. *Nat Genet* 2011; 44(1): 47–52.
99 Landau, D.A., et al. Evolution and impact of subclonal mutations in chronic lymphocytic leukemia. *Cell* 2013; 152(4): 714–726.
100 Rosati, E., et al. Constitutively activated Notch signaling is involved in survival and apoptosis resistance of B-CLL cells. *Blood* 2009; 113(4): 856–865.
101 Rosati, E., et al. γ-Secretase inhibitor I induces apoptosis in chronic lymphocytic leukemia cells by proteasome inhibition, endoplasmic reticulum stress increase and Notch down-regulation. *Int J Cancer* 2013; 132(8): 1940–1953.
102 Di Ianni, M., et al. A new genetic lesion in B-CLL: a *NOTCH1* PEST domain mutation. *Br J Haematol* 2009; 146(6): 689–691.
103 Rossi, D., et al. Mutations of *NOTCH1* are an independent predictor of survival in chronic lymphocytic leukemia. *Blood* 2011; 119(2): 521–529.
104 Shedden, K., et al. Characteristics of chronic lymphocytic leukemia with somatically acquired mutations in 52 exon 34. *Leukemia* 2012; 26(5): 1108–1110.
105 Rasi, S., et al. Analysis of 52 mutations in monoclonal B cell lymphocytosis. *Haematologica* 2001; 97(1): 153–154.
106 Rossi, D., et al. Integrated mutational and cytogenetic analysis identifies new prognostic subgroups in chronic lymphocytic leukemia. *Blood* 2013; 121(8): 1403–1412.
107 Yoshida, K., et al. Frequent pathway mutations of splicing machinery in myelodysplasia. *Nature* 2011; 478(7367): 64–69.
108 Papaemmanuil, E., et al. Somatic *SF3B1* mutation in myelodysplasia with ring sideroblasts. *N Engl J Med* 2011; 365(15): 1384–1395.
109 Rossi, D., et al. Mutations of the *SF3B1* splicing factor in chronic lymphocytic leukemia: association with progression and fludarabine-refractoriness. *Blood* 2011; 118(26): 6904–6908.
110 Gibney, G.T., Sondak V.K. Extending the reach of BRAF-targeted cancer therapy. *Lancet* 2012; 379(9829): 1858–1859.
111 Herman, S.E., et al. Phosphatidylinositol 3-kinase-δ inhibitor CAL-101 shows promising preclinical activity in chronic lymphocytic leukemia by antagonizing intrinsic and extrinsic cellular survival signals. *Blood* 2010; 116(12): 2078–2088.
112 Armitage, J.O., Weisenburger D.D. New approach to classifying non-Hodgkin's lymphomas: clinical features of the major histologic subtypes. Non-Hodgkin's Lymphoma Classification Project. *J Clin Oncol* 1998; 16(8): 2780–2795.
113 Johnson, P.W., et al. Patterns of survival in patients with recurrent follicular lymphoma: a 20-year study from a single center. *J Clin Oncol* 1995; 13(1): 140–147.
114 Montoto, S., et al. Risk and clinical implications of transformation of follicular lymphoma to diffuse large B-cell lymphoma. *J Clin Oncol* 2007; 25(17): 2426–2433.

115 Haioun, C., et al. Failure of combination chemotherapy of advanced follicular (low-grade) non-Hodgkin's lymphoma. *Am J Clin Oncol* 1987; 10(3): 196–198.
116 Bastion, Y., et al. Incidence, predictive factors and outcome of lymphoma transformation in follicular lymphoma patients. *J Clin Oncol* 1997; 15(4): 1587–1594.
117 Marcus, R., et al. Phase III study of R-CVP compared with cyclophosphamide, vincristine and prednisone alone in patients with previously untreated advanced follicular lymphoma. *J Clin Oncol* 2008; 26(28): 4579–4586.
118 Salles, G., et al. Rituximab combined with chemotherapy and interferon in follicular lymphoma patients: results of the GELA-GOELAMS FL2000 study. *Blood* 2008; 112(13): 4824–4831.
119 Herold, M., et al. Rituximab added to first-line mitoxantrone, chlorambucil and prednisolone chemotherapy followed by interferon maintenance prolongs survival in patients with advanced follicular lymphoma: an East German Study Group Hematology and Oncology study. *J Clin Oncol* 2007; 25(15): 1986–1992.
120 Tsujimoto, Y., et al. Clustering of breakpoints on chromosome 11 in human B-cell neoplasms with the t(11;14) chromosome translocation. *Nature* 1985; 315(6017): 340–343.
121 Tsujimoto, Y., et al. Involvement of the bcl-2 gene in human follicular lymphoma. *Science* 1985; 228(4706): 1440–1443.
122 Pasqualucci, L., et al. AID is required for germinal center-derived lymphomagenesis. *Nat Genet* 2008; 40(1): 108–112.
123 Limpens, J., et al. Lymphoma-associated translocation t(14;18) in blood B cells of normal individuals. *Blood* 1995; 85(9): 2528–2256.
124 McDonnell, T.J., et al. *bcl*-2–immunoglobulin transgenic mice demonstrate extended B cell survival and follicular lymphoproliferation. *Cell* 1989; 57(1): 79–88.
125 Weiss, L.M., et al. Molecular analysis of the t(14;18) chromosomal translocation in malignant lymphomas. *N Engl J Med* 1987; 317(19): 1185–1189.
126 Dolken, G., et al. BCL-2/JH rearrangements in circulating B cells of healthy blood donors and patients with nonmalignant diseases. *J Clin Oncol* 1996; 14(4): 1333–1344.
127 Cheung, K.J., et al. Genome-wide profiling of follicular lymphoma by array comparative genomic hybridization reveals prognostically significant DNA copy number imbalances. *Blood* 2009; 113(1): 137–148.
128 Horsman, D.E., et al. Analysis of secondary chromosomal alterations in 165 cases of follicular lymphoma with t(14;18). *Genes Chromosomes Cancer* 2001; 30(4): 375–382.
129 Johnson, N.A., et al. Prognostic significance of secondary cytogenetic alterations in follicular lymphomas. *Genes Chromosomes Cancer* 2008; 47(12): 1038–1048.
130 Schwaenen, C., et al. Microarray-based genomic profiling reveals novel genomic aberrations in follicular lymphoma which associate with patient survival and gene expression status. *Genes Chromosomes Cancer* 2009; 48(1): 39–54.
131 Martinez-Climent, J.A., et al. Transformation of follicular lymphoma to diffuse large cell lymphoma is associated with a heterogeneous set of DNA copy number and gene expression alterations. *Blood* 2003; 101(8): 3109–3117.

132 Fitzgibbon, J., et al. Genome-wide detection of recurring sites of uniparental disomy in follicular and transformed follicular lymphoma. *Leukemia* 2007; 21(7): 1514–1520.
133 O'Shea, D., et al. Regions of acquired uniparental disomy at diagnosis of follicular lymphoma are associated with both overall survival and risk of transformation. *Blood* 2009; 113(10): 2298–2301.
134 Ross, C.W., et al. Comprehensive analysis of copy number and allele status identifies multiple chromosome defects underlying follicular lymphoma pathogenesis. *Clin Cancer Res* 2007; 13(16): 4777–4785.
135 Cheung, K.J., et al. High resolution analysis of follicular lymphoma genomes reveals somatic recurrent sites of copy-neutral loss of heterozygosity and copy number alterations that target single genes. *Genes Chromosomes Cancer* 2010; 49(8): 669–681.
136 Carlotti, E., et al. Transformation of follicular lymphoma to diffuse large B-cell lymphoma may occur by divergent evolution from a common progenitor cell or by direct evolution from the follicular lymphoma clone. *Blood* 2009; 113(15): 3553–3557.
137 Matolcsy, A., et al. Clonal evolution of B cells in transformation from low- to high-grade lymphoma. *Eur J Immunol* 1999; 29(4): 1253–1264.
138 Cheung, K.J., et al. Acquired TNFRSF14 mutations in follicular lymphoma are associated with worse prognosis. *Cancer Res* 2010; 70(22): 9166–74.
139 Launay, E., et al. High rate of TNFRSF14 gene alterations related to 1p36 region in de novo follicular lymphoma and impact on prognosis. *Leukemia* 2012; 26(3): 559–62.
140 Ware, C.F., Sedy, J.R. TNF superfamily networks: bidirectional and interference pathways of the herpesvirus entry mediator (TNFSF14). *Curr Opin Immunol* 2011; 23(5): 627–631.
141 Oricchio, E., et al. The Eph-receptor A7 is a soluble tumor suppressor for follicular lymphoma. *Cell* 2011; 147(3): 554–564.
142 Wertz, I.E., et al. De-ubiquitination and ubiquitin ligase domains of A20 downregulate NF-κB signalling. *Nature* 2004; 430(7000): 694–699.
143 Dawson, D.W., et al. Global DNA methylation profiling reveals silencing of a secreted form of Epha7 in mouse and human germinal center B-cell lymphomas. *Oncogene* 2007; 26(29): 4243–4252.
144 Mansour, M.R., Look, A.T. Discovery of a secreted tumor suppressor provides a promising therapeutic strategy for follicular lymphoma. *Cancer Cell* 2011; 20(5): 559–561.
145 Herman, J.G., Baylin, S.B., Gene silencing in cancer in association with promoter hypermethylation. *N Engl J Med* 2003; 349(21): 2042–2054.
146 Muller, C.I., et al. DNA hypermethylation of myeloid cells, a novel therapeutic target in MDS and AML. *Curr Pharm Biotechnol* 2006; 7(5): 315–321.
147 Hayslip, J., Montero, A. Tumor suppressor gene methylation in follicular lymphoma: a comprehensive review. *Mol Cancer* 2006; 5: 44.
148 Chim, C.S., et al. *SOCS1* and *SHP1* hypermethylation in mantle cell lymphoma and follicular lymphoma: implications for epigenetic activation of the Jak/STAT pathway. *Leukemia* 2004; 18(2): 356–358.

149 Koyama, M., et al. Activated proliferation of B-cell lymphomas/leukemias with the *SHP1* gene silencing by aberrant CpG methylation. *Lab Invest* 2003; 83(12): 1849–1858.
150 Nakatsuka, S., et al. Hypermethylation of death-associated protein (DAP) kinase CpG island is frequent not only in B-cell but also in T- and natural killer (NK)/T-cell malignancies. *Cancer Sci* 2003; 94(1): 87–91.
151 Rossi, D., et al. Aberrant promoter methylation of multiple genes throughout the clinico-pathologic spectrum of B-cell neoplasia. *Haematologica* 2004; 89(2): 154–164.
152 Baur, A.S., et al. Frequent methylation silencing of $p15^{INK4b}$ (*MTS2*) and $p16^{INK4a}$ (*MTS1*) in B-cell and T-cell lymphomas. *Blood* 1999; 94(5): 1773–1781.
153 Li, Y., et al. Aberrant DNA methylation of $p57^{KIP2}$ gene in the promoter region in lymphoid malignancies of B-cell phenotype. *Blood* 2002; 100(7): 2572–2577.
154 Villuendas, R., et al. Loss of p16/INK4A protein expression in non-Hodgkin's lymphomas is a frequent finding associated with tumor progression. *Am J Pathol* 1998; 153(3): 887–897.
155 O'Riain, C., et al. Array-based DNA methylation profiling in follicular lymphoma. *Leukemia* 2009; 23(10): 1858–1866.
156 Killian, J.K., et al. Large-scale profiling of archival lymph nodes reveals pervasive remodeling of the follicular lymphoma methylome. *Cancer Res* 2009; 69(3): 758–764.
157 Martin-Subero, J.I., et al. A comprehensive microarray-based DNA methylation study of 367 hematological neoplasms. *PLoS One* 2009; 4(9): e6986.
158 O'Riain, C., et al. Array-based DNA methylation profiling in follicular lymphoma. *Leukemia* 2009; 23(10): 1858–1866.
159 Bennett, L.B., et al. DNA hypermethylation accompanied by transcriptional repression in follicular lymphoma. *Genes Chromosomes Cancer* 2009; 48(9): 828–841.
160 Martin-Subero, J.I., et al. New insights into the biology and origin of mature aggressive B-cell lymphomas by combined epigenomic, genomic and transcriptional profiling. *Blood* 2009; 113(11): 2488–2497.
161 Widschwendter, M., et al. Epigenetic stem cell signature in cancer. *Nat Genet* 2007; 39(2): 157–158.
162 Bödör, C., et al. EZH2 Y641 mutations in follicular lymphoma. *Leukemia* 2011; 25(4): 726–729.
163 Morin, R.D., et al. Somatic mutations altering EZH2 (Tyr641) in follicular and diffuse large B-cell lymphomas of germinal-center origin. *Nat Genet* 2010; 42(2): 181–185.
164 Ryan, R.J., et al. EZH2 codon 641 mutations are common in BCL2-rearranged germinal center B cell lymphomas. *PLoS One* 2011; 6(12): e28585.
165 Morin, R.D., et al. Frequent mutation of histone-modifying genes in non-Hodgkin lymphoma. *Nature* 2011; 476(7360): 298–303.
166 Pasqualucci, L., et al. Inactivating mutations of acetyltransferase genes in B-cell lymphoma. *Nature* 2011; 471(7337): 189–195.
167 Cao, R., et al. Role of histone H3 lysine 27 methylation in Polycomb-group silencing. *Science* 2002; 298(5595): 1039–1043.

168 Su, I.H., et al. Ezh2 controls B cell development through histone H3 methylation and Igh rearrangement. *Nat Immunol* 2003; 4(2): 124–131.
169 Velichutina, I., et al. EZH2-mediated epigenetic silencing in germinal center B cells contributes to proliferation and lymphomagenesis. *Blood* 2010; 116(24): 5247–5255.
170 Sneeringer, C.J., et al. Coordinated activities of wild-type plus mutant EZH2 drive tumor-associated hypertrimethylation of lysine 27 on histone H3 (H3K27) in human B-cell lymphomas. *Proc Natl Acad Sci U S A* 2010; 107(49): 20980–20985.
171 Yap, D.B., et al. Somatic mutations at EZH2 Y641 act dominantly through a mechanism of selectively altered PRC2 catalytic activity, to increase H3K27 trimethylation. *Blood* 2011; 117(8): 2451–2459.
172 Knutson, S.K., et al. A selective inhibitor of EZH2 blocks H3K27 methylation and kills mutant lymphoma cells. *Nat Chem Biol* 2012; 8(11): 890–896.
173 McCabe, M.T., et al. EZH2 inhibition as a therapeutic strategy for lymphoma with EZH2-activating mutations. *Nature* 2012; 492(7427): 108–112.
174 Qi, W., et al. Selective inhibition of Ezh2 by a small molecule inhibitor blocks tumor cells proliferation. *Proc Natl Acad Sci U S A* 2012; 109(52): 21360–21365.
175 Shilatifard, A., Molecular implementation and physiological roles for histone H3 lysine 4 (H3K4) methylation. *Curr Opin Cell Biol* 2008; 20(3): 341–348.
176 Issaeva, I., et al. Knockdown of ALR (MLL2) reveals ALR target genes and leads to alterations in cell adhesion and growth. *Mol Cell Biol* 2007; 27(5): 1889–1903.
177 Campo, E., et al. The 2008 WHO classification of lymphoid neoplasms and beyond: evolving concepts and practical applications. *Blood* 2011; 117(19): 5019–5032.
178 Alizadeh, A.A., et al. Distinct types of diffuse large B-cell lymphoma identified by gene expression profiling. *Nature* 2000; 403(6769): 503–511.
179 Pasqualucci, L., et al. Analysis of the coding genome of diffuse large B-cell lymphoma. *Nat Genet* 2011; 43(9): 830–837.
180 Lohr, J.G., et al. Discovery and prioritization of somatic mutations in diffuse large B-cell lymphoma (DLBCL) by whole-exome sequencing. *Proc Natl Acad Sci U S A* 2012; 109(10): 3879–3884.
181 Zhang, J., et al. Genetic heterogeneity of diffuse large B-cell lymphoma. *Proc Natl Acad Sci U S A* 2013; 110(4): 1398–1403.
182 Gururajan, M., Jennings, C.D., Bondada, S. Cutting edge: constitutive B cell receptor signaling is critical for basal growth of B lymphoma. *J Immunol* 2006; 176(10): 5715–5719.
183 Davis, R.E., et al. Constitutive nuclear factor κB activity is required for survival of activated B cell-like diffuse large B cell lymphoma cells. *J Exp Med* 2001; 194(12): 1861–1874.
184 Dunleavy, K., et al. Differential efficacy of bortezomib plus chemotherapy within molecular subtypes of diffuse large B-cell lymphoma. *Blood* 2009; 113(24): 6069–6076.
185 Lenz, G., et al. Oncogenic CARD11 mutations in human diffuse large B cell lymphoma. *Science* 2008; 319(5870): 1676–1679.

186 Davis, R.E., et al. Chronic active B-cell-receptor signalling in diffuse large B-cell lymphoma. *Nature* 2010; 463(7277): 88–92.

187 Lin, S.C., Lo, Y.C., and Wu, H. Helical assembly in the MyD88–IRAK4–IRAK2 complex in TLR/IL-1R signalling. *Nature* 2010; 465(7300): 885–890.

188 Ngo, V.N., et al. Oncogenically active MYD88 mutations in human lymphoma. *Nature* 2011; 470(7332): 115–119.

189 Advani, R.H., et al. Bruton tyrosine kinase inhibitor ibrutinib (PCI-32765) has significant activity in patients with relapsed/refractory B-cell malignancies. *J Clin Oncol* 2013; 31(1): 88–94.

190 Rushworth, S.A., et al. BTK inhibitor ibrutinib is cytotoxic to myeloma and potently enhances bortezomib and lenalidomide activities through NF-κB. *Cell Signal* 2013; 25(1): 106–112.

191 Friedberg, J.W., et al. Inhibition of Syk with fostamatinib disodium has significant clinical activity in non-Hodgkin lymphoma and chronic lymphocytic leukemia. *Blood* 2010; 115(13): 2578–2585.

192 Young, R.M., Staudt, L.M., Targeting pathological B cell receptor signalling in lymphoid malignancies. *Nat Rev Drug Discov* 2013; 12(3): 229–243.

193 Kulis, M., et al. Epigenomic analysis detects widespread gene-body DNA hypomethylation in chronic lymphocytic leukemia. *Nat Genet* 2012; 44(11): 1236–1242.

194 Velusamy, T., et al. Recurrent reciprocal RNA chimera involving YPEL5 and PPP1CB in chronic lymphocytic leukemia. *Proc Natl Acad Sci U S A* 2013; 110(8): 3035–3040.

195 Döhner, H., et al. p53 gene deletion predicts for poor survival and non-response to therapy with purine analogs in chronic B-cell leukemias. *Blood* 1995; 85(6): 1580–1589.

196 De Paoli, L., et al. *MGA*, a suppressor of *MYC*, is recurrently inactivated in high risk chronic lymphocytic leukemia. *Leuk Lymphoma* 2013; 54(5): 1087–1090.

197 Gaidano, G., et al. p53 mutations in human lymphoid malignancies: association with Burkitt lymphoma and chronic lymphocytic leukemia. *Proc Natl Acad Sci U S A* 1991; 88(12): 5413–5417.

198 Rodríguez, D., et al. Functional analysis of sucrase–isomaltase mutations from chronic lymphocytic leukemia patients. *Hum Mol Genet* 2013; 22(11): 2273–2282.

199 Zhang, X., et al. Sequence analysis of 515 kinase genes in chronic lymphocytic leukemia. *Leukemia* 2011; 25(12): 1908–1910.

200 Havelange, V., et al. IRF4 mutations in chronic lymphocytic leukemia. *Blood* 2011; 118(10): 2827–2829.

201 Morin, R.D., et al. Somatic mutations altering EZH2 (Tyr641) in follicular and diffuse large B-cell lymphomas of germinal-center origin. *Nat Genet* 2010; 42(2): 181–185.

202 Pasqualucci, L., et al. Inactivating mutations of acetyltransferase genes in B-cell lymphoma. *Nature* 2011; 471(7337): 189–195.

203 Ngo, V.N., et al. Oncogenically active MYD88 mutations in human lymphoma. *Nature* 2011; 470(7332): 115–119.

204 Lohr, J.G., et al. Discovery and prioritization of somatic mutations in diffuse large B-cell lymphoma (DLBCL) by whole-exome sequencing. *Proc Natl Acad Sci U S A* 2012; 109(10): 3879–3884.
205 Richter, J., et al. Recurrent mutation of the ID3 gene in Burkitt lymphoma identified by integrated genome, exome and transcriptome sequencing. *Nat Genet* 2012; 44(12): 1316–1320.
206 Love, C., et al. The genetic landscape of mutations in Burkitt lymphoma. *Nat Genet* 2012; 44(12): 1321–1325.
207 Kridel, R., et al. Whole transcriptome sequencing reveals recurrent *NOTCH1* mutations in mantle cell lymphoma. *Blood* 2012; 119(9): 1963–1971.
208 Rossi, D., et al. The coding genome of splenic marginal zone lymphoma: activation of *NOTCH2* and other pathways regulating marginal zone development. *J Exp Med* 2012; 209(9): 1537–1551.
209 Kiel, M.J., et al. Whole-genome sequencing identifies recurrent somatic *NOTCH2* mutations in splenic marginal zone lymphoma. *J Exp Med* 2012; 209(9): 1553–1565.
210 Tiacci, E., et al. BRAF mutations in hairy-cell leukemia. *N Engl J Med* 2011; 364(24): 2305–2315.
211 Montoto, S., Fitzgibbon, J. Transformation of indolent B-cell lymphomas. *J Clin Oncol* 2011; 29(14): 1827–1834.

CHAPTER 7

The genetics of chronic myelogenous leukaemia

Philippa C. May, Jamshid S. Khorashad, Mary Alikian, Danilo Perrotti and Alistair G. Reid

Introduction

Chronic myelogenous leukaemia (CML) is probably the most extensively studied and best understood of all human cancers. It was the first leukaemia to be described, in almost simultaneous reports in the 1840s by Donné,[1] Virchow[2] and Bennett.[3] The association of CML with the Philadelphia (Ph) chromosome was made by Nowell and Hungerford[4] in 1960 and was the first consistent chromosome abnormality to be associated with malignancy. With improvements in chromosome banding techniques, the Ph chromosome was subsequently found to be the result of a translocation between the long arms of chromosomes 9 and 22.[5] A decade later, in the mid-1980s, the Ph translocation was shown to fuse part of the *ABL1* proto-oncogene on chromosome 9 with the breakpoint cluster region (*BCR*) gene on chromosome 22.[6,7] This rearrangement leads to a chimeric gene, *BCR-ABL1*, which codes for a novel fusion protein with dysregulated tyrosine kinase activity.[8,9] The expression of *BCR-ABL1* transforms cell lines[10] and in murine bone marrow cells has been shown to be necessary for both initiation and maintenance of the leukaemias, with some diseases phenotypically resembling CML.[11–14] This confirmation of *BCR-ABL1* as the pathogenic basis of CML fostered the development of ABL1-specific tyrosine kinase inhibitors[15–17] that specifically inhibit the growth of *BCR-ABL1* positive cells *in vitro*[18,19] and *in vivo*, dramatically improving the outcome of patients with this disease.[20,21] Historically, CML has provided a paradigm for many aspects of tumour biology and demonstrates that dissection of the specific molecular pathways that lead to oncogenic transformation will identify rational targets for therapy.

The Genetic Basis of Haematological Cancers, First Edition. Edited by Sabrina Tosi and Alistair G. Reid.

Clinical features

CML affects 1–2 per 100,000 adults and accounts for around 10–15% of all adult leukaemias.[22,23] The disease is characterized by progression through two, or sometimes three, stages. An initial chronic phase (CP) is relatively indolent and may be asymptomatic. Left untreated, patients inevitably progress through an ill-defined accelerated phase (AP) to blast crisis (BC), analogous to an acute leukaemia, with proliferation of leukaemic blasts in the peripheral blood and bone marrow. Blasts may either be myeloid or lymphoid in lineage, emphasizing the stem cell nature of the disease.

A proportion of CML patients are diagnosed coincidentally following routine blood tests. Symptomatic patients present with increased white blood cell (WBC) counts of over 100×10^9/L. Splenomegaly is the most common finding. When present, symptoms are mainly the consequence of splenomegaly and anaemias, such as fatigue, weight loss and malaise. During progression, symptoms worsen to include bleeding, fevers and infections. Some patients progress to BC with no warning signals, whereas others progress through the transitional AP with progressive headaches, bone pain, arthralgias or fever.[24]

The frontline therapy for CML is one of the targeted tyrosine kinase inhibitors such as imatinib, dasatinib or nilotinib, which, in the majority of patients, reduces the peripheral granulocyte count, restores the peripheral counts to normal and reduces the size of the spleen, eventually inducing a 3–5 log reduction in the burden of Ph-positive cells. The later stages of the disease are less amenable to treatment, however, with blast crisis more likely to be resistant to therapy and frequently fatal.[25–28]

CML is characterized by a consistent cytogenetic abnormality, the Ph chromosome.[4] This small marker is actually the derivative chromosome 22 resulting from the translocation t(9;22)(q34;q11)[5] or variants thereof (Fig. 7.1). Variants of the Ph chromosome translocation involve breakpoints in one or more chromosome regions in addition to 9q34 and 22q11.2, and are found in up to 10% of patients with CML.[29–31] In ~1–2% of CML patients, the translocation is cryptic at the level of G-banding (aka masked Ph), and is characterized by direct insertion of chromosome 9 sequences into chromosome 22 or vice versa. The molecular consequence of a Ph chromosome translocation is the fusion of the *BCR* and *ABL1* genes to form the chimeric *BCR-ABL1* oncogene. However, for a disease with a uniform pathogenetic molecular abnormality, CML demonstrates a paradoxical clinical heterogeneity. The duration of the

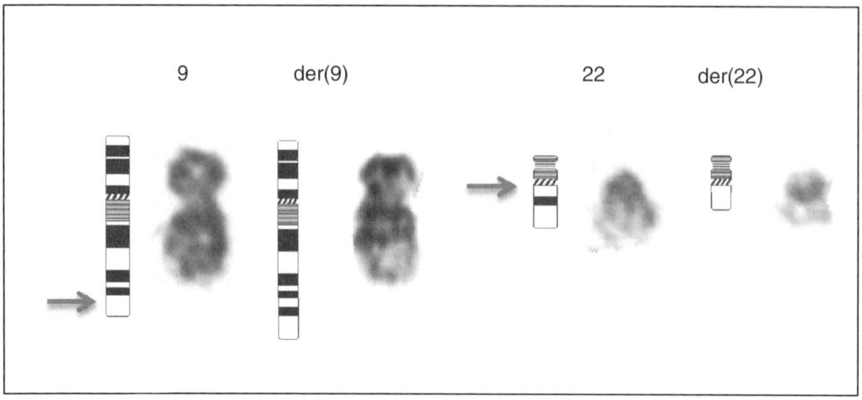

(a)

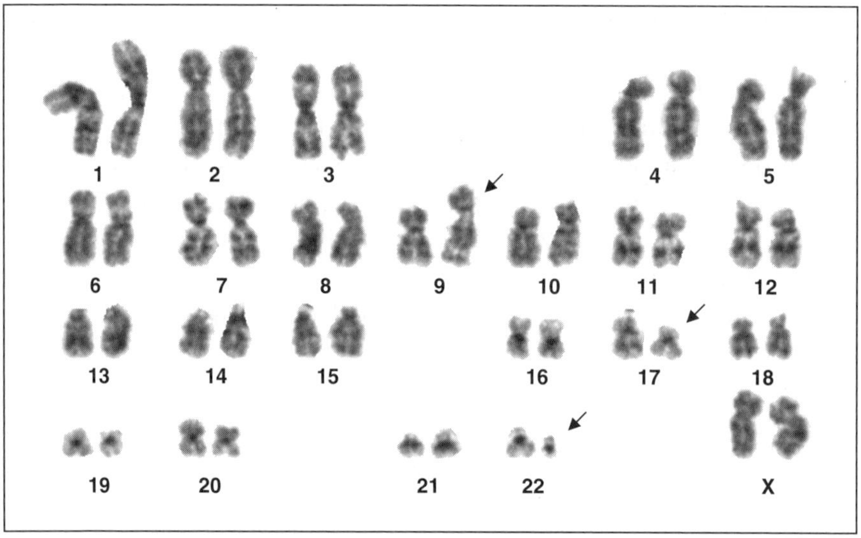

(b)

Figure 7.1 The Philadelphia (Ph) chromosome translocation. (a) Karyogram of the classical Ph chromosome translocation t(9;22)(q34.1;q11.2) from the bone marrow of a patient with CML, in which a reciprocal exchange of material is apparent between the terminal part of the long arm of chromosome 9 and the majority of the long arm of chromosome 22. This results in an elongated derivative chromosome 9 [der(9)] and a foreshortened derivative chromosome 22 [der(22)] or Philadelphia chromosome. The locations of the respective chromosome breakpoints are shown on the normal homologues by arrows. (b) An example of a variant Philadelphia chromosome translocation involving a third breakpoint on the long arm of chromosome 17, in addition to chromosomes 9 and 22.

chronic phase, and hence survival, in patients is variable even in the modern TKI era. Although most patients achieve a long-term response to therapy, a significant minority lose response and progress rapidly to blast crisis. The basis of this clinical heterogeneity and the molecular mechanisms of disease transformation remain largely obscure.

There are two commonly used prognostic scoring systems for patients with CML: the Sokal and Hasford systems.[32,33] These are mathematically derived calculations, designed and validated on large numbers of patients using a number of diagnostic, clinical and laboratory parameters. However, although they have general prognostic value, the Sokal and Hasford scoring systems have not proved robust enough to guide individual management decisions. Attempts have been made to identify individual genetic or biochemical variables that may allow improved risk stratification. A reduction in telomere length was shown to correlate with a more rapid onset of disease transformation,[34,35] and in a small series telomere length also appeared to predict for response to α-interferon,[36] although these methods were beyond the scope of most routine laboratories. The presence, in a proportion of patients, of submicroscopic deletions adjacent to the *BCR* and/or *ABL1* genomic breakpoints on the derivative chromosome 9 [aka 'der(9) deletions'] has been described extensively and was shown to correlate with a shorter survival in several large cohorts treated with α-interferon,[37–39] with one study suggesting that the size of deletion also affected prognosis.[40] However, the deletions appear to have no prognostic association in patients on TKI therapy.[41,42] A method has been proposed for measuring one of the surrogate targets for phosphorylation of *BCR-ABL1*, P-Crkl, to evaluate response to imatinib.[43] However, upon trial to validate the preliminary findings on CML patients at diagnosis, no predictive value of measuring P-Crkl could be demonstrated.[44] Other studies correlated the activity and/or level of expression of the human organic cation transporter 1 (*hOCT1*) gene, the main transporter for imatinib uptake, with subsequent response to imatinib.[45,46] However, a more recent study on presentation samples collected from CML patients before starting imatinib failed to confirm any correlation between *hOCT1* gene expression and imatinib response.[47] This gene remains of interest as several groups have also reported an association between SNPs within *hOCT1* and imatinib uptake and/or clinical outcome.[48–50] The emergence of *BCR-ABL1* kinase domain (KD) mutations (discussed later in this chapter) has also been found to be predictive of loss of response, irrespective of mutation type.[51] However, the logistical challenge of regular screening for KD mutations is considerable. Therefore, a method for prospectively distinguishing those patients who will progress rapidly to blast crisis from those whose disease will pursue an indolent course is still lacking and would be of great use to the clinician.

The structure and physiological function of *BCR* and *ABL1*

Both the 145 kilodalton (kDa) ABL1 protein and the 160 kDa BCR protein are ubiquitously expressed.[52] The *ABL1* gene is the human homologue of the *v-abl* oncogene encoded by the Abelson murine leukaemia virus (A-MuLV).[53] Human *ABL1* encodes a non-receptor tyrosine kinase with several distinct structural domains. Toward the N-terminus there are three SRC homology regions (SH1–SH3). The SH1 domain contains the tyrosine kinase function, whereas the SH2 and SH3 domains allow for interaction with other proteins.[54] Three proline-rich regions in the centre of the protein can interact with the SH3 domains of other proteins.[55] The C-terminus contains the nuclear-localization signals and also the DNA and actin-binding motifs.[56,57a,58] *ABL1* has two different isoforms, 1a and 1b, depending on which of the first two alternatively spliced exons are incorporated.[59,60] The longer and more prevalent type 1b has a myristoylation site at the N-terminus which can target the protein towards the membrane.[61] Like most protein kinases,[62] *ABL1* forms a latent conformation in the absence of cellular signals and therefore its regulation is usually tightly controlled. The SH2 and SH3 domains and the linker between them form a tight 'clamp' with the kinase domain, anchored in place by the N-terminal cap, which prevents *ABL1* activity.[63]

ABL1 appears to have many diverse functions, reviewed by several groups,[52,54,64–66] such as the regulation of the cell cycle,[57b,67] integrin signalling[68] and in the response of the cell to genotoxic stress.[69,70] ABL1 therefore appears to have a complex role in integrating diverse intracellular and extracellular signals that control the cell cycle and apoptosis.

Several structural motifs can be identified within the BCR protein. The N-terminus contains a serine–threonine kinase and the coiled-coil domain, which allows dimer formation *in vivo*.[71] The centre of the protein contains a motif with dbl-like and pleckstrin homology domains, which stimulates the exchange of guanidine triphosphate (GTP) for guanidine diphosphate (GDP) on Rho guanidine exchange factors, which may in turn activate transcription factors such as NF-κB.[72] The C-terminus of BCR encodes a domain with the ability to activate GTPases involved in the RAS signalling pathway.[73] The functional domains present within the BCR protein suggest a role in signal transduction, but the physiological role of BCR is not fully understood.

The structure of the *BCR-ABL1* fusion gene

Translocation between chromosomes 9 and 22 results in the fusion of 5′ *BCR* and 3′ *ABL1* on the Ph chromosome.[74] The fusion protein thus has chimeric features of both its wild-type counterparts (Fig. 7.2). The replacement of the first one or two exons of *ABL1* by at least *BCR* exon 1 – crucially, the only *BCR* exon common to all transforming *BCR-ABL1* fusion genes – prevents the auto-inhibition of the ABL1 kinase activity and instead allows oligomerization.[63,75] Oligomerized BCR-ABL1 subsequently phosphorylates both itself and other proteins in an uncontrolled manner, leading to proliferation and thus transformation.[76,77] Indeed, disruption of the *BCR* exon 1 oligomerization domain[71] abrogates the transforming ability of BCR-ABL1 and disruption of *ABL1* exon 1 transforms the wild-type protein to an oncoprotein,[63] demonstrating how crucial these changes are to deregulation of normal cellular control. The BCR-ABL1 protein is entirely cytoplasmic.[78]

The molecular architecture of the *ABL1* and *BCR* genes showing their common genomic breakpoints, alongside the resulting chimeric proteins, is shown in Fig. 7.2. The majority of breakpoints in the *ABL1* gene occur in the 150 kb region between the two alternate first exons, with rare cases occurring upstream of exon 1b or downstream of exon 1a.[79–81] However, regardless of the specific *ABL1* breakpoint, splicing of the chimeric transcript fuses *BCR* sequences to *ABL1* exon 2(a2). The breakpoints in *BCR* localize to three main breakpoint cluster regions, giving the gene its name. In the majority of CML patients and one-third of patients with Ph-positive acute lymphoblastic leukaemia (ALL), the *BCR* breakpoint occurs in a 6 kb region encompassing exons e12 to e16 (previously termed b1 to b5), referred to as the major breakpoint cluster region, or M-*bcr*. Following alternative splicing events, transcripts with either a e13a2 or e14a2 *BCR-ABL1* junction are formed, encoding a 210 kDa protein ($p210^{BCR\text{-}ABL1}$). The majority of patients with Ph-positive ALL and occasional patients with CML demonstrate a breakpoint upstream of the M-*bcr*, in the 54 kb region between the alternative

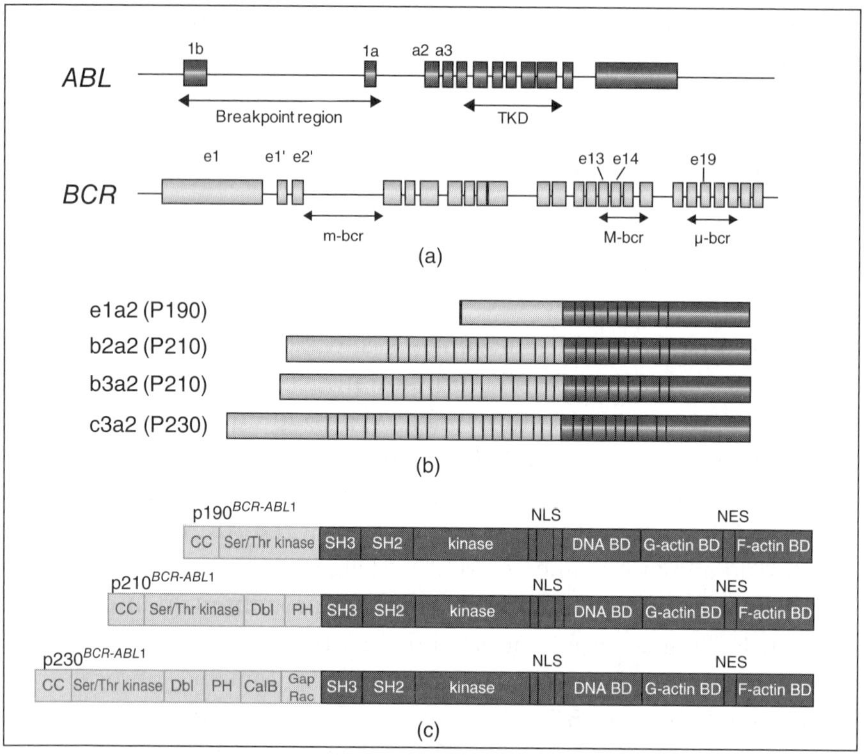

Figure 7.2 Anatomy of the Philadelphia translocation partners *ABL1* and *BCR*. (a) Genomic structure of the *ABL1* and *BCR* genes. The common breakpoint region in *ABL1* is indicated and different breakpoint cluster regions are shown within *BCR*. (b) Exon structures of the different BCR-ABL1 transcripts. The e1a2 transcript which results in the P190 BCR-ABL protein, the e13a2/e14a2 transcripts resulting in the P210 protein and the rare e19a2 transcript coding for the P230 protein are shown. (c) Structures of different BCR-ABL1 oncoproteins. Protein domains are indicated using the following abbreviations: NLS, nuclear localization signal; NES, nuclear export signal; BD, binding domain; SH, src homology region; CaLB, calcium-dependent lipase binding site; Dbl, diffuse B-cell lymphoma protein homology domain; PH, pleckstrin homology domain; Gap Rac, domain with the ability to activate GTPases involved in RAS signalling pathways.

BCR exons e2′ and e2, an area termed the minor breakpoint cluster region (m-*bcr*). The mRNA produced by this rearrangement encodes a protein of 190 kDa (p190$^{BCR\text{-}ABL1}$) with increased protein tyrosine kinase activity compared with p210$^{BCR\text{-}ABL1}$.[9] A third breakpoint cluster region exists downstream of *BCR* exon e19 (previously c3), the μ-*bcr* region, which encodes a 230 kDa fusion protein and has been associated with prominent neutrophilic maturation,[82] although it has also been described in typical CML.[83]

Whereas a large amount of data exists concerning the *BCR-ABL1* fusion, relatively little is known about the reciprocal translocation product *ABL1-BCR* and its resulting transcript. *ABL1-BCR* is expressed in around 60–70% of patients with CML and probably in a higher number of patients with Ph-positive ALL,[84–86] and stable expression of both p40$^{ABL1\text{-}BCR}$ and p96$^{ABL1\text{-}BCR}$ reciprocal fusion proteins has been demonstrated.[87] The primary reasons for abrogation of *ABL1-BCR* expression include the aforementioned der(9) deletions, which entirely remove one or both components of the fusion, and variant translocations which occur in 5–10% of patients and result in relocation of 3′ *BCR* sequences to a third partner chromosome.[88,89] In the pre-TKI era, it was speculated that the *ABL1-BCR1* transcript may modulate disease progression, a concept reminiscent of the proposed role of the reciprocal *RARA-PML* fusion gene in murine models of acute promyelocytic leukaemia;[90,91] however, expression of *ABL1-BCR* did not appear to correlate with cytogenetic response to α-interferon.[88,89,92,93] The role of *ABL1-BCR*, if any, remains unclear.

Mechanisms of *BCR-ABL1*-induced oncogenesis

Although *BCR-ABL1* is widely believed to be the single causal lesion for the initiation of CML, its expression instigates a complex network of deregulated downstream pathways that together give rise to the CML cellular phenotype. Most of the interactions are mediated by tyrosine phosphorylation and require the binding of BCR-ABL1 to adapter proteins such as growth factor receptor-bound protein 2 (GRB2), DOK, CRK, CRK-like proteins, SHC and casitas B lineage lymphoma protein.[94] Interaction with these proteins activates a range of signalling pathways that activate proteins such as RAS, PI3K, AKT, JNK, SRC family kinases and their respective downstream targets, and also transcription factors including the STATs, NF-κB and MYC. BCR-ABL1 also induces expression of cytokines such as interleukin-3 (IL3), granulocyte colony-stimulating factor (G-CSF) and granulocyte-macrophage colony-stimulating factor (GM-CSF).[13,95] The net result is deregulated cellular proliferation, decreased adherence of leukaemia cells to the bone marrow stroma, altered proteasome-mediated degradation and a reduced apoptotic response to mutagenic stimuli.[96,97] A summary of the many BCR-ABL1 downstream signalling pathways defined to date is shown in Fig. 7.3.

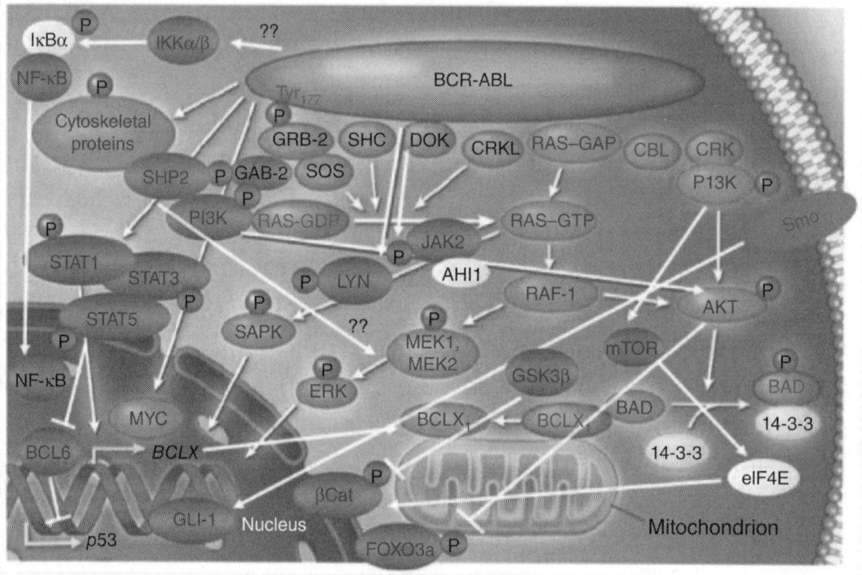

Figure 7.3 Known BCR-ABLl-mediated signal transduction pathways. Reproduced from Ahmed and Van Etten,[97] with permission.

Potential mechanisms underlying the genesis of CML

The underlying cause of the Ph translocation remains unknown. Exposure to ionizing radiation or benzene appears to be the only known external risk factor for developing CML.[98,99] A variable number tandem repeat polymorphism of *XRCC5* may be a predisposing genetic factor.[100] Various lifestyle factors, such as a lack of vigorous exercise and obesity, although not dietary factors *per se*,[101] have been reported, but a link with cigarette smoking remains uncertain.[102] It was also reported that the physical proximity of these two genomic regions in human haematopoietic cells may favour the translocation.[103,104] Related to this finding is the observation that a 76 kb duplicon maps close to the *BCR* gene on chromosome 22 and the *ABL1* gene on chromosome 9, leading to speculation that this region of homology between the two chromosomes may facilitate the genesis of the Ph translocation.[105]

The presence of the *BCR-ABL1* fusion within a haematopoietic cell at low levels does not appear to be sufficient to cause leukaemia, since *BCR-ABL1* fusion transcripts can be detected at very low levels in haematologically normal individuals.[106,107] It is unclear why these individuals do not develop leukaemia. One explanation may be the degree of differentiation of the cell in which the translocation occurs. If the target cell is committed to terminal differentiation without the potential for repopulation and renewal, it will therefore be eliminated. Alternatively there may be a threshold of *BCR-ABL1* expression below which the immune system is capable of suppressing or eliminating the neoplastic clone. It has also been suggested that *BCR-ABL1* is not the only genetic lesion required to induce chronic-phase CML.[108] This is supported by recent data suggesting that induction of *BCR-ABL1* expression via its native *BCR* promoter (as opposed to a more potent non-BCR promoter) fails to produce a leukaemic phenotype in a murine model.[109] Mathematical modelling of the chronic phase and subsequent transformation of the disease has also suggested that more than one mutation may be necessary to sustain the chronic phase of CML.[110] Nevertheless, evidence from X-inactivation studies has shown that imatinib restores a polyclonal haematopoiesis in the majority of patients, arguing against a clonal aberration pre-existing *BCR-ABL1*.[111]

CML blast crisis transformation

Left untreated, CML inevitably transforms into an acute leukaemia that is refractory to therapy. The time course over which this occurs is very variable and the overall incidence of progression has dramatically reduced with the introduction of TKI therapy. The transition from CP to BC likely results from acquired increase in *BCR-ABL1* expression at both the mRNA and protein levels and increased kinase activity,[112–114] which leads to altered mRNA processing,[115] genetic instability[116] and alteration of protein expression and activity in pathways regulating survival, proliferation and differentiation of myeloid- or lymphoid-committed $CD34^+$ progenitors.[117] Among these events, BCR-ABL1 dose- and kinase-dependent loss of protein phosphatase 2A (PP2A) activity[118,119] and ROS-induced genomic instability[119–121] appear to play a critical role in generating the epigenetic and genetic heterogeneity that characterizes the blastic phase of CML.[117] However, it is currently unclear whether a specific temporal sequence of molecular events induced by *BCR-ABL1* over-expression in Ph+ stem and/or progenitor cells is required for blastic transformation.

The blast crisis may be of either myeloid or lymphoid lineage, with a relative incidence of approximately 2:1, respectively. Transformation is associated with maturation arrest and secondary chromosome changes, observed in around 80% of patients.[122,123] The majority (75%) of these show the so-called 'major route' changes of an extra copy of the Ph chromosome, trisomy 8, trisomy 19 and/or an isochromosome of 17q. The less common 'minor route' secondary aberrations include five numerical changes (–Y, –7, –17, +17 and +21) as well as structural rearrangements involving the *MECOM* gene at 3q26, in particular the translocation, t(3;21)(q26;q22).

Some consistent cytogenetic abnormalities such as duplication of the Philadelphia chromosome, isochromosome 17q and t(3;21) are associated with obvious molecular mechanisms of transformation: an additional copy of the *BCR-ABL1* fusion, loss of the tumour suppressor gene *TP53* and a *MECOM-RUNX1* fusion gene, respectively. However, such abnormalities account for only a minority of patients. It is likely that in the remainder, more subtle genetic abnormalities – either those that have already been identified or hitherto unknown lesions – are responsible for disease progression. Mutations in cancer-associated genes such as *TP53*, *RB1* and *RAS* have frequently been reported in CML from targeted investigations, although occurrence rates have varied in different reports.[124–127] A more complete understanding of the contribution of genetic abnormalities to BC transformation will require exhaustive global genome screening for cooperating molecular lesions.

The value of the latter approach was evidenced when, using single nucleotide polymorphism (SNP) array analysis, Mullighan et al.[128] identified a near-obligate deletion of the *IKZF1* gene, encoding the lymphoid transcription factor Ikaros, in 84% of *BCR-ABL1*-positive B-ALL patients as well as a high proportion of patients in CML lymphoid BC (L-BC). The *IKZF1* deletions resulted in haploinsufficiency, expression of a dominant-negative Ikaros isoform or the complete loss of Ikaros expression, supporting their role in arresting lymphoid development in cooperation with *BCR-ABL1*. Subsequent investigations in larger cohorts of CML BC patients have confirmed both the preponderance of *IKZF1* deletions in lymphoid BC (at a frequency of at least 50%) and its virtual absence in myeloid BC, while analysis of paired samples taken at presentation and BC demonstrated that in CML the deletions are absent at diagnosis and thus acquired during the course of disease progression.[129] Interestingly, in *BCR-ABL1*-positive ALL, there is an association between *IKZF1* loss and deletions of *CDKN2A*.[130] Data demonstrating a similar relationship between aberrations of these two loci in CML are currently lacking, although prior to the discovery of *IKZF1* deletions, homozygous deletion of the *CDKN2A* locus was reported in a significant percentage (29–50%) of cases of CML BC of lymphoid, but not myeloid, phenotype.[131–133] The *CDKN2A* locus encodes two tumour-suppressor proteins, *INK4A*[134] and *ARF*,[135] loss of which results in aberrant mitogenic signalling.[136] In view of recent findings in ALL, it seems plausible that mutations of *IKZF1* and *CDKN2A* may coexist in CML L-BC; however, further studies are required to confirm this association.

In hindsight, however, global array CGH studies have failed to identify other recurrent copy-number aberrations (CNAs) associated with progression to blast crisis, although sporadic CNAs appear to be relatively common, in keeping with an increased level of genomic instability.[128,129,137,138] Although whole-genome studies of CML at the nucleotide level are currently lacking at any disease stage, one recent study employed targeted sequencing of likely CML-associated genes to shed some light on the frequency of point mutations in a number of key oncogenic drivers in BC.[129] The authors screened 39 BC samples for mutations in 11 candidate genes that play important roles in HSC differentiation and self-renewal, including *TP53*, *KRAS*, *RUNX1* and *WT1*. Mutations in some of these genes had been observed previously,[124–126] while the remainder were logical functional candidates.[139–142] The two most commonly mutated genes in this study were found to be *RUNX1* (33%) and *ASXL1* (21%), the latter being exclusive to CML-MBC.[129] Both *RUNX1* and *ASXL1* mutations were frequently seen in combination with additional mutations. In contrast to previous reports, mutations of *TP53* were found to be relatively infrequent.[124,143,144] In total, 76.9% of patients harboured point mutations in one of the 11 genes tested and, when combined with the findings of concurrent SNP array and conventional cytogenetic analyses, only 10.2% of patients lacked detectable secondary abnormalities. Although presentation (CP) material was unavailable for most patients in the study, evidence from paired samples from a minority of patients strongly supported the notion that most of the observed mutations had arisen after diagnosis and were therefore plausible cooperating lesions that may play a role in disease progression. In contrast, however, *ASXL1*, *DNMT3A*, *RUNX1* and *TET2* mutations have subsequently been reported in a substantial minority of CML patients at diagnosis and also in the Ph-negative cells of patients responding to TKI.[145] The role of point mutations in cancer-associated genes in CML initiation and progression therefore requires further investigation.

The contribution of the bone marrow microenvironment to enhanced genomic instability and induction and maintenance of the stem-like properties of the leukaemic progenitor cells during disease progression is currently unclear, but may prove to be important.[146–149] Nevertheless, it seems that *BCR-ABL1 per se* is sufficient to promote the genetic instability that ultimately leads to additional genetic aberrations that might serve as the 'second hit' required for progression into advanced phases. Given the spectrum of genetic changes already reported in CML-BC patients, it is unlikely that a single consistent secondary genetic aberration can be the cause of disease progression. Most likely, progression results from the accumulation of a critical number or combination of different mutations and from the pleiotropic effect of enhanced *BCR-ABL1* activity in CML AP.[117,150] Indeed, it has been shown that increased *BCR-ABL1* expression activates mitogenic and anti-apoptotic transduction pathways and facilitates the acquisition of self-renewal and differentiation arrest of the Ph+ blasts.[113,118,151–154] Expression studies have revealed that BCR-ABL1 dramatically perturbs the CML transcriptome,[155] resulting in altered expression of genes, some of which likely play a role in blastic transformation.[150,156–158] A plethora of studies have highlighted the post-transcriptional, translational and post-translational kinase-dependent effects of *BCR-ABL1* in CML-BC.[137,150,159,160] The molecular events leading to enhanced expression and activity of *BCR-ABL1* remain to be fully understood, although there is evidence indicating that different genetic and epigenetic events including *BCR-ABL1* gene amplification,[161,162] increased BCR promoter activity,[163] impaired PP2A activity[118] and inhibition of SHP1 phosphatase[118,164] might occur alone or in cooperation to increase *BCR-ABL1* expression and activity.[113]

The genetic targets of many secondary molecular lesions identified in CML BC to date are consistent with a model in which cooperating mutations facilitate the development of blast crisis by blocking differentiation. In experimental models, expression of *BCR-ABL1* and either *MECOM-RUNX1* or *NUP98/HOXA9*, or, alternatively, concomitant loss of *BCL11B* or over-expression of *NOTCH1*, has also led to the rapid development of acute myeloid leukaemia.[165–168] This suggests that cooperation between dysregulated tyrosine kinase activity and arrest of differentiation through the disruption of haematopoietic gene transcription may be a common theme in the development of the acute leukaemia phenotype. Nevertheless, inhibition of differentiation is also dependent on BCR-ABL1 dose and kinase activity in most myeloid CML-BC cases and furthermore relies on a marked reduction in miR-328 expression.[152,154,169] This microRNA not only negatively regulates survival of leukaemic progenitors upon interaction with PIM-1 kinase mRNA, but also exhibits decoy activity by interacting with the RNA binding protein hnRNP E2, thereby preventing the inhibitory effects of hnRNP E2 on the translation of C/EBPα,[154] a transcription factor essential for normal and leukaemic myeloid differentiation.[154,170,171] Other microRNAs (e.g. miR-130a/b, miR-486-5p) and miRNA-regulating factors (e.g. Lin 28/28b) may have a role in blastic transformation and acquisition of TKI resistance because of their BCR-ABL1 kinase-dependent aberrant expression in CML-BC.[172–181] Interestingly, miR-486–5p regulates survival and TKI sensitivity of CML-BC progenitors through restoration of the expression of *FOXO1* and *PTEN*,[181] two factors whose activity has been implicated in the regulation of survival of CML stem cells.[182,183]

Aside from several miRNAs, there are limited data concerning the role of alternative epigenetic changes in CML transformation, or, indeed, at any disease stage. Interestingly, methylation of one of the two alternative *ABL1* promoters has been proposed as a likely marker of CML transformation,[184,185] although its frequency in chronic phase remains controversial and its biological significance is currently unclear.[186] Methylation of the developmental transcription factors *TFAP2A* and *EBF2* has been observed more frequently in CML BC.[187] An increased incidence of methylation of the genes *Calcitonin, HIC1, ER, PDLIM4, HOXA4, HOPXA5, DDIT3 CDKN2B, OSCP1, PGRA, PGRB* and *TFAP2E* in the advanced phases of CML has also been reported by Jelinek et al.[185] (reviewed by Polakova et al.[186]). The functional relevance of these changes in disease progression is unclear and comprehensive comparative studies of the CML methylome in advanced or refractory disease are currently lacking.

Tyrosine kinase inhibitor (TKI) therapy

The discovery that the tyrosine kinase activity of ABL1 is essential for BCR-ABL1-mediated transformation made ABL1 kinase a logical therapeutic target. Imatinib mesylate (Glivec, previously known as STI571 and CGP 57148), a potent inhibitor of ABL1, ARG, PDGFR and KIT tyrosine kinases, was subsequently shown to induce apoptosis of *BCR-ABL1*-positive cells.[18] Imatinib works by binding close to the ATP binding site of BCR-ABL1, locking it in a closed, self-inhibited conformation and thereby blocking its enzymatic activity. Following successful early trials, imatinib was rapidly adopted as first-line therapy for CML patients.[188–190] In newly diagnosed CML in chronic phase, imatinib induces 'complete cytogenetic response' (CCyR; an absence of detectable Ph+ metaphases by bone marrow cytogenetics) in around 80% of patients with a 5-year overall survival of 90%.[191] Patients with more advanced phases of CML also respond to imatinib, albeit at a lower frequency and with less durable responses.[192–194] 'Second-generation' tyrosine kinase inhibitors were consequently developed and introduced, including nilotinib and dasatinib, which are effective in inducing a response in a high proportion of patients who fail to respond or lose their response to imatinib.[195,196] Although the molecular structure and potency of new-generation TKIs differ significantly from those of imatinib, their principal mode of action is the same, namely inhibition of the *BCR-ABL1* protein via blocking of its ATP binding site.[197–199]

Tyrosine kinase inhibitor therapy has therefore transformed the outlook of CML from a median survival of about 5 years to one in which perhaps 90% of patients are well after 9 years of therapy, with a proportion of these predicted to have a normal life expectancy.[200–202] Loss of response to TKI, however, and subsequent disease progression remain a significant clinical challenge.

The genetic basis of TKI resistance

The existence of a minority of patients who either failed to respond or lost their response to imatinib was apparent soon after its widespread introduction in the clinic.[203] Attempts were subsequently made to model the development of resistance in the laboratory. Exposure of cell lines to gradually increasing concentrations of imatinib successfully produced a number of resistant lines. Mechanisms deemed to be responsible for resistance in these lines included over-expression of *BCR-ABL1* associated with amplification of the fusion gene, over-expression of the multidrug-resistant P-glycoprotein (MDR-1)[204,205] and the presence of point mutations in the *ABL1*-kinase domain of the fusion gene.[206] Further *in vitro* experimentation combined with observations from primary material led to the implication of a number of other potential mechanisms, including oral bioavailability, deregulation of cellular drug transporters, aberrant plasma–protein binding and clonal evolution (reviewed by Apperley[207]). Aside from biological mechanisms of resistance, sub-optimal drug compliance has also been shown to be a major cause of loss of response to TKI.[208]

Resistance to imatinib and other TKIs is therefore likely to be a multifactorial process. Nevertheless, the acquisition of point mutations within the *BCR-ABL1* kinase domain has emerged as the best understood of the known potential biological mechanisms and is the only indicator that maintains widespread clinical utility in the management of CML patients showing sub-optimal response. In 2001, Gorre and co-workers[162] described 11 patients treated with imatinib for CML blast crisis or Ph-chromosome-positive acute lymphoblastic leukaemia (ALL) who relapsed on treatment. On sequencing the ATP-binding pocket and the activation loop of the kinase domain, an identical cytosine to thymidine mutation at *ABL1* nucleotide 944 was observed in six of nine assessable patients (two with Ph-chromosome-positive ALL, one with lymphoid blast crisis CML and three with myeloid blast crisis CML). The mutation resulted in an amino acid change from threonine to isoleucine at position 315, designated T315I, thus preventing crucial hydrogen bond formation with imatinib. In addition, the larger isoleucine molecule was predicted to induce steric hindrance, which led to the designation of the 315 residue as the so-called gatekeeper for imatinib.

To date, more than 80 mutants have been described, at varying frequencies, in association with resistance to imatinib (Figure 7.4).[207,209,210] Fifteen amino acid substitutions account for over 85% of the mutations[211] and the mutations responsible for 66% of reported cases of resistance occur at only seven positions (G250, Y253, E255, T315, M351, F359, H396). In addition, different substitutions can occur at the same amino acid residue, e.g. F317C, F317L and F317V, and can confer different sensitivities to imatinib. Interestingly, certain mutations seem to be associated with particular disease phases. Substitutions at residues M244, L248, F317, H396 and S417 are more likely to emerge in patients with chronic-phase disease, whereas those at Q252, Y253, E255, T315, E459 and F486 are predominantly associated with advanced-phase disease. However, it is currently unclear whether the latter mutations are responsible for disease progression (perhaps by conferring a growth advantage on affected cells) or whether they simply act as a surrogate marker of the increased genetic instability associated with advanced-phase disease. Nevertheless, marked differences have been shown to exist in the tyrosine phosphorylation patterns of cells expressing different *BCR-ABL1* mutants, which would be consistent with differences in substrate use and signalling pathway activation.[212]

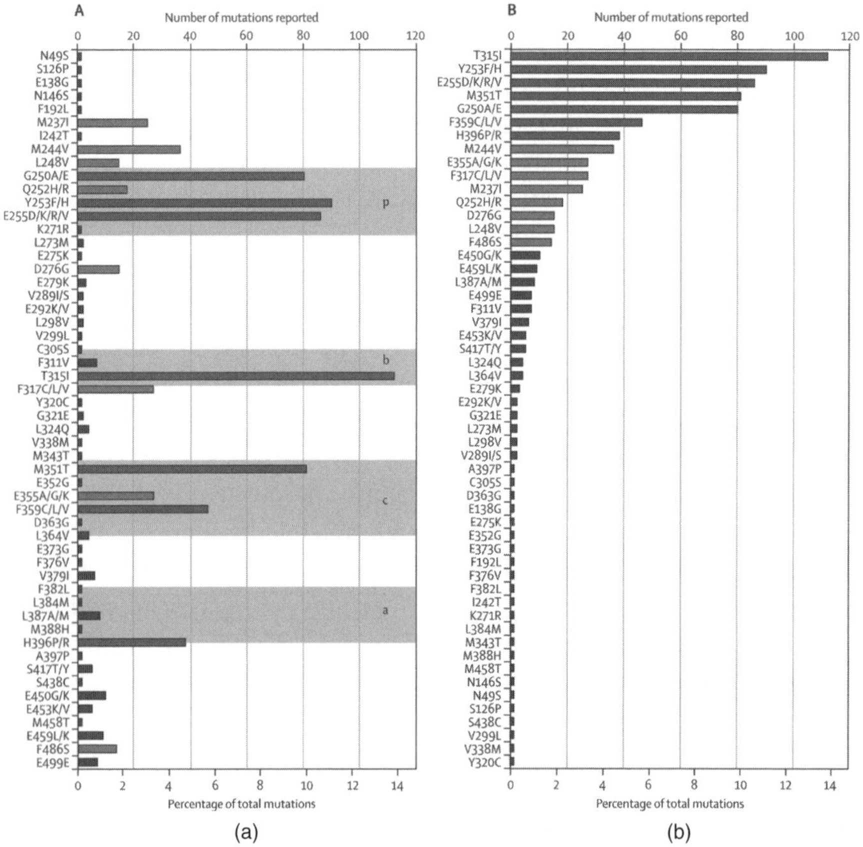

Figure 7.4 Relative frequency of different *BCR-ABL1* point mutations. Note that the most common 15 substitutions account for over 85% of cases. Reproduced from Apperley,[207] with permission. *(See plate section for color representation of this figure.)*

Nilotinib and dasatinib have significantly smaller spectra of resistance-conferring mutations than imatinib and are therefore effective in many cases of imatinib resistance; however, neither inhibits the T315I.[213] Patients who relapse on nilotinib are most frequently found to have acquired Y253H, E255K/V, F359V/C/I or T315I mutations, whereas those relapsing while taking dasatinib are most likely to acquire V299L, F317L/V/I/C, T315A or T315I mutations.[209,210,214–217] T315I also confers resistance to the second-generation TKI bosutinib,[199,218] whereas the third-generation drug ponatinib inhibits T315I *in vitro* and is effective in patients with T315 *in vivo*, restoring complete haematological (100%) and major cytogenetic (92%) response in patients on the PACE trial whose disease harboured a T315I mutation.[219–223]

BCR-ABL1 mutation frequency varies according to the method of detection, the disease phase and the definition of resistance.[224] In general, they account for around 40% of patients with imatinib resistance.[188–190,217] There are four categories of kinase domain mutation found to be associated with clinical resistance to imatinib affecting (i) the phosphate (P)-loop, (ii) the imatinib binding site, (iii) the catalytic domain and (iv) the activation (A) loop. Imatinib recognizes the inactive conformation of the *ABL1* moiety of *BCR-ABL1*, in which the A loop (amino acids 381–402 of ABL1) blocks the catalytic centre. Mutations within the A loop will destabilize the kinase or prevent it from adopting the inactive conformation, thus preventing interaction with imatinib. Dasatinib is able to bind and inhibit both the active and inactive conformations of *ABL1*,[225,226] which may explain, in part, the increased efficacy of dasatinib in cases of imatinib-resistant CML with A loop mutations.

While the impact of individual mutations on therapeutic response appears to be variable, depending on the precise amino acid change, the emergence of a detectable mutation *per se* has been associated with an inferior long-term outcome.[51,227] Khorashad et al.[51] studied kinase domain mutations in over 300 CML patients and correlated the occurrence of KD mutation with cytogenetic response after starting imatinib therapy. The presence of a KD mutation was found to be predictive of loss of complete cytogenetic response, regardless of the resistance-conferring properties of individual mutations or the size of the mutant clone. However, the logistical challenge of regular screening for KD mutations in patients responding to imatinib has thus far precluded the adoption of routine monitoring to aid patient management.

Shah et al.[228] reported that the sequential treatment of patients with various TKI was associated with development of 'compound' mutations, referring to the acquisition of two or more mutations in the same *BCR-ABL1* kinase molecule. More recently, it was reported that compound mutations are common in patients with sequencing evidence for two *BCR-ABL1* mutations that may previously have been misinterpreted as single polyclonal mutations. The detection of compound mutations frequently reflects a dynamic and highly complex clonal network whose evolution may be limited only by the negative impact of missense mutations on kinase function.[51,217]

Clonal cytogenetic evolution refers to the acquisition of cytogenetic abnormalities in addition to the Ph chromosome after presentation and has been linked with poor response to imatinib and increased haematological relapse[229] with a subsequent reduction in overall survival.[230] In addition to gain of the Ph chromosome (resulting in extra copies of *BCR-ABL1*), additional abnormalities include those whose contribution to pathogenesis is relatively ill-defined, such as chromosomal aneuploidy, and also the rare abnormalities that give rise to well-characterized oncogenic drivers. Clonal evolution is generally associated with CML progression and is thought to reflect the genetic instability of the CML proliferative progenitors.[231] It is therefore unlikely that clonal cytogenetic evolution is the primary cause of imatinib resistance in all cases, but it is plausible that the activation of *BCR-ABL1*-independent signal transduction pathways that underlie certain additional chromosome abnormalities (e.g. the rearrangement of the *MECOM* gene at 3q26) might be linked more causally to TKI resistance. The frequent acquisition of point mutations in cancer-associated genes including *RUNX1*, *AXL1* and *WT1*[129] is also likely to have a further negative effect on response to imatinib in advanced-stage disease, although this has yet to be explored in functional experiments.

One universal feature of targeted therapy that is intrinsically linked to loss of response is the inability of imatinib and other TKIs to eradicate the primitive leukaemic stem cell. Bhatia et al.[232] were the first to identify a population of quiescent *BCR-ABL1*-positive cells that were insensitive to concentrations of imatinib that were sufficient to eliminate CML progenitor cells. These primitive $CD34^+/CD38^-$ leukaemic cells, which account for less than 1% of total $CD34^+$ cells present at diagnosis,[233] have the ability to sustain the disease with the constant potential for re-expansion and progression.[234] Quiescent stem cells (QSCs) have also been shown to persist in patients despite the achievement of cytogenetic and molecular remission with imatinib.[232,235] It is therefore likely that this quiescent cell population provides a potential reservoir for the acquisition of TKD mutations and other resistance-inducing intracellular mechanisms. However, the biological basis governing the likelihood of an individual patient acquiring such changes remains unknown.

Novel therapeutic approaches

The identification of a population of TKI-resistant CML stem cells that may represent a source of future relapse has resulted in the intense development of potential strategies for targeting the signalling pathways involved in maintenance and survival of this cellular compartment. These include the WNT–β-catenin, Hedgehog, PML, SDF-1/CXCR4, BMP and Notch signalling pathways,[236] and also those regulating autophagy.[237–239] Many of these small molecules, peptides and blocking antibodies remain in early development and further research is required to optimize their use in targeting LSCs, either alone or in combination, and to reduce unwanted effects on normal stem cells. Studies aimed at identifying the biomarkers that could be used to predict the responses of individual patients to these treatments may also prove beneficial in this regard.

The identification of leukaemia-associated antigens (LAAs), such as proteinase 3 (PR3) and Wilms' tumour antigen 1 (WT1), has been a further area of therapeutic interest in CML, leading to the development of peptide vaccines for myeloid leukaemia.[240,241] However, although theoretically feasible, LAA-based vaccination to treat residual *BCR-ABL1*-positive leukaemia remains a challenging undertaking. Efforts have also been focused on the development of adoptive immunotherapy of leukaemia based on the administration of antibodies specific for leukaemia antigens.[242–244]

In addition to the kinase domain, other domains exist in BCR-ABL1 that may have a role in leukaemogenesis and, thus, could explain the resistance of CML LSCs to long-term TKI and provide rational targets for treatment.[245] Although ABL1 kinase activity has been shown to be critical for CML development *in vivo*, when expressed alone it is insufficient to produce the full CML phenotype, implying that other domains within the oncoprotein are necessary to reproduce a CML-like disease.[13,246,247] Recently, it was demonstrated that uncoupling the oncoprotein Abelson helper integration site-1 (AHI-1) from BCR-ABL1 and JAK2 resulted in enhanced sensitivity of CML LSCs to TKI,[248] suggesting that BCR-ABL1 kinase inhibition plus inhibition of AHI-1-mediated interactions might warrant further investigation as a combination treatment for eliminating CML LSCs. Similarly, inhibition of the tumour suppressor protein phosphatase 2A (PP2A) is caused by BCR-ABL1 expression, but not kinase activity, through recruitment of JAK2, resulting in the persistence of CML stem cells. Drugs targeting PP2A are able to eradicate CML stem cells, thus providing an additional therapeutic strategy for future investigation.[249] One further pathway activated by BCR-ABL1 in a kinase-independent manner is mediated by the arachidonate 5-lipoxygenase (5-LO) gene (*Alox5*), which appears to be a critical regulator CML stem cells[250] and inhibition of which impairs CML stem cell function in a mouse model.

Following their efficacy in the treatment of MDS and AML, demethylating agents targeting DNA methyl transferase 1 are also being considered for use in advanced or refractory CML. 5-Aza-2′′-deoxycitidine (decitabine), hydralazine and valorate showed early promise when administered in combination with imatinib to patients with acquired imatinib resistance.[251,252] Further understanding of the role of epigenetic changes in all disease phases may help to tailor this approach for future use.

Genetics in patient management

Despite the success of TKI drugs, a substantial minority of patients fail to respond, or lose their response, to individual agents, requiring prompt administration of one or more alternative TKIs and, if these prove unsuccessful, salvage chemotherapy or bone marrow transplantation. For this reason, regular and accurate measurement of residual *BCR-ABL1*-positive disease burden is critical in the management of CML patients on TKIs, both to assess that rate of early disease reduction is optimal and to detect any expansion of residual disease that may indicate imminent loss of response.

Cytogenetic and molecular cytogenetic monitoring

Traditionally, the degree of response to therapy in CML was monitored by conventional cytogenetic analysis (G-banding) of a cultured bone marrow specimen, and this remains a valuable investigation at diagnosis. In post-treatment samples, 30 metaphase cells are typically examined, allowing the exclusion of a Ph-positive clone present at a level of 10% with 95% confidence.[253] The accuracy of disease quantification by G-banding is therefore limited, especially when Ph-positive cells constitute less than 10% of the total. Nevertheless, the number of Ph-positive metaphases and the presence or absence of additional clonal cytogenetic abnormalities (CCAs) are robust indicators of response and the achievement of complete cytogenetic remission (CCyR; an absence of detectable Ph-positive metaphases) continues to be an important early treatment milestone in the TKI era.[254]

One intriguing side-effect of regular cytogenetic screening of CML patients has been the observation of clonal cytogenetic abnormalities in Ph-negative bone marrow cells (CCA/Ph–) of a minority of patients responding to therapy. The phenomenon was reported anecdotally in the pre-TKI era,[255–257] but rose in prominence following the introduction of TKIs.[258] The abnormalities are generally imbalances that would commonly be associated with myeloid malignancy, particularly trisomy 8 and abnormalities of chromosome 7, although rearrangements of specific genes such as *MECOM* have been reported in rare cases.[259] Surprisingly, in the majority of cases the clones appear to be transient, with no clinical significance, but occasionally such abnormalities are found in patients who subsequently progress to high-risk MDS or AML. The latter scenario is almost exclusively associated with clones harbouring deletion or monosomy of chromosome 7, suggesting that such patients require closer monitoring.[260] The frequency of CCA/Ph– is estimated at around 5% of patients on TKI, although the proportion of patients harbouring trisomy 8 or deletion/monosomy of chromosome 7 in a Ph-negative clone increased to 21% when the $CD34^+/CD38^-$ primitive stem cell compartment was specifically targeted by fluorescence *in situ* hybridization (FISH),[111] suggesting that the phenomenon may be more common than indicated by karyotype analysis.

Although the precise mechanism by which these Ph-negative clones arise remains to be elucidated, one explanation might be the hindering of normal cellular response to DNA damage via TKI-induced inhibition of wild-type kinase activity of ABL1.[261] It has also been suggested that CCA/Ph– provide evidence of a two-step process of CML pathogenesis, in which a monoclonal pre-leukaemic stage exists that favours the acquisition of either the Philadelphia rearrangement or other chromosome aberrations.[111,259,262] This notion has gained support from the observation that some patients responding to imatinib retain specific cytogenetic abnormalities other than the Ph chromosome that were present originally in Ph-positive cells[263] and also by the observation that point mutations in cancer-associated genes are frequently detectable in CCA/Ph– clones.[145] Alternatively, it is possible that Ph– clones arise as a result of increased pressure on normal haematopoietic stem cells to expand rapidly to replace the *BCR-ABL1*-positive population. Such an environment would favour any *BCR-ABL1*-negative cell that acquired a mechanism of selective advantage such as one that might be conferred by a cytogenetic abnormality. It is tempting to speculate that the latter effect might be compounded in those patients experiencing cytopenias, a side effect observed in around 50% of CML patients undergoing TKI therapy.[264] However, a consistent link between TKI-induced myelosuppression and Ph-negative clonal chromosome abnormalities has yet to be demonstrated.

FISH has been a useful adjunct to conventional cytogenetic analysis, with predesigned *BCR-ABL1*-specific assays commercially available since the early 1990s.[265] The strategy for identification of the *BCR-ABL1* fusion by locus-specific FISH involves the simultaneous hybridization of a DNA probe specific for the *ABL1* gene, fluorescently labelled (for example) in red, together with a second probe specific for the *BCR* gene labelled in another colour (e.g. green). In *BCR-ABL1*-positive metaphases, this results in a red–green doublet on the Ph chromosome, marking the fusion of 5′*BCR* and 3′*ABL1* sequences, a second red–green doublet marking *ABL1-BCR* on the der(9) and also single red and green signals on the normal chromosome 9 and 22 homologues, respectively (Figure 7.5). In addition to providing rapid confirmation of a *BCR-ABL1* fusion at diagnosis, FISH is essential in 1–2% of CML patients with a 'masked' (cryptic) *BCR-ABL1* rearrangement[266–268] and also in cases in which metaphase quality is too poor for successful G-band analysis. *BCR-ABL1* FISH analysis of large numbers of interphase nuclei provides a marginal improvement in sensitivity over G-banding in the detection and quantification of residual disease following TKI therapy or bone marrow transplantation. Current FISH probe systems were specifically designed to reduce the incidence of false positivity and enable the detection of residual disease as low as 0.1%[269,270] or one abnormal cell present in 1000 cells.

Since the introduction of TKI therapy, however, disease levels in the majority of patients are expected to fall below the threshold of detection of both conventional and molecular cytogenetic techniques 3–12 months after diagnosis. Therefore, although both approaches maintain an important role at diagnosis and during the first few months of therapy, they have little value in assessing therapeutic response in patients who respond optimally to TKI therapy.

Quantitative reverse transcriptase polymerase chain reaction (RT-qPCR)

The current gold standard technique for the serial quantification of disease burden CML patients on TKI therapy is quantitative measurement of *BCR-ABL1* RNA transcript by reverse transcriptase PCR (RT-qPCR).[271–273] RT-qPCR uses cDNA, rather than DNA, as a substrate, which means that the range of possible chimeric sequences is neatly reduced to the few common exon-to-exon junctions. Despite the advantages of using a common set of assays to quantify the few recurrent transcript types,

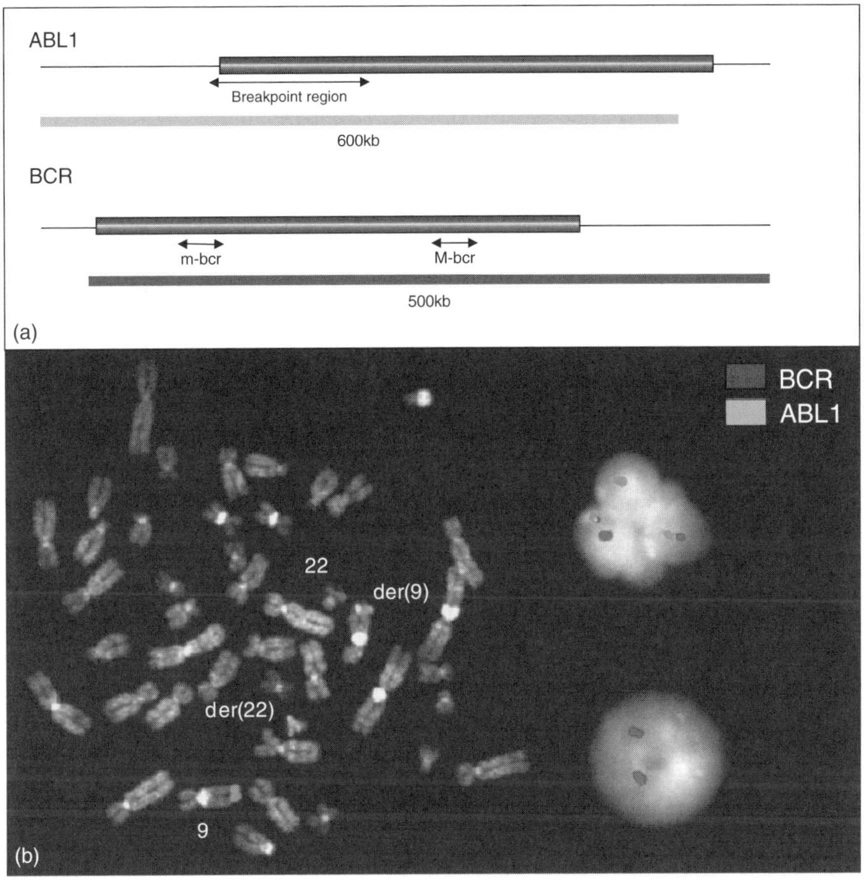

Figure 7.5 FISH detection of the *BCR-ABL1* gene fusion. (a) Coverage of the probes used in a standard dual-colour, dual-fusion fluorescence *in situ* hybridization (FISH) probe system with regard to the *ABL1* and *BCR* loci. At the *ABL1* locus the probe typically consists of a single large red-labelled contig of at least 3–400 kb, spanning the *ABL* breakpoint region, covering the majority of the gene and extending beyond ABL in a 5′ (centromeric) direction. At the *BCR* locus, the probe is represented by a second large contig specific for the majority of the *BCR* gene, usually extending beyond the 3′ (telomeric) end of the gene. The *BCR* component is designed to span all common breakpoints, with both probes producing two hybridization signals of roughly equal size when split by a *BCR-ABL1* rearrangement. Note that the colour scheme may differ between manufacturers. (b) Application of a dual-fusion *BCR-ABL1* FISH probe to a bone marrow metaphase and interphase cells from a patient with CML. Single green and red signals signal mark the unrearranged *ABL1* and *BCR* genes, respectively. Fused red–green doublets are present on the der(9) and der(22), marking the presence of *ABL1-BCR* and *BCR-ABL1* genes, respectively. The same hybridization pattern of one red, one green and two fusion signals (1R1G2F) is apparent on interphase cells (LHS, upper cell) and is readily distinguishable from a normal cell showing two red and two green signals (2R2G) only (LHS, lower cell). *(See plate section for color representation of this figure.)*

there are complications inherent to RT-qPCR that require careful deployment of this method, particularly when communicating results between laboratories. Areas of RT-qPCR in which variation might be introduced that could affect assay sensitivity include the RT step and the absence of universal reference material to quantify transcript numbers accurately. Notwithstanding, there has been extensive effort over recent years to standardize the routine measurement of *BCR-ABL1* transcripts.[274–276]

The number of *BCR-ABL1* transcripts detected by RT-qPCR is articulated as a ratio, relative to that of a reference gene such as *GUSB, B2M, G6PD* or *ABL1* itself.[277–279] Attempts to align results from different laboratories were first made during the multicentre imatinib versus interferon IRIS trial in 2003,[280] which calculated the disease level, or 'minimal residual disease' (MRD), as $\log_{10}$ reduction compared with each laboratory's standardized baseline, determined by analysis of 30 common untreated patients. The introduction of the International Scale (IS) developed this idea further; an individual laboratory's results could be converted to the IS via a unique conversion factor defined through a process of exchange of material and locally obtained measurements with a designated reference laboratory.[272,281] In the IRIS study, a major molecular response (MMR) was defined as a three log reduction (0.1%) in *BCR-ABL1* transcripts from the standardized baseline. A deeper category of response, termed complete molecular remission (CMR), was subsequently introduced, referring to the cases where no transcripts were detectable by RT-qPCR. This definition was later revised and redefined into three sub-categories, MR^4, $MR^{4.5}$ and MR^5, determined by the number of *ABL1* control molecules amplified in the reaction (10,000, 32,000 or 100,000, respectively), thereby defining the sensitivity achieved.[282]

Serial monitoring of *BCR-ABL1* transcripts therefore provides an accurate account of treatment response over time, with treatment failure for both first- and second-generation TKIs defined in international guidelines for the management of CML patients as a failure to reach key landmarks of disease burden reduction at defined time points.[254] The rate at which certain landmarks of molecular response are achieved on first-line therapy also has prognostic value. Achievement of $<10\%$ at 3 months, $<1\%$ at 6 months and ≥ 3 log decrease in *BCR-ABL1* RNA by 12 months has been shown to predict outcome strongly.[254,283]

With second- and third-generation TKI therapies allowing deeper and faster disease reduction,[21,195,284] there is an increasing possibility of a drug-free 'operational cure' for the subset of CML patients who achieve and maintain a deep molecular response.[285] This prospect is being explored in several clinical studies evaluating the consequences of TKI cessation in patients who achieve durable molecular responses of MR^4 or even MR^3.[285–288] The optimal criteria for drug withdrawal have yet to be determined, particularly regarding the required depth of MR prior to stopping treatment and whether cessation should be preceded by a period of therapy reduction.[283] Nevertheless, evidence suggests that at least 40% of patients in sustained MR^4, and a smaller minority of patients in MR^3, might maintain their response off-therapy.

Accurate and standardized definitions of deep molecular response will be critical for the safe introduction of TKI withdrawal into routine clinical practice. The probability of relapse post-withdrawal could be related to the precise level of persistent disease, including transcriptionally quiescent TKI-resistant leukaemic stem cells. A means of detecting and quantifying these cells that does not depend on oncogene transcription may therefore be clinically valuable, either in conjunction with, or as an alternative to, a gene expression-based method. Previous studies using real-time technology suggest that a quantitative PCR based on genomic DNA might be more sensitive for the detection of residual disease than one that relies on RNA.[289–291] Although traditionally laborious, recent advances in high-throughput sequencing provide a means of identifying patient-specific DNA junctions that is better suited to a clinical diagnostic setting.[292,293] Undertaking quantification by digital PCR, rather than on a real-time platform, is likely to enhance assay sensitivity further by facilitating absolute quantification without the need for common reference material.[294–296] The benefit of these techniques, in the context of therapy withdrawal and disease management in general, is currently under evaluation.

BCR-ABL1 mutation analysis

One further aspect of the routine genetic management of CML patients in the TKI era involves screening for mutations in the *BCR-ABL1* kinase domain. A nested PCR approach is required to exclude the normal *ABL1* allele and to allow a realistic estimation of mutation burden as a proportion of total disease. Sequencing is usually undertaken using the Sanger method, with a sensitivity of 10–20%, coupled with pyrosequencing for clinically relevant mutations to obtain improved sensitivity (5%) with quantitative capability.[297] Best practice recommendations are that mutation screening is carried out in the following scenarios: (i) at diagnosis: only in AP or BC patients; (ii) while on first-line imatinib: in case of failure, increase in *BCR-ABL1* transcript levels leading to MMR loss or in any other case described as suboptimal response or warning, before changing TKI therapy; and (iii) during second-line treatment with second-generation TKI: in case of loss of previously achieved response, before changing TKI therapy.[254,298] Therapy change is thus guided by aforementioned data concerning the degree of resistance of any detected mutations to the various TKIs, with ponatinib currently the only TKI recommended for patients with the T315I mutation, although there are some safety concerns.[222,223]

As previously mentioned, the emergence of any KD mutation is predictive of loss of response, but the logistical challenge of regular mutation screening in TKI responders has precluded its routine introduction. Next-generation sequencing of PCR-generated amplicons provides a potential method whereby regular screening for the early emergence of low-level mutations may feasibly be performed in a clinical setting.[299] In addition, analysis of long sequence reads allows the determination of whether multiple mutations are compound or polyclonal.[51,217,300] While each of multiple mutant clones retains its individual sensitivity to a given TKI, compound mutations can dramatically reduce TKI sensitivity, with several compound mutations even conferring resistance to ponatinib,[219,228,298,301] hence the distinction between compound and polyclonal mutations is clinically important.

Conclusion

Despite a consistent causal genetic aberration and a clinically effective targeted therapeutic approach, significant challenges persist in the understanding and management of chronic myeloid leukaemia. Key areas for future research that remain incompletely understood include pathways governing initiation of the CML clone, LSC maintenance, therapy resistance, disease progression and the suppression of low levels of residual disease in patients off TKI. Improved knowledge and understanding of additional genetic changes to *BCR-ABL1*, both pre-existing and secondary, are a prerequisite for these investigations. There is therefore an urgent need for a comprehensive summary, coupled with functional investigations, of the aetiology of recurrent genetic lesions associated with CML pathogenesis and progression.

References

1 Donne, A. De l'origine des globules du sang de leur mode de formation et leur fin. *C R Acad Sci* 1842; 14: 366–368.
2 Virchow, R. *Weisses Blut. Froriep's Notizen* 1845; 36: 151–156.
3 Bennett, J. Case of hypertrophy of the spleen and liver, in which death took place from suppuration of the blood *Edinb Med Surg J 1845*; 64: 413–423.
4 Nowell, P.C., Hungerford, D.A. A minute chromosome in human chronic granulocytic leukemia. *Science* 1960; 132: 1497.
5 Rowley, J.D. New consistent chromosomal abnormality in chronic myelogenous leukemia identified by quinacrine fluorescence and Giemsa staining. *Nature* 1973; 243(5405): 290–293.
6 Bartram, C.R., Deklein, A., Hagemeijer, A., et al. Translocation of c-abl oncogene correlates with the presence of a Philadelphia-chromosome in chronic myelocytic-leukemia. *Nature* 1983; 306(5940): 277–280.
7 Groffen, J., Stephenson, J.R., Heisterkamp, N., Deklein, A., Bartram, C.R., Grosveld, G. Philadelphia chromosomal breakpoints are clustered within a limited region, bcr, on chromosome-22. *Cell* 1984; 36(1): 93–99.
8 Konopka, J.B., Witte, O.N. Activation of the abl oncogene in murine and human leukemias. *Biochim Biophys Acta* 1985; 823(1): 1–17.
9 Clark, S.S., McLaughlin, J., Timmons, M., et al. Expression of a distinctive BCR-ABL oncogene in Ph1-positive acute lymphocytic leukemia (ALL). *Science* 1988; 239(4841): 775–777.
10 Lugo, T.G., Pendergast, A.M., Muller, A.J., Witte, O.N. Tyrosine kinase activity and transformation potency of bcr-abl oncogene products. *Science* 1990; 247(4946): 1079–1082.
11 Daley, G.Q., Van Etten, R.A., Baltimore, D. Induction of chronic myelogenous leukemia in mice by the P210bcr/abl gene of the Philadelphia chromosome. *Science* 1990; 247(4944): 824–830.
12 Pear, W.S., Miller, J.P., Xu, L.W., et al. Efficient and rapid induction of a chronic myelogenous leukemia-like myeloproliferative disease in mice receiving P210 bcr/abl-transduced bone marrow. *Blood* 1998; 92(10): 3780–3792.
13 Zhang, X.W., Ren, R.B. Bcr-Abl efficiently induces a myeloproliferative disease and production of excess interleukin-3 and granulocyte–macrophage colony-stimulating factor in mice: a novel model for chronic myelogenous leukemia. *Blood* 1998; 92(10): 3829–3840.
14 Huettner, C.S., Zhang, P., Van Etten, R.A., Tenen, D.G. Reversibility of acute B-cell leukaemia induced by BCR-ABL1. *Nat Genet* 2000; 24(1): 57–60.
15 Druker, B.J., Tamura, S., Buchdunger, E., et al. Effects of a selective inhibitor of the Abl tyrosine kinase on the growth of Bcr-Abl positive cells. *Nat Med* 1996; 2(5): 561–566.
16 Golemovic, M., Verstovsek, S., Giles, F., et al. AMN107, a novel aminopyrimidine inhibitor of Bcr-Abl, has in vitro activity against imatinib-resistant chronic myeloid leukemia. *Clin Cancer Res* 2005; 11(13): 4941–4947.

17 Schittenhelm, M.M., Shiraga, S., Schroeder, A., et al. Dasatinib (BMS-354825), a dual SRC/ABL kinase inhibitor, inhibits the kinase activity of wild-type, juxtamembrane and activation loop mutant KIT isoforms associated with human malignancies. *Cancer Res* 2006; 66(1): 473–481.
18 Druker, B.J., Talpaz, M., Resta, D.J., et al. Efficacy and safety of a specific inhibitor of the BCR-ABL tyrosine kinase in chronic myeloid leukemia. *N Engl J Med* 2001; 344(14): 1031–1037.
19 Deininger, M.W.N., Goldman, J.M., Lydon, N., Melo, J.V. The tyrosine kinase inhibitor CGP57148B selectively inhibits the growth of BCR-ABL-positive cells. *Blood* 1997; 90(9): 3691–3698.
20 Kantarjian, H., Giles, F., Wunderle, L., et al. Nilotinib in imatinib-resistant CML and Philadelphia chromosome-positive ALL. *N Engl J Med* 2006; 354(24): 2542–2551.
21 Kantarjian, H., Shah, N.P., Hochhaus, A., et al. Dasatinib versus imatinib in newly diagnosed chronic-phase chronic myeloid leukemia. *N Engl J Med* 2010; 362(24): 2260–2270.
22 Sawyers, C.L. Chronic myeloid leukemia. *N Engl J Med* 1999; 340(17): 1330–1340.
23 Siegel, R., Naishadham, D., Jemal, A. Cancer statistics, 2012. *CA Cancer J Clin* 2012; 62(1): 10–29.
24 Hehlmann, R. How I treat CML blast crisis. *Blood* 2012; 120(4): 737–747.
25 Kantarjian, H.M., Cortes, J., O'Brien, S., et al. Imatinib mesylate (ST1571) therapy for Philadelphia chromosome-positive chronic myelogenous leukemia in blast phase. *Blood* 2002; 99(10): 3547–3553.
26 Kantarjian, H.M., Giles, F., Gattermann, N., et al. Nilotinib (formerly AMN107), a highly selective BCR-ABL tyrosine kinase inhibitor, is effective in patients with Philadelphia chromosome-positive chronic myelogenous leukemia in chronic phase following imatinib resistance and intolerance. *Blood* 2007; 110(10): 3540–3546.
27 Goldman, J.M., Marin, D., Olavarria, E., Apperley, J.F. Clinical decisions for chronic myeloid leukemia in the imatinib era. *Semin Hematol* 2003; 40(2): 98–103.
28 Cortes, J., Rousselot, P., Kim, D.W., et al. Dasatinib induces complete hematologic and cytogenetic responses in patients with imatinib-resistant or -intolerant chronic myeloid leukemia in blast crisis. *Blood* 2007; 109(8): 3207–3213.
29 Potter, A.M., Watmore, A.E., Cooke, P., Lilleyman, J.S., Sokol, R.J. Significance of nonstandard Philadelphia chromosomes in chronic granulocytic-leukemia. *Br J Cancer* 1981; 44(1): 51–54.
30 Heim, S., Billstrom, R., Kristoffersson, U., Mandahl, N., Strombeck, B., Mitelman, F. Variant Ph translocations in chronic myeloid leukemia. *Cancer Genet Cytogenet* 1985; 18(3): 215–227.
31 Marzocchi, G., Castagnetti, F., Luatti, S., et al. Variant Philadelphia translocations: molecular-cytogenetic characterization and prognostic influence on frontline imatinib therapy, a GIMEMA Working Party on CML analysis. *Blood* 2011; 117(25): 6793–6800.

32 Sokal, J.E., Cox, E.B., Baccarani, M., et al. Prognostic discrimination in good-risk chronic granulocytic leukemia. *Blood* 1984; 63(4): 789–799.

33 Hasford, J., Pfirrmann, M., Hehlmann, R., et al. A new prognostic score for survival of patients with chronic myeloid leukemia treated with interferon alfa. *J Natl Cancer Inst* 1998; 90(11): 850–858.

34 Boultwood, J., Fidler, C., Shepherd, P., et al. Telomere length shortening is associated with disease evolution in chronic myelogenous leukemia. *Am J Hematol* 1999; 61(1): 5–9.

35 Brummendorf, T.H., Holyoake, T.L., Rufer, N., et al. Prognostic implications of differences in telomere length between normal and malignant cells from patients with chronic myeloid leukemia measured by flow cytometry. *Blood* 2000; 95(6): 1883–1890.

36 Ohyashiki, K., Iwama, H., Hayashi, S., et al. Sequential alterations of telomeric DNA length correlated with cytogenetic response in chronic myeloid leukemia treated with interferon alpha. *Blood* 1998; 92(10): 606A–607A.

37 Sinclair, P.B., Nacheva, E.P., Leversha, M., et al. Large deletions at the t(9;22) breakpoint are common and may identify a poor-prognosis subgroup of patients with chronic myeloid leukemia. *Blood* 2000; 95(3): 738–744.

38 Cohen, N., Rozenfeld-Granot, G., Hardan, I., et al. Subgroup of patients with Philadelphia-positive chronic myelogenous leukemia characterized by a deletion of 9q proximal to ABL gene: expression profiling, resistance to interferon therapy and poor prognosis. *Cancer Genet Cytogenet* 2001; 128(2): 114–119.

39 Huntly, B.J.P., Reid, A.G., Bench, A.J., et al. Deletions of the derivative chromosome 9 occur at the time of the Philadelphia translocation and provide a powerful and independent prognostic indicator in chronic myeloid leukemia. *Blood* 2001; 98(6): 1732–1738.

40 Fourouclas, N., Campbell, P.J., Bench, A.J., et al. Size matters: the prognostic implications of large and small deletions of the derivative 9 chromosome in chronic myeloid leukemia. *Haematologica* 2006; 91(7): 952–955.

41 Castagnetti, F., Testoni, N., Luatti, S., et al. Deletions of the derivative chromosome 9 do not influence the response and the outcome of chronic myeloid leukemia in early chronic phase treated with imatinib mesylate: GIMEMA CML Working Party analysis. *J Clin Oncol* 2010; 28(16): 2748–2754.

42 Quintas-Cardama, A., Cortes, J.E., Kantarjian, H.M. Early cytogenetic and molecular response during first-line treatment of chronic myeloid leukemia in chronic phase. *Cancer* 2011; 117(23): 5261–5270.

43 Hamilton, A., Elrick, L., Myssina, S., et al. BCR-ABL activity and its response to drugs can be determined in $CD34^+$ CML stem cells by CrkL phosphorylation status using flow cytometry. *Leukemia* 2006; 20(6): 1035–1039.

44 Khorashad, J.S., Wagner, S., Greener, L., et al. The level of BCR-ABL1 kinase activity before treatment does not identify chronic myeloid leukemia patients who fail to achieve a complete cytogenetic response on imatinib. *Haematologica* 2009; 94(6): 861–864.

45 White, D.L., Saunders, V.A., Dang, P., et al. Most CML patients who have a suboptimal response to imatinib have low OCT-1 activity: higher doses of imatinib

may overcome the negative impact of low OCT-1 activity. *Blood* 2007; 110(12): 4064–4072.
46 Wang, L., Giannoudis, A., Lane, S., Williamson, P., Pirmohamed, M., Clark, R.E. Expression of the uptake drug transporter hOCT1 is an important clinical determinant of the response to imatinib in chronic myeloid leukemia. *Clin Pharmacol Ther* 2008; 83(2): 258–264.
47 Zhang, W.W., Cortes, J.E., Yao, H., et al. Predictors of primary imatinib resistance in chronic myelogenous leukemia are distinct from those in secondary imatinib resistance. *J Clin Oncol* 2009; 27(22): 3642–3649.
48 Engler, J.R., Frede, A., Saunders, V.A., Zannettino, A.C.W., Hughes, T.P., White, D.L. Chronic myeloid leukemia $CD34^+$ cells have reduced uptake of imatinib due to low OCT-1 activity. *Leukemia* 2010; 24(4): 765–770.
49 White, D.L., Dang, P., Engler, J., et al. Functional activity of the OCT-1 protein is predictive of long-term outcome in patients with chronic-phase chronic myeloid leukemia treated with imatinib. *J Clin Oncol* 2010; 28(16): 2761–2767.
50 Grinfeld, J., Gerrard, G., Alikian, M., et al. A common novel splice variant of SLC22A1 (OCT1) is associated with impaired responses to imatinib in patients with chronic myeloid leukaemia. *Br J Haematol* 2013; 163(5): 631–639.
51 Khorashad, J.S., de Lavallade, H., Apperley, J.F., et al. Finding of kinase domain mutations in patients with chronic phase chronic myeloid leukemia responding to imatinib may identify those at high risk of disease progression. *J Clin Oncol* 2008; 26(29): 4806–4813.
52 Greuber, E.K., Smith-Pearson, P., Wang, J.X., Pendergast, A.M. Role of ABL family kinases in cancer: from leukaemia to solid tumours. *Nat Rev Cancer* 2013; 13: 559–571.
53 Abelson, H.T., Rabstein, L.S. Influence of prednisolone on Moloney leukemogenic virus in BALB/c mice. *Cancer Res* 1970; 30(8): 2208–2212.
54 Panjarian, S., Iacob, R.E., Chen, S.G., Engen, J.R., Smithgall, T.E. Structure and dynamic regulation of Abl kinases. *J Biol Chem* 2013; 288(8): 5443–5450.
55 Feller, S.M., Ren, R.B., Hanafusa, H., Baltimore, D. SH2 and SH3 domains as molecular adhesives – the interactions of Crk and Abl. *Trends Biochem Sci* 1994; 19(11): 453–458.
56 Van Etten, R.A., Jackson, P., Baltimore, D. The mouse type IV c-abl gene product is a nuclear protein, and activation of transforming ability is associated with cytoplasmic localization. *Cell* 1989; 58(4): 669–678.
57a Kipreos, E.T., Wang, J.Y.J. Cell cycle-regulated binding of c-Abl tyrosine kinase to DNA. *Science* 1992; 256(5055): 382–385.
57b Kipreos, E.T., Wang, J.Y.J. Differential phosphorylation of c-Abl in cell-cycle determined by cdc2 kinase and phosphatase activity. *Science* 1990; 248(4952): 217–220.
58 McWhirter, J.R., Wang, J.Y.J. An actin-binding function contributes to transformation by the Bcr-Abl oncoprotein of Philadelphia chromosome-positive human leukemias. *EMBO J* 1993; 12(4): 1533–1546.
59 Ben-Neriah, Y., Bernards, A., Paskind, M., Daley, G.Q., Baltimore, D. Alternative 5′ exons in c-Abl messenger-RNA. *Cell* 1986; 44(4): 577–586.

60 Franz, W.M., Berger, P., Wang, J.Y.J. Deletion of an N-terminal regulatory domain of the c-Abl tyrosine kinase activates its oncogenic potential. *EMBO J* 1989; 8(1): 137–147.

61 Hantschel, O., Superti-Furga, G. Regulation of the c-Abl and Bcr-Abl tyrosine kinases. *Nat Rev Mol Cell Biol* 2004; 5(1): 33–44.

62 Huse, M., Kuriyan, J. The conformational plasticity of protein kinases. *Cell* 2002; 109(3): 275–282.

63 Pluk, H., Dorey, K., Superti-Furga, G. Autoinhibition of c-Abl. *Cell* 2002; 108(2): 247–259.

64 Van Etten, R.A. Cycling, stressed-out and nervous: cellular functions of c-Abl. *Trends Cell Biol* 1999; 9(5): 179–186.

65 Levav-Cohen, Y., Goldberg, Z., Zuckerman, V., Grossman, T., Haupt, S., Haupt, Y. c-Abl as a modulator of p53. *Biochem Biophys Res Commun* 2005; 331(3): 737–749.

66 Sirvent, A., Benistant, C., Roche, S. Cytoplasmic signalling by the c-Abl tyrosine kinase in normal and cancer cells. *Biol Cell* 2008; 100(11): 617–631.

67 Sawyers, C.L., McLaughlin, J., Goga, A., Havlik, M., Witte, O. The nuclear tyrosine kinase c-Abl negatively regulates cell growth. *Cell* 1994; 77(1): 121–131.

68 Lewis, J.M., Schwartz, M.A. Integrins regulate the association and phosphorylation of paxillin by c-Abl. *J Biol Chem* 1998; 273(23): 14225–14230.

69 Yuan, Z.M., Shioya, H., Ishiko, T., et al. P73 is regulated by tyrosine kinase c-Abl in the apoptotic response to DNA damage. *Nature* 1999; 399(6738): 814–817.

70 Gonfloni, S. DNA damage stress response in germ cells: role of c-Abl and clinical implications. *Oncogene* 2010; 29(47): 6193–6202.

71 McWhirter, J.R., Galasso, D.L., Wang, J.Y.J. A coiled-coil oligomerization domain of Bcr is essential for the transforming function of Bcr-Abl oncoproteins. *Mol Cell Biol* 1993; 13(12): 7587–7595.

72 Montaner, S., Perona, R., Saniger, L., Lacal, J.C. Multiple signalling pathways lead to the activation of the nuclear factor κB by the Rho family of GTPases. *J Biol Chem* 1998; 273(21): 12779–12785.

73 Diekmann, D., Brill, S., Garrett, M.D., et al. Bcr encodes a GTPase-activating protein for p21rac. *Nature* 1991; 351(6325): 400–402.

74 Heisterkamp, N., Stam, K., Groffen, J., Deklein, A., Grosveld, G. Structural organization of the bcr gene and its role in the Ph′ translocation. *Nature* 1985; 315(6022): 758–761.

75 Nagar, B., Hantschel, O., Young, M.A., et al. Structural basis for the autoinhibition of c-Abl tyrosine kinase. *Cell* 2003; 112(6): 859–871.

76 Zhang, X.W., Subrahmanyam, R., Wong, R., Gross, A.W., Ren, R.B. The NH_2-terminal coiled-coil domain and tyrosine 177 play important roles in induction of a myeloproliferative disease in mice by Bcr-Abl. *Mol Cell Biol* 2001; 21(3): 840–853.

77 Arlinghaus, R.B. Bcr: a negative regulator of the Bcr-Abi oncoprotein in leukemia. *Oncogene* 2002; 21(56): 8560–8567.

78 Lifshitz, B., Fainstein, E., Marcelle, C., et al. Bcr genes and transcripts. *Oncogene* 1988; 2(2): 113–117.

79 Jiang, X.Y., Trujillo, J.M., Liang, J.C. Chromosomal breakpoints within the first intron of the ABL gene are nonrandom in patients with chronic myelogenous leukemia. *Blood* 1990; 76(3): 597–601.
80 Morris, C.M., Heisterkamp, N., Groffen, J., Fitzgerald, P.H. Entire ABL gene is joined with 5′-BCR in some patients with Philadelphia-positive leukemia. *Blood* 1991; 78(4): 1078–1084.
81 Chissoe, S.L., Bodenteich, A., Wang, Y.F., et al. Sequence and analysis of the human ABL gene, the BCR gene, and regions involved in the Philadelphia chromosomal translocation. *Genomics* 1995; 27(1): 67–82.
82 Pane, F., Frigeri, F., Sindona, M., et al. Neutrophilic-chronic myeloid leukemia: a distinct disease with a specific molecular marker (BCR/ABL with C3/A2 junction). *Blood* 1996; 88(7): 2410–2414.
83 Wilson, G., Frost, L., Goodeve, A., Vandenberghe, E., Peake, I., Reilly, J. BCR-ABL transcript with an e19a2 (c3a2) junction in classical chronic myeloid leukemia. *Blood* 1997; 89(8): 3064–3064.
84 Melo, J.V., Gordon, D.E., Cross, N.C.P., Goldman, J.M. The ABL-BCR fusion gene is expressed in chronic myeloid leukemia. *Blood* 1993; 81(1): 158–165.
85 Melo, J.V., Gordon, D.E., Tuszynski, A., Dhut, S., Young, B.D., Goldman, J.M. Expression of the ABL-BCR fusion gene in Philadelphia-positive acute lymphoblastic leukemia. *Blood* 1993; 81(10): 2488–2491.
86 Melo, J.V. The diversity of BCR-ABL fusion proteins and their relationship to leukemia phenotype. *Blood* 1996; 88(7): 2375–2384.
87 Zheng, X.M., Oancea, C., Henschler, R., Moore, M.A.S., Ruthardt, M. Reciprocal t(9;22) ABL/BCR fusion proteins: leukemogenic potential and effects on B cell commitment. *PLoS One* 2009; 4(10): e7661.
88 Huntly, B.J.P., Bench, A.J., Delabesse, E., et al. Derivative chromosome 9 deletions in chronic myeloid leukemia: poor prognosis is not associated with loss of ABL-BCR expression, elevated BCR-ABL levels or karyotypic instability. *Blood* 2002; 99(12): 4547–4553.
89 Reid, A.G., Huntly, B.J.P., Grace, C., Green, A.R., Nacheva, E.P. Survival implications of molecular heterogeneity in variant Philadelphia-positive chronic myeloid leukaemia. *Br J Haematol* 2003; 121(3): 419–427.
90 Pollock, J.L., Westervelt, P., Kurichety, A.K., Pelicci, P.G., Grisolano, J.L., Ley, T.J. A bcr-3 isoform of RAR alpha-PML potentiates the development of PML-RAR alpha-driven acute promyelocytic leukemia. *Proc Natl Acad Sci U S A* 1999; 96(26): 15103–15108.
91 He, L.Z., Bhaumik, M., Tribioli, C., et al. Two critical hits for promyelocytic leukemia. *Mol Cell* 2000; 6(5): 1131–1141.
92 Melo, J.V., Hochhaus, A., Yan, X.H., Goldman, J.M. Lack of correlation between ABL-BCR expression and response to interferon-alpha in chronic myeloid leukaemia. *Br J Haematol* 1996; 92(3): 684–686.
93 Huntly, B.J.P., Bench, A.J., Reid, A.G., et al. Lack of expression of the ABL-BCR transcript is not always associated with deletion of the derivative chromosome 9 and does not explain the poor prognosis associated with these deletions in CML. *Blood* 2001; 98(11): 260B.
94 Jones, D., Luthra, R., Cortes, J., et al. BCR-ABL fusion transcript types and levels and their interaction with secondary genetic changes in determining the

phenotype of Philadelphia chromosome-positive leukemias. *Blood* 2008; 112(13): 5190–5192.
95 Jiang, X.Y., Lopez, A., Holyoake, T., Eaves, A., Eaves, C. Autocrine production and action of IL-3 and granulocyte colony-stimulating factor in chronic myeloid leukemia. *Proc Natl Acad Sci U S A* 1999; 96(22): 12804–12809.
96 Fujisawa, S., Nakamura, S., Naito, K., Kobayashi, M., Ohnishi, K. A variant transcript, e1a3, of the minor BCR-ABL fusion gene in acute lymphoblastic leukemia: case report and review of the literature. *Int J Hematol* 2008; 87(2): 184–188.
97 Ahmed, W., Van Etten, R.A. Signal transduction in the chronic leukemias: implications for targeted therapies. *Curr Hematol Malig Rep* 2013; 8(1): 71–80.
98 Tanaka, K., Takechi, M., Hong, J., et al. 9;22 translocation and bcr rearrangements in chronic myelocytic leukemia patients among atomic bomb survivors. *J Radiat Res* 1989; 30(4): 352–358.
99 Corso, A., Lazzarino, M., Morra, E., et al. Chronic myelogenous leukemia and exposure to ionizing radiation – a retrospective study of 443 patients. *Annals of Hematology* 1995; 70(2): 79–82.
100 Gorre, M., Mohandas, P.E., Kagita, S., Annamaneni, S., Digumarti, R., Satti, V. Association of XRCC5 VNTR polymorphism with the development of chronic myeloid leukemia. *Tumour Biol* 2014; 35(2): 923–927.
101 Kabat, G.C., Wu, J.W., Moore, S.C., et al. Lifestyle and dietary factors in relation to risk of chronic myeloid leukemia in the NIH-AARP Diet and Health Study. *Cancer Epidemiol Biomarkers Prev* 2013; 22(5): 848–854.
102 Musselman, J.R.B., Blair, C.K., Cerhan, J.R., Nguyen, P., Hirsch, B., Ross, J.A. Risk of adult acute and chronic myeloid leukemia with cigarette smoking and cessation. *Cancer Epidemiol* 2013; 37(4): 410–416.
103 Kozubek, S., Lukasova, E., Ryznar, L., et al. Distribution of ABL and BCR genes in cell nuclei of normal and irradiated lymphocytes. *Blood* 1997; 89(12): 4537–4545.
104 Neves, H., Ramos, C., da Silva, M.G., Parreira, A., Parreira, L. The nuclear topography of ABL, BCR, PML and RAR alpha genes: Evidence for gene proximity in specific phases of the cell cycle and stages of hematopoietic differentiation. *Blood* 1999; 93(4): 1197–1207.
105 Saglio, G., Storlazzi, C.T., Giugliano, E., et al. A 76-kb duplicon maps close to the BCR gene on chromosome 22 and the ABL gene on chromosome 9: possible involvement in the genesis of the Philadelphia chromosome translocation. *Proc Natl Acad Sci U S A* 2002; 99(15): 9882–9887.
106 Biernaux, C., Loos, M., Sels, A., Huez, G., Stryckmans, P. Detection of major bcr-abl gene-expression at a very-low level in blood cells of some healthy individuals. *Blood* 1995; 86(8): 3118–3122.
107 Bose, S., Deininger, M., Gora-Tybor, J., Goldman, J.M., Melo, J.V. The presence of typical and atypical BCR-ABL fusion genes in leukocytes of normal individuals: biologic significance and implications for the assessment of minimal residual disease. *Blood* 1998; 92(9): 3362–3367.
108 Deininger, M.W.N., Goldman, J.M., Melo, J.V. The molecular biology of chronic myeloid leukemia. *Blood* 2000; 96(10): 3343–3356.

109 Foley, S.B., Hildenbrand, Z.L., Soyombo, A.A., et al. Expression of BCR/ABL p210 from a knockin allele enhances bone marrow engraftment without inducing neoplasia. *Cell Rep* 2013; 5(1): 51–60.
110 Vickers, M. Estimation of the number of mutations necessary to cause chronic myeloid leukaemia from epidemiological data. *Br J Haematol* 1996; 94(1): 1–4.
111 Bumm, T., Deininger, J., Newell, A.H., et al. Clonal chromosomal abnormalities in $CD34^+/CD38^-$ hematopoietic cells from cytogenetically normal chronic myeloid leukemia patients with a complete cytogenetic response to tyrosine kinase inhibitors. *Leukemia* 2010; 24(8): 1525–1528.
112 Barnes, D.J., Palaiologou, D., Panousopoulou, E., et al. Bcr-Abl expression levels determine the rate of development of resistance to imatinib mesylate in chronic myeloid leukemia. *Cancer Res* 2005; 65(19): 8912–8919.
113 Barnes, D.J., Schultheis, B., Adedeji, S., Melo, J.V. Dose-dependent effects of Bcr-Abl in cell line models of different stages of chronic myeloid leukemia. *Oncogene* 2005; 24(42): 6432–6440.
114 Modi, H., McDonald, T., Chu, S., Yee, J.K., Forman, S.J., Bhatia, R. Role of BCR/ABL gene-expression levels in determining the phenotype and imatinib sensitivity of transformed human hematopoietic cells. *Blood* 2007; 109(12): 5411–5421.
115 Perrotti, D., Harb, J.G. BCR-ABL1 kinase-dependent alteration of mRNA metabolism: potential alternatives for therapeutic intervention. *Leuk Lymphoma* 2011; 52: 30–44.
116 Skorski, T. Oncogenic tyrosine kinases and the DNA damage response. *Nat Rev Cancer* 2002; 2(5): 351–360.
117 Perrotti, D., Jamieson, C., Goldman, J., Skorski, T. Chronic myeloid leukemia: mechanisms of blastic transformation. *J Clin Invest* 2010; 120(7): 2254–2264.
118 Neviani, P., Santhanam, R., Trotta, R., et al. The tumor suppressor PP2A is functionally inactivated in blast crisis CML through the inhibitory activity of the BCR/ABL-regulated SET protein. *Cancer Cell* 2005; 8(5): 355–368.
119 Flis, K., Irvine, D., Copland, M., Bhatia, R., Skorski, T. Chronic myeloid leukemia stem cells display alterations in expression of genes involved in oxidative phosphorylation. *Leuk Lymphoma* 2012; 53(12): 2474–2478.
120 Koptyra, M., Cramer, K., Slupianek, A., Richardson, C., Skorski, T. BCR/ABL promotes accumulation of chromosomal aberrations induced by oxidative and genotoxic stress. *Leukemia* 2008; 22(10): 1969–1972.
121 Stoklosa, T., Poplawski, T., Koptyra, M., et al. BCR/ABL inhibits mismatch repair to protect from apoptosis and induce point mutations. *Cancer Res* 2008; 68(8): 2576–2580.
122 Mitelman, F. The cytogenetic scenario of chronic myeloid-leukemia. *Leuk Lymphoma* 1993; 11(Suppl 1): 11–15.
123 Chase, A., Huntly, B.J.P., Cross, N.C.P. Cytogenetics of chronic myeloid leukaemia. *Best Pract Res Clin Haematol* 2001; 14(3): 553–571.
124 Ahuja, H., Bareli, M., Advani, S.H., Benchimol, S., Cline, M.J. Alterations in the p53 gene and the clonal evolution of the blast crisis of chronic myelocytic leukemia. *Proc Natl Acad Sci U S A* 1989; 86(17): 6783–6787.

125 Collins, S.J., Howard, M., Andrews, D.F., Agura, E., Radich, J. Rare occurrence of N-ras point mutations in Philadelphia chromosome positive chronic myeloid leukemia. *Blood* 1989; 73(4): 1028–1032.

126 Lemaistre, A., Lee, M.S., Talpaz, M., et al. Ras oncogene mutations are rare late stage events in chronic myelogenous leukemia. *Blood* 1989; 73(4): 889–891.

127 Beck, Z., Kiss, A., Toth, F.D., et al. Alterations of P53 and RB genes and the evolution of the accelerated phase of chronic myeloid leukemia. *Leuk Lymphoma* 2000; 38(5–6): 587–597.

128 Mullighan, C.G., Goorha, S., Radtke, I., et al. Genome-wide analysis of genetic alterations in acute lymphoblastic leukaemia. *Nature* 2007; 446(7137): 758–764.

129 Grossmann, V., Kohlmann, A., Zenger, M., et al. A deep-sequencing study of chronic myeloid leukemia patients in blast crisis (BC-CML) detects mutations in 76.9% of cases. *Leukemia* 2011; 25(3): 557–560.

130 Iacobucci, I., Ferrari, A., Lonetti, A., et al. CDKN2A/B alterations impair prognosis in adult BCR-ABL1-positive acute lymphoblastic leukemia patients. *Clin Cancer Res* 2011; 17(23): 7413–7423.

131 Serra, A., Gottardi, E., Dellaragione, F., Saglio, G., Iolascon, A. Involvement of the cyclin-dependent kinase-4 inhibitor (CDKN2) gene in the pathogenesis of lymphoid blast crisis of chronic myelogenous leukemia. *Br J Haematol* 1995; 91(3): 625–629.

132 Sill, H., Goldman, J.M., Cross, N.C.P. Homozygous deletions of the p16 tumor-suppressor gene are associated with lymphoid transformation of chronic myeloid leukemia. *Blood* 1995; 85(8): 2013–2016.

133 Hernandez-Boluda, J.C., Cervantes, F., Colomer, D., et al. Genomic p16 abnormalities in the progression of chronic myeloid leukemia into blast crisis: a sequential study in 42 patients. *Exp Hematol* 2003; 31(3): 204–210.

134 Serrano, M., Hannon, G.J., Beach, D. A new regulatory motif in cell-cycle control causing specific inhibition of cyclin D/CDK4. *Nature* 1993; 366(6456): 704–707.

135 Stott, F.J., Bates, S., James, M.C., et al. The alternative product from the human CDKN2A locus, p14(ARF), participates in a regulatory feedback loop with p53 and MDM2. *EMBO J* 1998; 17(17): 5001–5014.

136 Gil, J., Peters, G. Regulation of the INK4b–ARF–INK4a tumour suppressor locus: all for one or one for all. *Nat Rev Mol Cell Biol* 2006; 7(9): 667–677.

137 Brazma, D., Grace, C., Howard, J., et al. Genomic profile of chronic myelogenous leukemia: imbalances associated with disease progression. *Genes Chromosomes Cancer* 2007; 46(11): 1039–1050.

138 Menezes, J., Salgado, R.N., Acquadro, F., et al. ASXL1, TP53 and IKZF3 mutations are present in the chronic phase and blast crisis of chronic myeloid leukemia. *Blood Cancer J* 2013; 3: e157.

139 Yamagami, T., Sugiyama, H., Inoue, K., et al. Growth inhibition of human leukemic cells by WT1 (Wilms tumor gene) antisense oligodeoxynucleotides: implications for the involvement of WT1 in leukemogenesis. *Blood* 1996; 87(7): 2878–2884.

140 Miething, C., Grundler, R., Mugler, C., et al. Retroviral insertional mutagenesis identifies RUNX genes involved in chronic myeloid leukemia disease persistence under imatinib treatment. *Proc Natl Acad Sci U S A* 2007; 104(11): 4594–4599.

141 Abdel-Wahab, O., Pardanani, A., Patel, J., et al. Concomitant analysis of EZH2 and ASXL1 mutations in myelofibrosis, chronic myelomonocytic leukemia and blast-phase myeloproliferative neoplasms. *Leukemia* 2011; 25(7): 1200–1202.

142 Mughal, T.I., Vannucchi, A.M., Soverini, S., et al. Current pre-clinical and clinical advances in the BCR-ABL1-positive and -negative chronic myeloproliferative neoplasms. *Haematologica* 2014; 99(5): 797–801.

143 Feinstein, E., Cimino, G., Gale, R.P., et al. P53 in chronic myelogenous leukemia in acute phase. *Proc Natl Acad Sci U S A* 1991; 88(14): 6293–6297.

144 Nakai, H., Misawa, S., Toguchida, J., Yandell, D.W., Ishizaki, K. Frequent p53 gene mutations in blast crisis of chronic myelogenous leukemia, especially in myeloid crisis harboring loss of a chromosome-17p. *Cancer Res* 1992; 52(23): 6588–6593.

145 Schmidt, M., Rinke, J., Schaefer, V., et al. Molecular-defined clonal evolution in patients with chronic myeloid leukemia independent of the BCR-ABL status. *Leukemia* 2014; 28(12): 2292–2299.

146 Jamieson, C.H., Barroga, C.F., Vainchenker, W.P. Miscreant myeloproliferative disorder stem cells. *Leukemia* 2008; 22(11): 2011–2019.

147 Sengupta, A., Arnett, J., Dunn, S., Williams, D.A., Cancelas, J.A. Rac2 GTPase deficiency depletes BCR-ABL$^+$ leukemic stem cells and progenitors in vivo. *Blood* 2010; 116(1): 81–84.

148 Schemionek, M., Spieker, T., Kerstiens, L., et al. Leukemic spleen cells are more potent than bone marrow-derived cells in a transgenic mouse model of CML. *Leukemia* 2012; 26(5): 1030–1037.

149 Tabe, Y., Jin, L., Iwabuchi, K., et al. Role of stromal microenvironment in non-pharmacological resistance of CML to imatinib through Lyn/CXCR4 interactions in lipid rafts. *Leukemia* 2012; 26(5): 883–892.

150 Calabretta, B., Perrotti, D. The biology of CML blast crisis. *Blood* 2004; 103(11): 4010–4022.

151 Jamieson, C.H., Ailles, L.E., Dylla, S.J., et al. Granulocyte–macrophage progenitors as candidate leukemic stem cells in blast-crisis CML. *N Engl J Med* 2004; 351(7): 657–667.

152 Perrotti, D., Cesi, V., Trotta, R., et al. BCR-ABL suppresses C/EBPalpha expression through inhibitory action of hnRNP E2. *Nat Genet* 2002; 30(1): 48–58.

153 Trotta, R., Vignudelli, T., Candini, O., et al. BCR/ABL activates mdm2 mRNA translation via the La antigen. *Cancer Cell* 2003; 3(2): 145–160.

154 Eiring, A.M., Harb, J.G., Neviani, P., et al. miR-328 functions as an RNA decoy to modulate hnRNP E2 regulation of mRNA translation in leukemic blasts. *Cell* 2010; 140(5): 652–665.

155 Yong, A.S., Melo, J.V. The impact of gene profiling in chronic myeloid leukaemia. *Best Pract Res Clin Haematol* 2009; 22(2): 181–190.

156 Radich, J.P., Dai, H., Mao, M., et al. Gene expression changes associated with progression and response in chronic myeloid leukemia. *Proc Natl Acad Sci U S A* 2006; 103(8): 2794–2799.

157 Terragna, C., Durante, S., Astolfi, A., et al. Gene expression profile (GEP) of chronic myeloid leukemia (CML) patients at diagnosis: two distinguished subgroups of CML patients identified, based on a molecular signature, irrespective of their Sokal risk score. *Blood 2008*; 112 (Suppl.): 3190.
158 Oehler, V.G., Guthrie, K.A., Cummings, C.L., et al. The preferentially expressed antigen in melanoma (PRAME) inhibits myeloid differentiation in normal hematopoietic and leukemic progenitor cells. *Blood* 2009; 114(15): 3299–3308.
159 Jamieson, C.H. Chronic myeloid leukemia stem cells. *Hematology Am Soc Hematol Educ Program* 2008; 436–442.
160 Quintas-Cardama, A., Cortes, J. Molecular biology of bcr-abl1-positive chronic myeloid leukemia. *Blood* 2009; 113(8): 1619–1630.
161 le Coutre, P., Tassi, E., Varella-Garcia, M., et al. Induction of resistance to the Abelson inhibitor STI571 in human leukemic cells through gene amplification. *Blood* 2000; 95(5): 1758–1766.
162 Gorre, M.E., Mohammed, M., Ellwood, K., et al. Clinical resistance to STI-571 cancer therapy caused by BCR-ABL gene mutation or amplification. *Science* 2001; 293(5531): 876–880.
163 Marega, M., Piazza, R., Meneghetti, I., Redaelli, S., Mogavero, A., Gambacorti, C. BCR and BCR/ABL regulation during myeloid differentiation in healthy donors and in chronic phase/blast crisis CML patients. *Blood* 2008; 112(Suppl): 3204.
164 Amin, H.M., Hoshino, K., Yang, H., Lin, Q., Lai, R., Garcia-Manero, G. Decreased expression level of SH2 domain-containing protein tyrosine phosphatase-1 (Shp1) is associated with progression of chronic myeloid leukaemia. *J Pathol* 2007; 212(4): 402–410.
165 Cuenco, G.M., Ren, R.B. Cooperation of BCR-ABL and AML1/MDS1/EVI1 in blocking myeloid differentiation and rapid induction of an acute myelogenous leukemia. *Oncogene* 2001; 20(57): 8236–8248.
166 Dash, A.B., Williams, I.R., Kutok, J.L., et al. BCR/ABL cooperates with NUP98/HOXA9 to cause CML blast crisis. *Blood* 2001; 98(11): 117A–118A.
167 Mizuno, T., Yamasaki, N., Miyazaki, K., et al. Overexpression/enhanced kinase activity of BCR/ABL and altered expression of Notch1 induced acute leukemia in p210BCR/ABL transgenic mice. *Oncogene* 2008; 27(24): 3465–3474.
168 Nagamachi, A., Yamasaki, N., Miyazaki, K., et al. Haploinsufficiency and acquired loss of Bcl11b and H2AX induces blast crisis of chronic myelogenous leukemia in a transgenic mouse model. *Cancer Sci* 2009; 100(7): 1219–1226.
169 Chang, J.S., Santhanam, R., Trotta, R., et al. High levels of the BCR/ABL oncoprotein are required for the MAPK-hnRNP-E2-dependent suppression of C/EBPα-driven myeloid differentiation. *Blood* 2007; 110(3): 994–1003.
170 Tenen, D.G., Hromas, R., Licht, J.D., Zhang, D.E. Transcription factors, normal myeloid development and leukemia. *Blood* 1997; 90(2): 489–519.
171 Tenen, D.G. Disruption of differentiation in human cancer: AML shows the way. *Nat Rev Cancer* 2003; 3(2): 89–101.
172 McCallum, L., Price, S., Planque, N., et al. A novel mechanism for BCR-ABL action: stimulated secretion of CCN3 is involved in growth and differentiation regulation. *Blood* 2006; 108(5): 1716–1723.

173 McCallum, L., Lu, W., Price, S., Lazar, N., Perbal, B., Irvine, A.E. CCN3 suppresses mitogenic signalling and reinstates growth control mechanisms in chronic myeloid leukaemia. *J Cell Commun Signal* 2012; 6(1): 27–35.

174 Suresh, S., McCallum, L., Lu, W., Lazar, N., Perbal, B., Irvine, A.E. MicroRNAs 130a/b are regulated by BCR-ABL and downregulate expression of CCN3 in CML. *J Cell Commun Signal* 2011; 5(3): 183–191.

175 Venturini, L., Battmer, K., Castoldi, M., et al. Expression of the miR-17-92 polycistron in chronic myeloid leukemia (CML) $CD34^+$ cells. *Blood* 2007; 109(10): 4399–4405.

176 Bueno, M.J., de Castro, I.P., de Cedron, M.G., et al. Genetic and epigenetic silencing of microRNA-203 enhances ABL1 and BCR-ABL1 oncogene expression. *Cancer Cell* 2008; 13(6): 496–506.

177 Viswanathan, S.R., Powers, J.T., Einhorn, W., et al. Lin28 promotes transformation and is associated with advanced human malignancies. *Nat Genet* 2009; 41(7): 843–848.

178 Chim, C.S., Wong, K.Y., Leung, C.Y., et al. Epigenetic inactivation of the hsa-miR-203 in haematological malignancies. *J Cell Mol Med* 2011; 15(12): 2760–2767.

179 Gao, S.M., Yang, J., Chen, C., et al. miR-15a/16-1 enhances retinoic acid-mediated differentiation of leukemic cells and is up-regulated by retinoic acid. *Leuk Lymphoma* 2011; 52(12): 2365–2371.

180 Lopotova, T., Zackova, M., Klamova, H., Moravcova, J. MicroRNA-451 in chronic myeloid leukemia: miR-451–BCR-ABL regulatory loop? *Leuk Res* 2011; 35(7): 974–977.

181 Wang, L.-S., Li, L., Li, L., et al. Role of microRNA-486-5p overexpression in CML $CD34^+$ cells in modulating BCR-ABL mediated hematopoietic stem/progenitor cell transformation and imatinib sensitivity. *Blood 2011*; 188(ASH Annual Meeting Abstracts): Abstr. 1667(21).

182 Ito, K., Bernardi, R., Morotti, A., Matsuoka, S., et al. PML targeting eradicates quiescent leukaemia-initiating cells. *Nature* 2008; 453(7198): 1072–1078.

183 Naka, K., Hoshii, T., Muraguchi, T., et al. TGF-β-FOXO signalling maintains leukaemia-initiating cells in chronic myeloid leukaemia. *Nature* 2010; 463(7281): 676–680.

184 Issa, J.P.J., Kantarjian, H., Mohan, A., et al. Methylation of the ABL1 promoter in chronic myelogenous leukemia: lack of prognostic significance. *Blood* 1999; 93(6): 2075–2080.

185 Jelinek, J., Gharibyan, V., Estecio, M.R.H., et al. Aberrant DNA methylation is associated with disease progression, resistance to imatinib and shortened survival in chronic myelogenous leukemia. *PLoS One* 2011; 6(7): e22110.

186 Polakova, K.M., Koblihova, J., Stopka, T. Role of epigenetics in chronic myeloid leukemia. *Curr Hematol Malig Rep* 2013; 8(1): 28–36.

187 Dunwell, T., Hesson, L., Rauch, T.A., et al. A genome-wide screen identifies frequently methylated genes in haematological and epithelial cancers. *Mol Cancer* 2010; 9.

188 O'Brien, S.G., Guilhot, F., Larson, R.A., et al. Imatinib compared with interferon and low-dose cytarabine for newly diagnosed chronic-phase chronic myeloid leukemia. *N Engl J Med* 2003; 348(11): 994–1004.

189 Druker, B.J., Guilhot, F., O'Brien, S., Larson, R.A. Long-term benefits of imatinib (IM) for patients newly diagnosed with chronic myelogenous leukemia in chronic phase (CML-CP): the 5-year update from the IRIS study. *J Clin Oncol* 2006; 24(18): 338S.

190 Druker, B.J., Guilhot, F., O'Brien, S.G., et al. Five-year follow-up of patients receiving imatinib for chronic myeloid leukemia. *N Engl J Med* 2006; 355(23): 2408–2417.

191 Hehlmann, R., Müller, M.C., Lauseker, M., et al. Deep molecular response is reached by the majority of patients treated with imatinib, predicts survival and is achieved more quickly by optimized high-dose imatinib: results from the randomized CML-Study IV. *J Clin Oncol* 2014; 32(5): 415–423.

192 Druker, B.J., Ohno, S., Buchdunger, E., Tamura, S., Zimmermann, J., Lydon, N.B. Selective killing of BCR-ABL positive cells with a specific inhibitor of the ABL tyrosine kinase. In *Cancer Genes: Functional Aspects*, ed. Mihich, E., Housman, D. New York, Plenum Press, 1996, pp. 255–267.

193 Fruehauf, S., Topaly, J., Buss, E.C., et al. Imatinib combined with mitoxantrone/etoposide and cytarabine is an effective induction therapy for patients with chronic myeloid leukemia in myeloid blast crisis. *Cancer* 2007; 109(8): 1543–1549.

194 Ohanian, M., Kantarjian, H.M., Quintas-Cardama, A., et al. Tyrosine kinase inhibitors as initial therapy for patients with chronic myeloid leukemia in accelerated phase. *Clin Lymphoma Myeloma Leuk* 2014; 14(2): 155–162.

195 Saglio, G., Kim, D.W., Issaragrisil, S., et al. Nilotinib versus imatinib for newly diagnosed chronic myeloid leukemia. *N Engl J Med* 2010; 362(24): 2251–2259.

196 Kantarjian, H.M., Shah, N.P., Cortes, J.E., et al. Dasatinib or imatinib in newly diagnosed chronic-phase chronic myeloid leukemia: 2-year follow-up from a randomized phase 3 trial (DASISION). *Blood* 2012; 119(5): 1123–1129.

197 Rix, U., Hantschel, O., Duernberger, G., et al. Chemical proteomic profiles of the BCR-ABL inhibitors imatinib, nilotinib and dasatinib reveal novel kinase and nonkinase targets. *Blood* 2007; 110(12): 4055–4063.

198 Karaman, M.W., Herrgard, S., Treiber, D.K., et al. A quantitative analysis of kinase inhibitor selectivity. *Nat Biotechnol* 2008; 26(1): 127–132.

199 Redaelli, S., Piazza, R., Rostagno, R., et al. Activity of bosutinib, dasatinib and nilotinib against 18 imatinib-resistant BCR/ABL mutants. *J Clin Oncol* 2009; 27(3): 469–471.

200 Cortes, J.E., Kim, D.W., Pinilla-Ibarz, J., et al. A phase 2 trial of ponatinib in Philadelphia chromosome-positive leukemias. *N Engl J Med* 2013; 369(19): 1783–1796.

201 Giles, F.J., le Coutre, P.D., Pinilla-Ibarz, J., et al. Nilotinib in imatinib-resistant or imatinib-intolerant patients with chronic myeloid leukemia in chronic phase: 48-month follow-up results of a phase II study. *Leukemia* 2013; 27(1): 107–112.

202 Jabbour, E., Kantarjian, H.M., Saglio, G., et al. Early response with dasatinib or imatinib in chronic myeloid leukemia: 3-year follow-up from a randomized phase 3 trial (DASISION). *Blood* 2014; 123(4): 494–500.
203 Druker, B.J., Sawyers, C.L., Kantarjian, H., et al. Activity of a specific inhibitor of the BCR-ABL tyrosine kinase in the blast crisis of chronic myeloid leukemia and acute lymphoblastic leukemia with the Philadelphia chromosome. *N Engl J Med* 2001; 344(14): 1038–1042.
204 Campbell, L.J., Patsouris, C., Rayeroux, K.C., Somana, K., Januszewicz, E.H., Szer, J. BCR/ABL amplification in chronic myelocytic leukemia blast crisis following imatinib mesylate administration. *Cancer Genet Cytogenet* 2002; 139(1): 30–33.
205 Burger, H., van Tol, H., Brok, M., et al. Chronic imatinib mesylate exposure leads to reduced intracellular drug accumulation by induction of the ABCG2 (BCRP) and ABCB1 (MDR1) drug transport pumps. *Cancer Biol Ther* 2005; 4(7): 747–752.
206 Jiang, X., Delaney, A., Eaves, A., Eaves, C. Leukemic stem cells from CML patients have uniquely elevated BCR-ABL activity explaining their selective resistance to imatinib mesylate but also contain subpopulations with kinase mutations. *Exp Hematol* 2005; 33(7): 50.
207 Apperley, J.F. Part I. Mechanisms of resistance to imatinib in chronic myeloid leukaemia. *Lancet Oncol* 2007; 8(11): 1018–1029.
208 Ibrahim, A.R., Eliasson, L., Apperley, J.F. Poor adherence is the main reason for loss of CCyR and imatinib failure for chronic myeloid leukemia patients on long-term therapy. Erratum for Ibrahim et al. *Blood 2011*; 117(14), 3733–3766. *Blood 2012*; 120(24): 4903.
209 Bixby, D., Talpaz, M. Seeking the causes and solutions to imatinib resistance in chronic myeloid leukemia. *Leukemia* 2011; 25(1): 7–22.
210 Soverini, S., Hochhaus, A., Nicolini, F.E., et al. BCR-ABL kinase domain mutation analysis in chronic myeloid leukemia patients treated with tyrosine kinase inhibitors: recommendations from an expert panel on behalf of European LeukemiaNet. *Blood* 2011; 118(5): 1208–1215.
211 Soverini, S., Colarossi, S., Gnani, A., et al. Contribution of ABL kinase domain mutations to imatinib resistance in different subsets of Philadelphia-positive patients: by the GIMEMA Working Party on Chronic Myeloid Leukemia. *Clin Cancer Res* 2006; 12(24): 7374–7379.
212 Picard, S., Titier, K., Etienne, G., et al. Trough imatinib plasma levels are associated with both cytogenetic and molecular responses to standard-dose imatinib in chronic myeloid leukemia. *Blood* 2007; 109(8): 3496–3499.
213 Muller, M.C., Cortes, J.E., Kim, D.W., et al. Dasatinib treatment of chronic-phase chronic myeloid leukemia: analysis of responses according to preexisting BCR-ABL mutations. *Blood* 2009; 114(24): 4944–4953.
214 Soverini, S., Gnani, A., Colarossi, S., et al. Philadelphia-positive patients who already harbor imatinib-resistant Bcr-Abl kinase domain mutations have a higher likelihood of developing additional mutations associated with resistance to second- or third-line tyrosine kinase inhibitors. *Blood* 2009; 114(10): 2168–2171.

215 Gruber, F.X., Ernst, T., Porkka, K., et al. Dynamics of the emergence of dasatinib and nilotinib resistance in imatinib-resistant CML patients. *Leukemia* 2012; 26(1): 172–177.
216 Hochhaus, A., Saglio, G., Larson, R.A., et al. Nilotinib is associated with a reduced incidence of BCR-ABL mutations vs imatinib in patients with newly diagnosed chronic myeloid leukemia in chronic phase. *Blood* 2013; 121(18): 3703–3708.
217 Khorashad, J.S., Kelley, T.W., Szankasi, P., et al. BCR-ABL1 compound mutations in tyrosine kinase inhibitor-resistant CML: frequency and clonal relationships. *Blood* 2013; 121(3): 489–498.
218 Cortes, J.E., Kim, D.-W., Kantarjian, H.M., et al. Bosutinib versus imatinib in newly diagnosed chronic-phase chronic myeloid leukemia: results from the BELA trial. *J Clin Oncol* 2012; 30(28): 3486–3492.
219 O'Hare, T., Shakespeare, W.C., Zhu, X., et al. AP24534, a pan-BCR-ABL inhibitor for chronic myeloid leukemia, potently inhibits the T315I mutant and overcomes mutation-based resistance. *Cancer Cell* 2009; 16(5): 401–412.
220 Cortes, J.E., Kantarjian, H., Shah, N.P., et al. Ponatinib in refractory Philadelphia chromosome-positive leukemias. *N Engl J Med* 2012; 367(22): 2075–2088.
221 Cortes, J.E., Kim, D.-W., Pinilla-Ibarz, J., et al. PACE: a pivotal phase II trial of ponatinib in patients with CML and Ph plus ALL resistant or intolerant to dasatinib or nilotinib or with the T315I mutation. *J Clin Oncol 2012*; 30(15 Suppl): Abstr. 6503.
222 Quintas-Cardama, A. Ponatinib in Philadelphia Chromosome-positive leukemias. *N Engl J Med* 2014; 370(6): 577.
223 Cortes, J.E., Talpaz, M., Kantarjian, H. Ponatinib in Philadelphia chromosome-positive leukemias: reply. *N Engl J Med* 2014; 370(6): 577.
224 Quintas-Cardama, A., Kantarjian, H.M., Cortes, J.E. Mechanisms of primary and secondary resistance to imatinib in chronic myeloid leukemia. *Cancer Control* 2009; 16(2): 122–131.
225 Shah, N.P., Tran, C., Lee, F.Y., Chen, P., Norris, D., Sawyers, C.L. Overriding imatinib resistance with a novel ABL kinase inhibitor. *Science* 2004; 305(5682): 399–401.
226 Tokarski, J.S., Newitt, J.A., Chang, C.Y.J., et al. The structure of dasatinib (BMS-354825) bound to activated ABL kinase domain elucidates its inhibitory activity against imatinib-resistant ABL mutants. *Cancer Res* 2006; 66(11): 5790–5797.
227 Parker, W.T., Ho, M., Scott, H.S., Hughes, T.P., Branford, S. Brief report. Poor response to second-line kinase inhibitors in chronic myeloid leukemia patients with multiple low-level mutations, irrespective of their resistance profile. *Blood* 2012; 119(10): 2234–2238.
228 Shah, N.P., Skaggs, B.J., Branford, S., et al. Sequential ABL kinase inhibitor therapy selects for compound drug-resistant BCR-ABL mutations with altered oncogenic potency. *J Clin Invest* 2007; 117(9): 2562–2569.
229 O'Dwyer, M.E., Mauro, M.J., Blasdel, C., et al. Clonal evolution and lack of cytogenetic response are adverse prognostic factors for hematologic relapse of chronic phase CML patients treated with imatinib mesylate. *Blood* 2004; 103(2): 451–455.

230 Cortes, J.E., Talpaz, M., Giles, F., et al. Prognostic significance of cytogenetic clonal evolution in patients with chronic myelogenous leukemia on imatinib mesylate therapy. *Blood* 2003; 101(10): 3794–3800.
231 Cortes, J., O'Dwyer, M.E. Clonal evolution in chronic myelogenous leukemia. *Hematol Oncol Clin North Am* 2004; 18(3): 671–684.
232 Bhatia, R., Holtz, M., Niu, N., et al. Persistence of malignant hematopoietic progenitors in chronic myelogenous leukemia patients in complete cytogenetic remission following imatinib mesylate treatment. *Blood* 2003; 101(12): 4701–4707.
233 Copland, M., Hamilton, A., Eirick, L.J., et al. Dasatinib (BMS-354825) targets an earlier progenitor population than imatinib in primary CML but does not eliminate the quiescent fraction. *Blood* 2006; 107(11): 4532–4539.
234 Nagar, B., Bornmann, W.G., Pellicena, P., et al. Crystal structures of the kinase domain of c-Abl in complex with the small molecule inhibitors PD173955 and imatinib (STI-571). *Cancer Res* 2002; 62(15): 4236–4243.
235 Chu, S., Xu, H., Shah, N.P., et al. Detection of BCR-ABL kinase mutations in $CD34^+$ cells from chronic myelogenous leukemia patients in complete cytogenetic remission on imatinib mesylate treatment. *Blood* 2005; 105(5): 2093–2098.
236 Helgason, G.V., Young, G.A., Holyoake, T.L. Targeting chronic myeloid leukemia stem cells. *Curr Hematol Malig Rep* 2010; 5(2): 81–87.
237 Bellodi, C., Lidonnici, M.R., Hamilton, A., et al. Targeting autophagy potentiates tyrosine kinase inhibitor-induced cell death in Philadelphia chromosome-positive cells, including primary CML stem cells. *J Clin Invest* 2009; 119(5): 1109–1123.
238 Carella, A.M., Beltrami, G., Pica, G., Carella, A., Catania, G. Clarithromycin potentiates tyrosine kinase inhibitor treatment in patients with resistant chronic myeloid leukemia. *Leuk Lymphoma* 2012; 53(7): 1409–1411.
239 Schafranek, L., Leclercq, T.M., White, D.L., Hughes, T.P. Clarithromycin enhances dasatinib-induced cell death in chronic myeloid leukemia cells, by inhibition of late stage autophagy. *Leuk Lymphoma* 2013; 54(1): 198–201.
240 Quintarelli, C., Dotti, G., Hasan, S.T., et al. High-avidity cytotoxic T lymphocytes specific for a new PRAME-derived peptide can target leukemic and leukemic-precursor cells. *Blood* 2011; 117(12): 3353–3362.
241 Dao, T., Liu, C., Scheinberg, D.A. Approaching untargetable tumor-associated antigens with antibodies. *Oncoimmunology* 2013; 2(7): e24678.
242 Bocchia, M., Korontsvit, T., Xu, Q., et al. Specific human cellular immunity to bcr-abl oncogene-derived peptides. *Blood* 1996; 87(9): 3587–3592.
243 Bocchia, M., Gentili, S., Abruzzese, E., et al. Effect of a p210 multipeptide vaccine associated with imatinib or interferon in patients with chronic myeloid leukaemia and persistent residual disease: a multicentre observational trial. *Lancet* 2005; 365(9460): 657–662.
244 Rojas, J.M., Knight, K., Wang, L., Clark, R.E. Clinical evaluation of BCR-ABL peptide immunisation in chronic myeloid leukaemia: results of the EPIC study. *Leukemia* 2007; 21(11): 2287–2295.
245 Ren, R. Mechanisms of BCR-ABL in the pathogenesis of chronic myelogenous leukaemia. *Nat Rev Cancer* 2005; 5(3): 172–183.
246 Gross, A.W., Zhang, X.W., Ren, R.B. Bcr-Abl with an SH3 deletion retains the ability to induce a myeloproliferative disease in mice, yet c-Abl activated by

an SH3 deletion induces only lymphoid malignancy. *Mol Cell Biol* 1999; 19(10): 6918–6928.

247 Johnson, K.J., Griswold, I.J., O'Hare, T., et al. A BCR-ABL mutant lacking direct binding sites for the GRB2, CBL and CRKL adapter proteins fails to induce leukemia in mice. *PLoS One* 2009; 4(10): e7439.

248 Chen, M., Gallipoli, P., DeGeer, D., et al. Targeting primitive chronic myeloid leukemia cells by effective inhibition of a new AHI-1–BCR-ABL–JAK2 complex. *J Natl Cancer Inst* 2013; 105(6): 405–423.

249 Neviani, P., Harb, J.G., Oaks, J.J., et al. PP2A-activating drugs selectively eradicate TKI-resistant chronic myeloid leukemic stem cells. *J Clin Invest* 2013; 123(10): 4144–4157.

250 Chen, Y., Hu, Y., Zhang, H., Peng, C., Li, S. Loss of the Alox5 gene impairs leukemia stem cells and prevents chronic myeloid leukemia. *Nat Genet* 2009; 41(7): 783–792.

251 Oki, Y., Kantarjian, H.M., Gharibyan, V., et al. Phase II study of low-dose decitabine in combination with imatinib mesylate in patients with accelerated or myeloid blastic phase of chronic myelogenous leukemia. *Cancer* 2007; 109(5): 899–906.

252 Cervera, E., Candelaria, M., Lopez-Navarro, O., et al. Epigenetic therapy with hydralazine and magnesium valproate reverses imatinib resistance in patients with chronic myeloid leukemia. *Clin Lymphoma Myeloma Leuk* 2012; 12(3): 207–212.

253 Hook, E.B. Exclusion of chromosomal mosaicism – tables of 90%, 95%, and 99% confidence-limits and comments on use. *Am J Hum Genet* 1977; 29(1): 94–97.

254 Baccarani, M., Deininger, M.W., Rosti, G., et al. European LeukemiaNet recommendations for the management of chronic myeloid leukemia: 2013. *Blood* 2013; 122(6): 872–884.

255 Casali, M., Truglio, F., Milone, G., et al. Trisomy-8 in Philadelphia chromosome (Ph1)-negative cells in the course of Ph1-positive chronic myelocytic leukemia. *Genes Chromosomes Cancer* 1992; 4(3): 269–270.

256 Izumi, T., Imagawa, S., Hatake, K., et al. Philadelphia chromosome-negative cells with trisomy 8 after busulfan and interferon treatment of Ph(1)-positive chronic myelogenous leukemia. *Int J Hematol* 1996; 64(1): 73–77.

257 Fayad, L., Kantarjian, H., O'Brien, S., et al. Emergence of new clonal abnormalities following interferon-alpha induced complete cytogenetic response in patients with chronic myeloid leukemia: report of three cases. *Leukemia* 1997; 11(5): 767–771.

258 Jabbour, E., Kantarjian, H.M., Abruzzo, L.V., et al. Chromosomal abnormalities in Philadelphia chromosome-negative metaphases appearing during imatinib mesylate therapy in patients with newly diagnosed chronic myeloid leukemia in chronic phase. *Blood* 2007; 110(8): 2991–2995.

259 De Melo, V.A.S., Milojkovic, D., Khorashad, J.S., et al. Philadelphia-negative clonal hematopoiesis is a significant feature of dasatinib therapy for chronic myeloid leukemia. *Blood* 2007; 110(8): 3086–3087.

260 Pitini, V., Arrigo, C., Sauta, M.G., Altavilla, G. Myelodysplastic syndrome appearing during imatinib mesylate therapy in a patient with GIST. *Leuk Res* 2009; 33(9): e143–e144.

261 van der Kuip, H., Moehring, A., Wohlbold, L., Miething, C., Duyster, J., Aulitzky, W.E. Imatinib mesylate (STI571) prevents the mutator phenotype of Bcr-Abl in hematopoietic cell lines. *Leuk Res* 2004; 28(4): 405–408.

262 Bumm, T., Muller, C., Al-Ali, H.K., et al. Emergence of clonal cytogenetic abnormalities in Ph– cells in some CML patients in cytogenetic remission to imatinib but restoration of polyclonal hematopoiesis in the majority. *Blood* 2003; 101(5): 1941–1949.

263 Zaccaria, A., Valenti, A.M., Donti, E., Gozzetti, A., Ronconi, S., Spedicato, F. Persistence of chromosomal abnormalities additional to the Philadelphia chromosome after Philadelphia chromosome disappearance during imatinib therapy for chronic myeloid leukemia. *Haematologica* 2007; 92(4): 564–565.

264 Sneed, T.B., Kantarjian, H.M., Talpaz, M., et al. The significance of myelosuppression during therapy with imatinib mesylate in patients with chronic myelogenous leukemia in chronic phase. *Cancer* 2004; 100(1): 116–121.

265 Dewald, G.W., Schad, C.R., Christensen, E.R., et al. The application of fluorescent in-situ hybridization to detect Mbcr/ABL fusion in variant Ph chromosomes in CML and ALL. *Cancer Genet Cytogenet* 1993; 71(1): 7–14.

266 Calabrese, G., Stuppia, L., Franchi, P.G., et al. Complex translocations of the Ph chromosome and Ph negative CML arise from similar mechanisms, as evidenced by FISH analysis. *Cancer Genet Cytogenet* 1994; 78(2): 153–159.

267 Nacheva, E., Holloway, T., Brown, K., Bloxham, D., Green, A.R. Philadelphia-negative chronic myeloid leukemia – detection by FISH of BCR-ABL fusion gene localized either to chromosome-9 or chromosome-22. *Br J Haematol* 1994; 87(2): 409–412.

268 De Melo, V.A.S., Milojkovic, D., Marin, D., Apperley, J.F., Nacheva, E.P., Reid, A.G. Deletions adjacent to BCR and ABL1 breakpoints occur in a substantial minority of chronic myeloid leukemia patients with masked Philadelphia rearrangements. *Cancer Genet Cytogenet* 2008; 182(2): 111–115.

269 Sinclair, P.B., Green, A.R., Grace, C., Nacheva, E.P. Improved sensitivity of BCR-ABL detection: a triple-probe three-color fluorescence in situ hybridization system. *Blood* 1997; 90(4): 1395–1402.

270 Dewald, G.W., Wyatt, W.A., Juneau, A.L., et al. Highly sensitive fluorescence in situ hybridization method to detect double BCR/ABL fusion and monitor response to therapy in chronic myeloid leukemia. *Blood* 1998; 91(9): 3357–3365.

271 Hughes, T., Branford, S. Molecular monitoring of chronic myeloid leukemia. *Semin Hematol* 2003; 40(2): 62–68.

272 Hughes, T., Deininger, M., Hochhaus, A., et al. Monitoring CML patients responding to treatment with tyrosine kinase inhibitors: review and recommendations for harmonizing current methodology for detecting BCR-ABL transcripts and kinase domain mutations and for expressing results. *Blood* 2006; 108(1): 28–37.

273 Branford, S., Cross, N.C.P., Hochhaus, A., et al. Rationale for the recommendations for harmonizing current methodology for detecting BCR-ABL transcripts in patients with chronic myeloid leukaemia. *Leukemia* 2006; 20(11): 1925–1930.

274 Branford, S., Fletcher, L., Cross, N.C., et al. Desirable performance characteristics for BCR-ABL measurement on an international reporting scale to allow consistent interpretation of individual patient response and comparison of response rates between clinical trials. *Blood* 2008; 112(8): 3330–3338.
275 Muller, M.C., Erben, P., Saglio, G., et al. Harmonization of BCR-ABL mRNA quantification using a uniform multifunctional control plasmid in 37 international laboratories. *Leukemia* 2008; 22(1): 96–102.
276 Cross, N.C.P. Standardisation of molecular monitoring for chronic myeloid leukaemia. *Best Pract Res Clin Haematol* 2009; 22(3): 355–365.
277 Beillard, E., Pallisgaard, N., van der Velden, V.H.J., et al. Evaluation of candidate control genes for diagnosis and residual disease detection in leukemic patients using 'real-time' quantitative reverse-transcriptase polymerase chain reaction (RQ-PCR) – a Europe Against Cancer program. *Leukemia* 2003; 17(12): 2474–2486.
278 Wang, Y.L., Lee, J.W., Cesarman, E., Jin, D.K., Csernus, B. Molecular monitoring of chronic myelogenous leukemia – Identification of the most suitable internal control gene for real-time quantification of BCR-ABL transcripts. *J Mol Diagn* 2006; 8(2): 231–239.
279 Rulcova, J., Zmekova, V., Zemanova, Z., Klamova, H., Moravcova, J. The effect of total-ABL, GUS and B2M control genes on BCR-ABL monitoring by real-time RT-PCR. *Leuk Res* 2007; 31(4): 483–491.
280 Hughes, T.P., Kaeda, J., Branford, S., et al. Frequency of major molecular responses to imatinib or interferon alfa plus cytarabine in newly diagnosed chronic myeloid leukemia. *N Engl J Med* 2003; 349(15): 1423–1432.
281 Muller, M.C., Saglio, G., Lin, F., et al. An international study to standardize the detection and quantitation of BCR-ABL transcripts from stabilized peripheral blood preparations by quantitative RT-PCR. *Haematologica* 2007; 92(7): 970–973.
282 Cross, N.C.P., White, H.E., Mueller, M.C., Saglio, G., Hochhaus, A. Standardized definitions of molecular response in chronic myeloid leukemia. *Leukemia* 2012; 26(10): 2172–2175.
283 Marin, D., Ibrahim, A.R., Lucas, C., et al. Assessment of BCR-ABL1 transcript levels at 3 months is the only requirement for predicting outcome for patients with chronic myeloid leukemia treated with tyrosine kinase inhibitors. *J Clin Oncol* 2012; 30(3): 232–238.
284 Saglio, G., Baccarani, M. First-line therapy for chronic myeloid leukemia: new horizons and an update. *Clin Lymphoma Myeloma Leuk* 2010; 10(3): 169–176.
285 Mahon, F.X., Rea, D., Guilhot, J., et al. Discontinuation of imatinib in patients with chronic myeloid leukaemia who have maintained complete molecular remission for at least 2 years: the prospective, multicentre Stop Imatinib (STIM) trial. *Lancet Oncol* 2010; 11(11): 1029–1035.
286 Rousselot, P., Huguet, F., Rea, D., et al. Imatinib mesylate discontinuation in patients with chronic myelogenous leukemia in complete molecular remission for more than 2 years. *Blood* 2007; 109(1): 58–60.
287 Rousselot, P., Charbonnier, A., Cony-Makhoul, P., et al. Loss of major molecular response as a trigger for restarting tyrosine kinase inhibitor therapy in patients

with chronic-phase chronic myelogenous leukemia who have stopped imatinib after durable undetectable disease. *J Clin Oncol* 2014; 32(5): 424–430.

288 Ross, D.M., Branford, S., Seymour, J.F., et al. Safety and efficacy of imatinib cessation for CML patients with stable undetectable minimal residual disease: results from the TWISTER study. *Blood* 2013; 122(4): 515–522.

289 Mattarucchi, E., Spinelli, O., Rambaldi, A., et al. Molecular monitoring of residual disease in chronic myeloid leukemia by genomic DNA compared with conventional mRNA analysis. *J Mol Diagn* 2009; 11(5): 482–487.

290 Ross, D.M., Branford, S., Seymour, J.F., et al. Patients with chronic myeloid leukemia who maintain a complete molecular response after stopping imatinib treatment have evidence of persistent leukemia by DNA PCR. *Leukemia* 2010; 24(10): 1719–1724.

291 Sobrinho-Simoes, M., Wilczek, V., Score, J., Cross, N.C.P., Apperley, J.F., Melo, J.V. In search of the original leukemic clone in chronic myeloid leukemia patients in complete molecular remission after stem cell transplantation or imatinib. *Blood* 2010; 116(8): 1329–1335.

292 Shibata, Y., Malhotra, A., Dutta, A. Detection of DNA fusion junctions for BCR-ABL translocations by Anchored ChromPET. *Genome Med* 2010; 2(9): 70.

293 Alikian, M., Ellery, P., Forbes, M., et al. NGS-assisted DNA-based digital qPCR facilitates stratification of CML patients in long-term molecular remission based on the presence of detectable BCR-ABL1 DNA. *Blood* 2013; 122(21): 4006.

294 Goh, H.G., Lin, M., Fukushima, T., et al. Sensitive quantitation of minimal residual disease in chronic myeloid leukemia using nanofluidic digital polymerase chain reaction assay. *Leuk Lymphoma* 2011; 52(5): 896–904.

295 Sanders, R., Huggett, J.F., Bushell, C.A., Cowen, S., Scott, D.J., Foy, C.A. Evaluation of digital PCR for absolute DNA quantification. *Anal Chem* 2011; 83(17): 6474–6484.

296 Sanders, R., Mason, D.J., Foy, C.A., Huggett, J.F. Evaluation of digital PCR for absolute RNA quantification. *PLoS One* 2013; 8(9): e75296.

297 Alikian, M., Gerrard, G., Subramanian, P.G., et al. BCR-ABL1 kinase domain mutations: methodology and clinical evaluation. *Am J Hematol* 2012; 87(3): 298–304.

298 Soverini, S., Branford, S., Nicolini, F.E., et al. Implications of BCR-ABL1 kinase domain-mediated resistance in chronic myeloid leukemia. *Leuk Res* 2014; 38(1): 10–20.

299 Machova Polakova, K., Kulvait, V., Benesova, A., et al. Next-generation deep sequencing improves detection of BCR-ABL1 kinase domain mutations emerging under tyrosine kinase inhibitor treatment of chronic myeloid leukemia patients in chronic phase. *J Cancer Res Clin Oncol* 2015; 141(5): 887–899.

300 Soverini, S., De Benedittis, C., Polakova, K.M., et al. Unraveling the complexity of tyrosine kinase inhibitor-resistant populations by ultra-deep sequencing of the BCR-ABL kinase domain. *Blood* 2013; 122(9): 1634–1648.

301 Zabriskie, M.S., Eide, C.A., Tantravahi, S.K., et al. BCR-ABL1 compound mutations combining key kinase domain positions confer clinical resistance to ponatinib in Ph chromosome-positive leukemia. *Cancer Cell* 2014; 26(3): 428–442.

Index

The Genetic Basis of Haematological Cancers, First Edition. Edited by Sabrina Tosi and Alistair G. Reid.